Analysis and Synthesis of Linear Control Systems

CHI-TSONG CHEN

Department of Electrical Sciences
State University of New York at Stony Brook

Holt, Rinehart and Winston, Inc.
New York   Chicago   San Francisco   Atlanta
Dallas   Montreal   Toronto   London   Sydney

To Beatrice
Janet, Pauline, and Stanley

Copyright © 1975 by Holt, Rinehart and Winston, Inc.
All rights reserved.
Printed in the United States of America

Chen, Chi-Tsong.
Analysis and synthesis of linear control systems.

(HRW series in electrical engineering, electronics, and systems)
Includes bibliographies.
1. Automatic control. I. Title.
TJ213.C476   629.8'32   74-12304
ISBN: 0-03-003511-2

5 6 7 8 9   038   9 8 7 6 5 4 3 2 1

Preface

This is an introductory book on the design of linear control systems. It is developed from the lecture notes used in a one-semester senior course taught at the State University of New York at Stony Brook. The notes, with some additional material, were also used in a one-year junior course at the National Chiao-Tung University, Taiwan. The background assumed is the knowledge of Newton's laws and Kirchhoff's laws. The familiarity of the Laplace transformation and elementary linear algebra is helpful but not necessary; they are reviewed in the appendixes to the extent needed in this book.

The main goal of the book is to enable the reader to carry out complete designs of control systems, starting from the formulations of mathematical equations for control components to the last stage of synthesizing the required compensators. Four different design techniques are introduced, and their relative merits are discussed. These techniques are applied to the theme example, a tracking antenna system with a transfer function of degree four; the design results are then simulated and compared. Since these techniques are developed for linear models of control systems, in practical applications the problems of saturation, loading, and noise have to be examined. They are repeatedly discussed in the book.

The designs of linear time-invariant control systems can be carried out by using either transfer functions or state-variable equations. The state-variable equations are often identified with modern or optimal control theory, whereas classical control theory uses transfer functions exclusively. In this book the two mathematical approaches are introduced. Optimal design is discussed using both transfer functions and state-variable equations. State-variable feedback problems are solved by the use of transfer functions. A unification of the two approaches is made.

The mathematics required in the proofs of optimal control theories is beyond the level of this book. This however should not prevent us from the employment of optimal design techniques. It is this author's opinion that some optimal design techniques are conceptually and computationally as simple as, if not simpler than, the root-locus and frequency-domain techniques. Therefore in this book the former is introduced before the latter. However the order of the studies of these techniques can be reversed. (See the logical dependence of chapters listed at the end of this preface.)

The root-locus method in most control texts is introduced together with the unit feedback system. This text begins with the formulation of a one-parameter problem. By so doing, the method can be applied to any configuration of control systems. The presentation of the frequency-domain technique also is different from other texts. Effort is made to derive the frequency-domain specifications from the time-domain specifications. Although these derivations are not rigorous, it is hoped that this approach will shed a new light on the study of the frequency-domain technique.

The text stresses syntheses, and analyses are kept at a minimum or are developed as tools for syntheses. Most results are carefully stated as theorems, but no proofs are given. To help the reader grasp the main ideas, review questions and problems are provided at the end of each chapter. A solutions manual for the problems can be obtained from the publisher.

Because of the advance of integrated circuits and wide availability of computers, the discrete-time techniques become increasingly important. A serious attempt is made to incorporate discrete-time techniques in the book. After all, many results in the continuous-time method, such as realization, optimal design, and root-locus method, can be directly applied to the discrete-time case. However the attempt was discarded for the following two reasons: First, the material in the text is more than that which can be covered in a one-semester course. Second, to give a meaningful introduction of discrete-time systems, discussions of coding, sampling, analog-to-digital conversion, and some digital circuit implementation cannot be avoided. This would enormously increase the size of the text.

I owe a great deal to many people in writing the book. To Professor J. G. Truxal, Dean of the College of Engineering at Stony Brook, and Professor S. S. L. Chang for their leadership and encouragement, to Professor V. A. Marsocci, Chairman of the Department of Electrical Sciences, for providing the departmental assistant, to Professor P. E. Barry for critical scrutiny of the manuscript and for many suggestions in improving the presentation, to Mrs. V. Donahue, Mrs. R. Bouleris, and Mrs. J. Logiovane

Books are to be returned on or before
the last date below

Analysis and Synthesis
of Linear Control Systems

Holt, Rinehart and Winston Series
in Electrical Engineering, Electronics, and Systems

Shu-Park Chan, Introductory Topological Analysis of Electrical Networks
Chi-Tsong Chen, Analysis and Synthesis of Linear Control Systems
Chi-Tsong Chen, Introduction to Linear System Theory
George R. Cooper and Clare D. McGillem, Methods of Signal and System Analysis
George R. Cooper and Clare D. McGillem, Probabilistic Methods of Signal and System Analysis
Woodrow W. Everett, Jr., editor, Topics in Intersystem Electromagnetic Compatibility
D. J. Hamilton, F. A. Lindholm, A. H. Marshak, Principles and Applications of Semiconductor Device Modeling
Benjamin J. Leon and Paul A. Wintz, Basic Linear Networks for Electrical and Electronics Engineers
Clare D. McGillem and George R. Cooper, Continuous and Discrete Signal and System Analysis
Richard Saeks, Generalized Networks
Amnon Yariv, Introduction to Optical Electronics

for typing various drafts of the book. I also wish to thank the National Chiao-Tung University, Taiwan, where the final draft was completed.

Logical dependence of chapters:

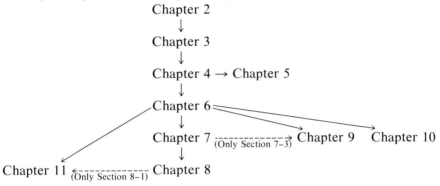

Chi-Tsong Chen

Stony Brook, New York

Contents

Preface   v

Chapter 1   **Introduction**   1
- 1-1   Control Systems   1
- 1-2   Problem Formulation   2
- 1-3   Scope of the Text   5
- References   6

Chapter 2   **Mathematical Descriptions of Linear Time-Invariant Systems**   7
- 2-1   Physical Systems and Models   7
- 2-2   Inputs, Outputs, and States of Lumped Systems   9
- 2-3   Linear Time-Invariant Systems   12
- 2-4   The Input-Output Description of Single-Variable Systems   15
- 2-5   The State-Variable Description of Single-Variable Systems   20
- 2-6   Mathematical Descriptions of Multivariable Systems   24
- 2-7   Complete Characterization   27
- 2-8   Remarks and Review Questions   32
- References   33
- Problems   33

Chapter 3   Control Components and Systems   38

- 3-1   Introduction   38
- 3-2   AC and DC Devices   39
- 3-3   Servomotors   40
- 3-4   Gear Trains   47
- 3-5   Transducers   51
- 3-6   Error Detectors   56
- 3-7   Amplifiers   59
- 3-8   Compensation Networks   60
- 3-9   Examples of Control Systems   63
- 3-10   Remarks and Review Questions   70
- References   72
- Problems   73

Chapter 4   Analysis: Quantitative and Qualitative   81

- 4-1   Mathematical Descriptions of Composite Systems   81
- 4-2   Transfer Function–Partial Fraction Expansion   85
- 4-3   Solutions of Dynamical Equations   90
- 4-4   BIBO, Asymptotic, and Total Stability   95
- 4-5   Stability and Complete Characterization   100
- 4-6   Routh-Hurwitz Criterion   104
- 4-7   Steady-State Responses of Stable Systems   109
- 4-8   Simplification   112
- 4-9   Remarks and Review Questions   115
- References   116
- Problems   117

Chapter 5   Computer Simulations and Realizations   122

- 5-1   Analog Computer Simulations of Dynamical Equations   122
- 5-2   Digital Computer Simulations of Dynamical Equations   132
- 5-3   Realization Problem   134
- 5-4   Realizations of Proper Rational Functions   135
- 5-5   Realizations of Vector Proper Rational Transfer Functions   143
- 5-6   Remarks and Review Questions   148
- References   149
- Problems   150

Chapter 6		**Introduction to Design**	**152**
	6-1	The Choice of a Plant 152	
	6-2	Performance Criterion 154	
	6-3	Two Basic Approaches in the Design 159	
	6-4	Difficulties in the Design 160	
	6-5	Review Questions 164	
		References 165	
		Problems 165	

Chapter 7		**Synthesizable Transfer Functions and Feedback**	**168**
	7-1	Pole-Zero Excesses of Overall Transfer Function 168	
	7-2	Zeros and Poles of an Overall Transfer Function 172	
	7-3	Why Feedback? 173	
	7-4	Remarks and Review Questions 181	
		References 182	
		Problems 182	

Chapter 8		**Optimal Control Systems**	**185**
	8-1	Quadratic Performance Criterion 185	
	8-2	Optimal Control Systems: Transfer-Function Approach 188	
	8-3	Optimal Control Systems: State-Variable Approach 194	
	8-4	The Regulator Problem 197	
	8-5	Are All State Variables Available for Feedback? 199	
	8-6	Design of Compensators: Output Feedback 202	
	8-7	Design of Compensators: State Feedback 208	
	8-8	Design of Compensators: Partial State Feedback 213	
	8-9	Remarks and Review Questions 219	
		References 221	
		Problems 221	

Chapter 9		**The Root-Locus Method**	**224**
	9-1	Problem Formulation: One-Parameter Variation 224	
	9-2	Desired Pole Locations 226	
	9-3	The Plot of Root Loci 232	

9-4	Design by the Root-Locus Technique 239	
9-5	Remarks and Review Questions 248	
	References 249	
	Problems 250	

Chapter 10 **The Frequency-Domain Technique** — 255
- 10-1 Frequency-Domain Specifications 256
- 10-2 Frequency-Domain Plots: Polar Plot, Bode Plot, and Log Magnitude-Phase Plot 260
- 10-3 Stability Test in the Frequency Domain 268
- 10-4 Specifications in Terms of Open-Loop Transfer Functions 270
- 10-5 Design on the Bode Plots 279
- 10-6 Remarks and Review Questions 293
- References 295
- Problems 295

Chapter 11 **Parameter Optimization** — 299
- 11-1 Analytical Method 300
- 11-2 Numerical Method 308
- 11-3 Concluding Remarks 312
- References 313
- Problems 314

Chapter 12 **Epilogue** — 315
- 12-1 Comparisons of Various Design Results of the Tracking Antenna Problem 315
- 12-2 Further Topics 319
  - Multivariable Systems 319
  - Linear Discrete-Time Systems 321
  - Time-Varying Systems 324
  - Nonlinear Systems 324
  - Stochastic Systems 325
  - Adaptive and Learning Systems 325
- References 326

Appendix A **The Laplace Transformation** — 327
- Reference 331

Appendix B **Matrix Theory** — 332
- B-1 Matrices 332
- B-2 The Rank of a Matrix 333
- B-3 Linear Algebraic Equations 335
- References 336

Index — 337

CHAPTER 1

Introduction

1-1 Control Systems

Since this text is concerned with the analysis and design of control systems, it is pertinent at the very outset to discuss what a control system is. Roughly speaking, any connection of devices or any arrangement that strives to achieve a certain goal is a control system. This definition is so vague that a great number of systems can be considered as control systems. Let us give some examples.

The success of an economic system is measured by the gross national product (GNP), the rate of unemployment, average hourly wage, and consumer price index. If the annual increase of the consumer price index is too large, or if the rate of unemployment is not acceptable, then the economic policy has to be modified. This is done by the changes of interest rates, of government spending, and of monetary policy. This economic system can be looked upon as a control system. The goal of the system is to obtain a healthy annual increase of GNP, to keep the rate of unemployment below a certain percentage, and to hold the consumer price index constant. The controlling factors are government spending, interest rate, and monetary policy. This system has several interrelated factors, the cause-and-effect relationships of which are not exactly known. Furthermore there are many uncertainties (such as consumer spending, labor disputes, or international crises) involved in the system; hence an economic system is an extremely complicated control system.

2 INTRODUCTION

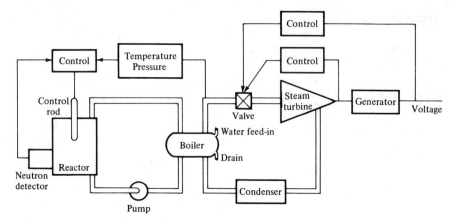

Figure 1-1 Simplified nuclear power plant.

Sending a space vehicle to the moon and bringing it back safely to the earth is another example of a complicated control system. If the trajectory of the space vehicle is not properly controlled, the space vehicle may stray into outer space. Another control problem is that of monitoring, from the earth, the television camera mounted on a rover on the moon surface. The camera is controlled remotely to scan the surface of the moon and to focus on interesting objects.

Not all control systems are as exotic as the ones just mentioned. A home heating system, a refrigerator, and an oven are all control systems; they control the temperatures of some closed spaces. A bathroom toilet tank is a control system; it maintains the water level at a preset height. Driving an automobile is also a control problem; we control the automobile to stay in a lane and to cruise at a desired speed.

Control systems are indispensable in modern manufacturing and industrial processes. For example, consider the nuclear power plant shown in Figure 1-1. In order to maintain the pressure and temperature of steam at fixed levels, a control system is required to set the position of the control rod. Another control system is required to regulate the generated voltage at, say, 120 volts. The regulation of the speed of the turbine shaft and the maintenance of the water level of the boiler also call for control systems. In a paper mill the entire process —from the first stage of moving raw materials to the last stage of controlling the weight, thickness, and strength of paper—can be accomplished automatically by the use of control systems. Hence control systems are indeed indispensable in industrial processes.

1-2 Problem Formulation

Although control systems may appear quite different in form, they are essentially the same type of design problem. This can be seen by drawing the schematic diagrams of control systems.

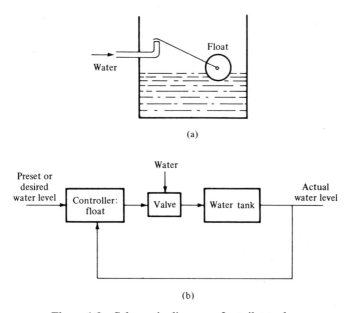

Figure 1-2 Schematic diagram of a toilet tank.

Consider the bathroom toilet tank shown in Figure 1-2(a). The mechanism is to close the valve automatically whenever the water level reaches a preset height. A schematic diagram of the system is shown in Figure 1-2(b). The float translates the water level into the valve position. Although a toilet tank is a very simple mechanism, it has all the ingredients of a control system.

A home heating system can be represented schematically as shown in Figure 1-3. The burner supplies the rooms with heat through either hot air or hot water. The thermostat measures the room temperature, compares it with the temperature desired, and then turns the burner on or off. We see that its schematic diagram is quite similar to that of the water tank.

The control of the descending stage of a lunar module is shown in Figure 1-4. The desired descending trajectory is precomputed. The goal of the control is to have the actual descending trajectory as close as possible to the desired one. This is achieved with the aid of an on-board computer, altimeters, and other devices. As a final example, the schematic diagram of the control of a clothes dryer is shown in Figure 1-5. Depending on the desired degree of dryness, and

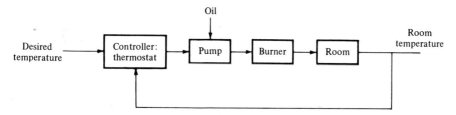

Figure 1-3 Schematic diagram of a home heating system.

4 INTRODUCTION

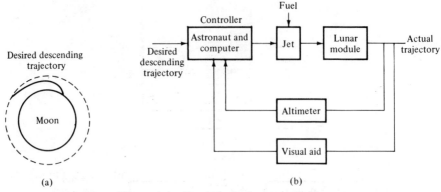

Figure 1-4 Control of a lunar module.

on the amount of clothes, we set the timer. After using the dryer for a number of times, we soon learn to set the timer properly.

We shall now extract the basic design problem from these examples. Every control system has a goal or mission, such as sending a lunar module to the moon or keeping a house warm. The lunar module and the air in the room are the objects to be controlled. In order to carry out the control, we need a set of devices, such as valves in the case of a gas jet or a burner. This set of devices is called an *actuator*. An actuator, together with the object to be controlled, will be called a *plant*. The output of a plant is the variable to be controlled. The input of a plant is called an *actuating signal*. In order to carry out the design, the mission is first translated into a quantitative entity, such as a desired trajectory or temperature. This quantitative entity is called the *desired signal*, or *command signal*. Hence the design problem can be stated as follows: Given a plant and a desired signal, as shown in Figure 1-6, design an overall system so that the output of the plant will be as close as possible to the desired signal. This will be achieved by the design of controllers, or compensators, and the employment of sensing devices. This text is devoted entirely to this problem.

There are basically two types of control systems: the open-loop system and the closed-loop, or feedback, system. In an open-loop system the actuating signal is predetermined and will not change, no matter what the actual output is. Clothes washers and simple dryers (those without sensing devices) are systems of this type. In a closed-loop system, the actuating signal will be a function of the desired signal as well as the actual output. The systems in Figure 1-1 through 1-4 are of this type. A properly designed feedback system can function better

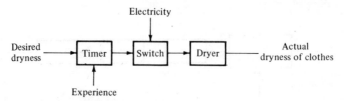

Figure 1-5 Schematic diagram of the control of a simple dryer.

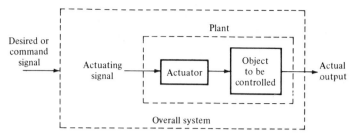

Figure 1-6 The basic design problem.

than an open-loop system, as will be discussed in detail. Hence the emphasis in this text will be on feedback systems.

1-3 Scope of the Text

This text is concerned exclusively with the design of the control problem posed in the previous section; however it does not intend to solve all the problems that can be so posed. In fact, we study only a very special class of control systems—namely linear, time-invariant, lumped, continuous-time, and deterministic systems. A control system can be classified dichotomously as linear or nonlinear, time-invariant or time-varying, lumped or distributed, continuous-time or discrete-time, deterministic or stochastic. Roughly speaking, a system is said to be linear if it satisfies the superposition property, time-invariant if its characteristics do not change with time, and lumped if it has a finite number of state variables (see Chapter 2). A system is a continuous-time system if its responses are defined for all time, and a deterministic system if it contains no noise. For this class of systems, it is possible to give a fairly complete treatment of the posed problem.

In Chapter 2 we introduce the mathematical equations used through the text. The concepts of linearity and time-invariance are first introduced. The input-output description, in particular the transfer function, and the state-variable description are then developed from these concepts. The concept of complete characterization is also introduced. This concept is important in the employment of transfer functions.

In Chapter 3 we study models and mathematical descriptions of control components. A number of control systems are then built from these components. This chapter illustrates how the mathematical equations that we shall use throughout the text are derived from physical systems. It also provides the motivation for the consideration of physical constraints in the design.

In Chapter 4 we study the quantitative and qualitative analyses of linear time-invariant systems. The analysis techniques are applicable to any system once its mathematical equation is found; hence we introduce first the mathematical descriptions of composite systems. The quantitative analyses are discussed only briefly, because they can be delegated to a digital or an analog computer. In the qualitative analyses, we introduce the concepts of BIBO (bounded-input bounded-output), asymptotic, and total stability. Steady-state responses of BIBO stable systems are then introduced. In Chapter 5 we study the analog and digital computer simulations. We also study the realization problem; the result is useful in the synthesis of a transfer function by using an operational-amplifier circuit.

Chapters 2 through 5 are more or less concerned with modeling and analysis problems; the remaining six chapters are concerned with the design problem. The design criteria, the basic approaches, and possible difficulties in the design are discussed in Chapter 6. The presence of noise in control systems restricts the class of compensators used in practice. This restriction inevitably imposes certain constraints on overall systems. This is discussed in Chapter 7. Comparisons of open-loop and closed-loop systems are also discussed in this chapter.

There are four methods available in the design of control systems. They are the optimal design, the root-locus method, the Bode-plot method, and the parameter optimization method. These methods are introduced respectively in Chapters 8, 9, 10, and 11. In the optimal design both the transfer function and the state-variable approaches are studied. A tracking antenna problem, which has a transfer function of degree 4, is designed by using these four methods. Comparisons of these methods are also made at the end of each chapter and in Chapter 12.

References

[1] Chang, S. S. L., "On the modeling and control of mega systems," *Proc. Princeton Conf. on Information Sciences and Systems*, March 1971.
[2] Chestnut, H., and R. W. Mayer, *Servomechanisms and Regulating System Design*. New York: Wiley, 1951.
[3] Newton, G. C., Jr., L. A. Gould, and J. F. Kaiser, *Analytical Design of Linear Feedback Controls*. New York: Wiley, 1957.

CHAPTER 2
Mathematical Descriptions of Linear Time-Invariant Systems

2-1 Physical Systems and Models

The ultimate goal of engineering is to build real physical systems to perform some specified jobs. For example, an engineer may be asked to design and install a heating system for a house, or a group of engineers may be asked to build a nuclear power plant or to send a space vehicle to the moon and bring it back safely to the earth. There are basically two approaches in which a design problem can be accomplished: the *ad hoc* approach and the analytical approach. In the *ad hoc* approach, an engineer combines his experience, know-how, and trial-and-error results to build the required system. This approach employs simple calculation and utilizes available components. It is often satisfactory for simple and routine tasks, such as the design of a heating system.

The *ad hoc* approach, unfortunately, is often found to be inadequate for complicated systems such as space vehicles or nuclear power plants because of the lack of experience and because of the cost and dangers involved in repeated testing. In these cases an engineer has to proceed differently—that is, through analytical methods. In the analytical study, an engineer proceeds to search for an idealized system that resembles the physical system in its salient features but is easier to study. This idealized system is called a *model*. As engineers we are all familiar with the distinction between a physical system and a model. A television set is a physical system; the circuit diagram in the manual

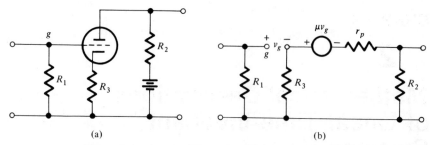

Figure 2-1 An amplifier. (a) Circuit. (b) Model.

is a model. In electrical engineering, an amplifier circuit, as shown in Figure 2-1(a), is often modeled as shown in Figure 2-1(b). In mechanical engineering, an automobile suspension system may be modeled as shown in Figure 2-2. In bioengineering, a human arm may be modeled as shown in Figure 2-3(b) or, more realistically, as in Figure 2-3(c). Modeling is a very important problem because the success of design depends upon whether or not a physical system is properly modeled.

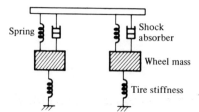

Figure 2-2
Model of an automobile suspension system.

Depending on different questions asked, or depending on different operating ranges, a physical system may have many different models. For example, an electronic amplifier has different models at high and low frequencies. A spaceship may be modeled as a particle in the study of trajectory; however it must be modeled as a rigid body in the study of maneuvering. In order to develop a suitable model for a physical system, we must understand thoroughly the physical system and its operating range. In this book we shall refer to models of physical systems as *systems*. Hence a physical system is a device or a collection of devices existing in the real world; a system is a model of a physical system.

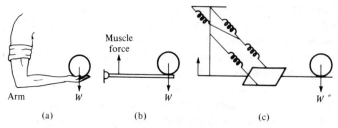

Figure 2-3 Models of an arm.

2-2 Inputs, Outputs, and States of Lumped Systems

The class of systems (models) studied in this book is assumed to have some input terminals and output terminals. The inputs (the causes or the excitations) are applied at the input terminals; the outputs (the effects or the responses) are measurable at the output terminals. The concepts of input and output are self-explanatory. For convenience, we shall classify systems into single-variable or multivariable systems according to the following definition.

Definition 2-1

A system is called a *single-variable system* if it has only one input terminal and only one output terminal. Otherwise it is called a *multivariable system*. ∎

In other words, a system with two or more input terminals and/or two or more output terminals is called a multivariable system. For single-variable systems, we use scalars u and y to denote the input and output, respectively. For multivariable systems, we use column vectors $\mathbf{u}$ and $\mathbf{y}$ to denote the inputs and outputs, respectively.

Consider the network (a model of some real circuit) shown in Figure 2-4. The input u is a voltage source. The output y is the voltage across the capacitor, as shown. If an input is applied to the circuit, for different initial current through the inductor and different initial voltage across the capacitor, we obtain different outputs. In other words, knowing the input but not knowing the initial conditions, we are not able to determine the output uniquely. Hence the function of the set of initial conditions is to help in uniquely determining the outputs. This set of initial conditions is called a *state* of the network.

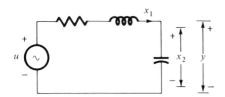

Figure 2-4
Network with two state variables.

Definition 2-2

A *state* of a system at time t_0 is the minimum amount of information[1] (minimum set of initial conditions) that, together with the input $\mathbf{u}_{[t_0,\infty)}$, determines uniquely the responses of the system for all $t \geq t_0$. ∎

---

[1] A state can also be defined without the qualification "minimum." See, for example, Reference [1]. We shall however adopt the qualification "minimum" for the convenience of the discussion of complete characterization in Section 2-7.

By $\mathbf{u}_{[t_0,\infty)}$ we mean the input functions defined over the time interval $[t_0, \infty)$. By the responses of a system we mean not only the responses at the output terminals but also the responses inside the system; for example, the response at *every* branch if the system is a network. In this definition, *how* the state actually attains the values at time t_0 is immaterial in determining the behavior of the system after time t_0. Hence the state summarizes, in a sense, the past information required in the determination of the future behavior of a system. The state of a system is often represented by a column vector $\mathbf{x}$ called the *state vector*. Each component of $\mathbf{x}$—that is, each initial condition—is called a *state variable*. The network shown in Figure 2-4 has two state variables: the current x_1 through the inductor and the voltage x_2 across the capacitor. The state vector of the network is

$$\mathbf{x} = \begin{bmatrix} x_1 \\ x_2 \end{bmatrix}$$

We shall now give several more examples to illustrate the concept of state.

Example 1

Consider the resistive network shown in Figure 2-5. Since there is no energy-storage element in the network, the output at time t_1 depends solely on the input applied at time t_1. There is no initial condition required to determine uniquely the output. Hence there is *no* state variable in the network. This kind of system is called an *instantaneous*, or *memoryless, system*.

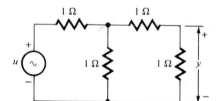

Figure 2-5
A memoryless system.

Example 2

Consider the mechanical system shown in Figure 2-6. If an external force u (input) is applied to the system at time t_0, the future position y (output) of the mass is not uniquely determinable unless the position and the velocity of the mass at time t_0 are known. Hence the set of two variables, the position and velocity of the mass, qualifies as a *state* of the system.

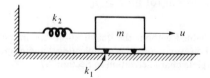

Figure 2-6
A mechanical system.

Example 3

Consider the circuit shown in Figure 2-7(a). The element denoted by T is a tunnel diode, with the characteristic shown in Figure 2-7(b). The tunnel diode is a memoryless element. Hence the network consists of two state variables: the current through the inductor and the voltage across the capacitor.

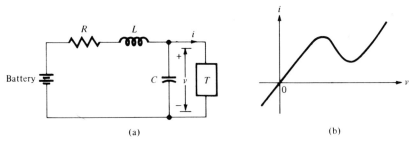

Figure 2-7 A nonlinear system.

Example 4

Consider the network shown in Figure 2-8. It is clear that if all the capacitor voltages are known, then for any applied voltage signal the responses of the network can be uniquely determined. The application of Kirchhoff's voltage law to the loop consisting of the three capacitors yields

$$x_1(t) - x_2(t) - x_3(t) = 0$$

for all t, which implies that if any two of x_1, x_2, and x_3 are known, then the remainder is also known. Hence the set of three capacitor voltages x_1, x_2, and x_3 is *not* a state of the network, for a state should, according to the definition, consist of the *minimum* set of initial conditions. However any two of x_1, x_2, and x_3 qualify as a state of the network. ∎

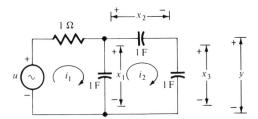

Figure 2-8 Network with two state variables.

From these examples we see that the way to choose a state is not unique. However because it is required that a state consist of the minimum set of initial conditions, *the number of state variables of a system is unique*. The state variables shown in these examples are associated with physical quantities. Although it is possible to look upon a state as a mathematical entity without having any physical meaning, it is convenient to associate a state with physical quantities.

Example 5

Consider a unit-time-delay system, defined as $y(t) = u(t - 1)$ for all t. In order to determine $y_{[t_0,\infty)}$ uniquely, we must know, in addition to $u_{[t_0,\infty)}$, $u_{[t_0-1,t_0]}$. Hence the state of the system at t_0 consists of all information stored in $u_{[t_0-1,t_0]}$. There are infinitely many points in the interval $[t_0 - 1, t_0]$; therefore the state of this system consists of infinitely many numbers. Hence the system has infinitely many state variables. ∎

A system may have, as shown in the above examples, a finite number of state variables or infinitely many state variables. The former is classified as a *lumped system*, and the latter as a *distributed system*. In this book we study only lumped systems—that is, the class of systems having a finite number of state variables.

2-3 Linear Time-Invariant Systems

In this section the concepts of linearity and time-invariance will be introduced. As discussed in the previous section, the responses of a system for all $t \geq t_0$ are uniquely determinable only if both the inputs $\mathbf{u}_{[t_0,\infty)}$ and the state at t_0, $\mathbf{x}(t_0)$ are known. In other words, the responses of a system are excited partly by $\mathbf{u}_{[t_0,\infty)}$ and partly by $\mathbf{x}(t_0)$. The effect of the input applied prior to time t_0 has been summarized in the state at t_0 and need not be considered further. By the responses of a system, we mean the output $\mathbf{y}_{[t_0,\infty)}$ as well as the responses inside the system—in particular the future state $\mathbf{x}(t)$, for $t > t_0$, of the system. We use the notation

$$\{\mathbf{u}_{[t_0,\infty)}, \mathbf{x}(t_0)\} \to \{\mathbf{x}_{[t_0,\infty)}, \mathbf{y}_{[t_0,\infty)}\}$$

to denote that the state $\mathbf{x}(t_0)$ and the input $\mathbf{u}_{[t_0,\infty)}$ excite the output $\mathbf{y}(t)$ and the state $\mathbf{x}(t)$, for $t \geq t_0$, and call it an input-state-output pair. An input-state-output pair of a system is called *admissible* if the system can generate such a pair.

Definition 2-3

A system is said to be *linear* if, for every two admissible pairs,

$$\{\mathbf{x}^1(t_0), \mathbf{u}^1_{[t_0,\infty)}\} \to \{\mathbf{x}^1_{[t_0,\infty)}, \mathbf{y}^1_{[t_0,\infty)}\} \quad (2\text{-}1)$$

$$\{\mathbf{x}^2(t_0), \mathbf{u}^2_{[t_0,\infty)}\} \to \{\mathbf{x}^2_{[t_0,\infty)}, \mathbf{y}^2_{[t_0,\infty)}\} \quad (2\text{-}2)$$

and every real number α, the following pairs,

$$\{\mathbf{x}^1(t_0) + \mathbf{x}^2(t_0), \mathbf{u}^1_{[t_0,\infty)} + \mathbf{u}^2_{[t_0,\infty)}\} \to \{\mathbf{x}^1_{[t_0,\infty)} + \mathbf{x}^2_{[t_0,\infty)}, \mathbf{y}^1_{[t_0,\infty)} + \mathbf{y}^2_{[t_0,\infty)}\} \quad (2\text{-}3)$$

and

$$\{\alpha \mathbf{x}^1(t_0), \alpha \mathbf{u}^1_{[t_0,\infty)}\} \to \{\alpha \mathbf{x}^1_{[t_0,\infty)}, \alpha \mathbf{y}^1_{[t_0,\infty)}\} \quad (2\text{-}4)$$

are also admissible. Otherwise the system is said to be *nonlinear*. ∎

2-3 LINEAR TIME-INVARIANT SYSTEMS

The property given in Equation (2-3) is often referred to as the *additivity* property. It asserts that the sum of the responses of any two excitations is equal to the responses of the sum of the two excitations. The property in Equation (2-4) is called the *homogeneity* property. It asserts that if the magnitudes of the excitations change by a factor, those of the responses will change by the same factor. The combination of these two properties is often referred to as the *superposition* property. Hence if all possible input-state-output pairs of a system satisfy the superposition property, then the system is linear.

In Definition 2-3, if

$$\mathbf{x}^1(t_0) = -\mathbf{x}^2(t_0) \quad \text{and} \quad \mathbf{u}^1_{[t_0,\infty)} = -\mathbf{u}^2_{[t_0,\infty)}$$

then the linearity implies that $\{0, 0\} \to \{0_{[t_0,\infty)}, 0_{[t_0,\infty)}\}$. Hence a necessary condition for a system to be linear is that if $\mathbf{x}(t_0) = 0$ and $\mathbf{u}_{[t_0,\infty)} \equiv 0$, then the responses of the system are identically zero. A very important property of any linear system is that the responses of the system can be decomposed into two parts, as

Responses due to $\{\mathbf{x}(t_0), \mathbf{u}_{[t_0,\infty)}\}$
= responses due to $\{\mathbf{x}(t_0), 0\}$ + responses due to $\{0, \mathbf{u}_{[t_0,\infty)}\}$ **(2-5)**

The responses due to $\{\mathbf{x}(t_0), 0\}$ are called *zero-input responses*; they are generated exclusively by the nonzero initial state $\mathbf{x}(t_0)$. The responses due to $\{0, \mathbf{u}_{[t_0,\infty)}\}$ are called *zero-state responses*; they are excited exclusively by the input $\mathbf{u}_{[t_0,\infty)}$. Equation (2-5) follows directly from (2-1) to (2-3) if we choose

$$\mathbf{x}^1(t_0) = \mathbf{x}(t_0) \quad \mathbf{u}^1 = 0 \quad \mathbf{x}^2(t_0) = 0 \quad \mathbf{u}^2_{[t_0,\infty)} = \mathbf{u}_{[t_0,\infty)}$$

We introduce next the concept of time-invariance.

Definition 2-4

A system is said to be time-invariant if, for every admissible pair,

$$\{\mathbf{x}(t_0), \mathbf{u}_{[t_0,\infty)}\} \to \{\mathbf{x}_{[t_0,\infty)}, \mathbf{y}_{[t_0,\infty)}\}$$

and every real number T, the pair

$$\{\mathbf{x}(t_0 + T), \mathbf{u}_{[t_0+T,\infty)}\} \to \{\mathbf{u}_{[t_0+T,\infty)}, \mathbf{y}_{[t_0+T,\infty)}\}$$

is also admissible. Otherwise the system is said to be time-varying. ∎

In words, for time-invariant systems, if the waveforms of excitations are the same, then the waveform of the responses will always be the same no matter at what instant the excitations are applied. This implies that the characteristics of a time-invariant system will not change with time. For example, the mass of a rocket decreases with time because of the consumption of the fuel; hence a flying rocket is a time-varying system. For time-invariant systems, because the responses are independent of the initial time it is always assumed, for convenience, that $t_0 = 0$ and that the time interval of interest is $[0, \infty)$. We note that $t_0 = 0$ is not absolute; rather it is the instant we start to study a system or to apply an input to a system.

A system may often consist of many subsystems or components. For example, the network in Figure 2-4 consists of three components: one resistor, one inductor, and one capacitor. It can be shown that if every component of a system is linear and time-invariant, then the overall system is also linear and time-invariant. (Note that the converse is not necessarily true. See Problem 2-6.) Hence in modeling a physical system, if every component is modeled as linear and time-invariant, then the resulting overall system will be linear and time-invariant.

Strictly speaking, no physical system is linear and time-invariant. A television set, an automobile, or a communication satellite cannot function forever; its performance will deteriorate with time because of aging or other factors. However if the changes of characteristics are very small in the time interval of interest—say, one year—then these physical systems can be considered as time-invariant. Hence over finite time intervals, a great number of physical systems can be modeled by time-invariant systems.

In order to obtain linear models for physical systems, we often have to employ the linearization technique. Consider the network shown in Figure 2-4. The characteristic of the inductor is shown in Figure 2-9 by the use of the solid line. If the current through the inductor is known to range between i_1 and $-i_2$, then the inductor can be modeled by a linear one with the characteristic shown in Figure 2-9 by the use of dashed line. If the current lies between i_3 and $-i_4$, then the inductor can be modeled by a linear one with the characteristic shown in Figure 2-9 by the use of dashed-and-dotted line. Similarly if the resistor and capacitor in Figure 2-4 are modeled by linear ones, then the circuit is a linear model. Consider next the mechanical system shown in Figure 2-6. The friction force between the floor and the mass generally consists of three components: static friction, Coulomb friction, and viscous friction as shown in Figure 2-10. If we neglect the static and Coulomb friction, then the friction and

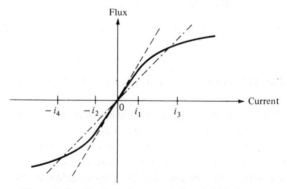

Figure 2-9 Typical characteristic of an inductor.

2-4 The Input-Output Description of Single-Variable Systems

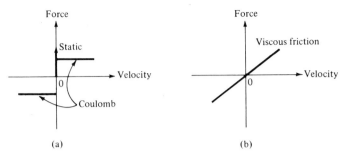

Figure 2-10 (a) Static and Coulomb friction. (b) Viscous friction.

the velocity are linearly related. Hence if we consider only the viscous friction and if the deflection of the spring does not exceed the elastic limit, the physical system in Figure 2-6 can be modeled as a linear system. This kind of linearization is commonly used in analytical study of physical systems.

The systems that will be used in this book to model physical systems are almost exclusively linear time-invariant systems. The reasons are as follows: (a) A great number of physical systems over the normal operating ranges, as discussed above, can be modeled by linear time-invariant systems. (b) The theory of linear time-invariant systems is complete and well organized. (c) The theory is the basis for the study of time-varying and/or nonlinear systems.

2-4 The Input-Output Description of Single-Variable Systems

Once a system (a model) is found for a physical system, the next step in the analytical study is the search of mathematical equations to describe the system. Depending on the question asked or the design techniques used, one form of equation might be preferable to another in describing the same system. In this section and subsequent sections, two different mathematical descriptions of linear time-invariant systems will be introduced. Since we study only time-invariant systems, we shall assume, without loss of generality, that the initial time is zero ($t_0 = 0$) and that the time interval of interest is $[0, \infty)$. We study first the single-variable case, namely the class of systems with only one input u and only one output y.

The responses of a system, as discussed earlier, are partly contributed by the input $u_{[0, \infty)}$ and partly by the initial state at $t_0 = 0$. If the initial state $\mathbf{x}(0)$ is $\mathbf{0}$, then the output $y_{[0, \infty)}$ will be excited exclusively by the applied input $u_{[0, \infty)}$. This part of the response is called the *zero-state response*. The input-output description of a system, as the name indicates, gives only the relationship between the input and output. Hence it describes only the zero-state response of the system. For convenience, we call a system *initially relaxed* if its initial state is zero. For example, if the initial conditions of the network in Figure 2-4 are zero, then the network is initially relaxed. Since the input-output description describes only the zero-state response, hence *whenever the input-output equation is used to describe a system, the system is implicitly assumed to be initially relaxed.*

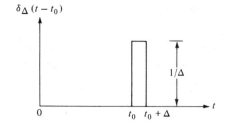

Figure 2-11
A pulse, $\delta_\Delta(t - t_0)$.

Impulse response. The input u and the output y of *any* linear time-invariant system that is initially relaxed at $t_0 = 0$ can be described by an equation of the form

$$y(t) = \int_0^t g(t - \tau)u(\tau)\, d\tau = \int_0^t g(\tau)u(t - \tau)\, d\tau \qquad (2\text{-}6)$$

This is called a *convolution integral*. The second integration is obtained from the first one by substituting $\tau' = t - \tau$ and then changing the dummy variable τ' to τ. The function $g(t)$ is defined only for $t \geq 0$ and is called the *impulse response* of the system. In order to derive Equation (2-6) and give an interpretation of g, we need the concept of impulse function.

Consider the unit pulse function $\delta_\Delta(t - t_0)$ defined in Figure 2-11; that is,

$$\delta_\Delta(t - t_0) \triangleq \begin{cases} \dfrac{1}{\Delta} & \text{for } 0 < (t - t_0) < \Delta \\ 0 & \text{otherwise} \end{cases} \qquad (2\text{-}7)$$

Note that $\delta_\Delta(t - t_0)$ has unit area for all $\Delta > 0$. As Δ approaches zero, the limiting "function"

$$\delta(t - t_0) \triangleq \lim_{\Delta \to 0} \delta_\Delta(t - t_0) \qquad (2\text{-}8)$$

is called the *impulse function*, or *δ-function*. Thus the δ-function $\delta(t - t_0)$ has the properties

$$\int_{-\infty}^{\infty} \delta(t - t_0)\, dt = \int_{t_0 - \varepsilon}^{t_0 + \varepsilon} \delta(t - t_0)\, dt = 1 \qquad (2\text{-}9)$$

for any $\varepsilon > 0$, and

$$\int_{-\infty}^{\infty} f(t)\delta(t - t_0)\, dt = f(t_0) \qquad (2\text{-}10)$$

for any function that is continuous at $t = t_0$.

Consider the input function $u(t)$ shown in Figure 2-12(a). Clearly it can be approximated by the summation of the pulse functions $u_i(t)$ shown in Figure 2-12(b). Note that every pulse function $u_i(t)$ can be expressed in terms of $\delta_\Delta(t)$ as

$$u_i(t) = \delta_\Delta(t - t_i)u(t_i)\Delta$$

2-4 THE INPUT-OUTPUT DESCRIPTION OF SINGLE-VARIABLE SYSTEMS   17

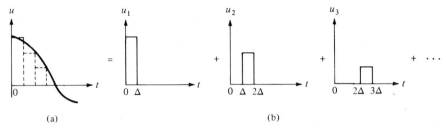

Figure 2-12 Decomposition of a function.

Hence we have

$$u(t) \doteq \sum_{i=0}^{\infty} \delta_\Delta(t - t_i)u(t_i)\Delta \tag{2-11}$$

Now if the output of the linear time-invariant system due to the input $\delta_\Delta(t)$ is denoted by $g_\Delta(t)$, then the properties of linearity and time-invariance imply that the output of the system due to $u(t)$ is given by

$$y(t) \doteq \sum_{i=0}^{\infty} g_\Delta(t - t_i)u(t_i)\Delta \tag{2-12}$$

As Δ approaches zero, the approximation tends to an exact equality, the summation becomes an integration, and the pulse function becomes a δ-function. Hence as $\Delta \to 0$, Equation (2-12) becomes

$$y(t) = \int_0^\infty g(t - \tau)u(\tau)\,d\tau \tag{2-13}$$

where $g(t) \triangleq \lim_{\Delta \to 0} g_\Delta(t)$; that is, $g(t)$ is the response of the impulse function applied at $t = 0$. Hence the function $g(t)$ is called the *impulse response* of the system. Since $g(t)$ is the response due to an input applied at $t = 0$, it is not defined for $t < 0$. By convention, we assume[2] that

$$g(t) = 0 \quad \text{for } t < 0 \tag{2-14}$$

On the basis of this property, Equation (2-13) reduces immediately to Equation (2-6).

A very important property of this input-output description is that it can be obtained from the input and output terminals of a system even if the system's internal structure is not known. If the system is known to be linear, time-invariant, and relaxed at $t = 0$, then by applying a δ-function at the input terminal, the response at the output terminal gives us immediately the impulse response of the system. If the impulse response is known, the output of the system due to any input can be computed from Equation (2-6).

---

[2] This, in fact, follows from the property of causality. Since causality is an intrinsic property of every physical system, this concept will not be introduced.

Transfer function. In the study of linear time-invariant systems, it is of great advantage to use the Laplace transform (see Appendix A). We use the circumflex (^) over a variable to denote the Laplace transform of the variable; for example,

$$\hat{g}(s) \triangleq \mathscr{L}[g(t)] \triangleq \int_0^\infty g(t)e^{-st}\,dt \qquad (2\text{-}15)$$

The application of the Laplace transform to (2-6) yields

$$\hat{y}(s) = \int_0^\infty y(t)e^{-st}\,dt = \int_0^\infty \int_0^t g(t-\tau)u(\tau)\,d\tau e^{-st}\,dt \qquad (2\text{-}16)$$

Since $g(t-\tau) = 0$ for $\tau > t$ as implied by (2-14), the upper limit t of the integration in (2-16) can be set at ∞. Hence Equation (2-16) becomes, after the change of the order of integrations,

$$\hat{y}(s) = \int_0^\infty \int_0^\infty g(t-\tau)e^{-s(t-\tau)}\,dt\,u(\tau)e^{-s\tau}\,d\tau \qquad (2\text{-}17)$$

which, if we substitute $v = t - \tau$, can be written

$$\hat{y}(s) = \int_0^\infty \int_{-\tau}^\infty g(v)e^{-sv}\,dv\,u(\tau)e^{-s\tau}\,d\tau \qquad (2\text{-}18)$$

We use again the fact that $g(v) = 0$ for $v < 0$ to reduce (2-18) to

$$\hat{y}(s) = \int_0^\infty \int_0^\infty g(v)e^{-sv}\,dv\,u(\tau)e^{-s\tau}\,d\tau$$

$$= \left[\int_0^\infty g(v)e^{-sv}\,dv\right]\left[\int_0^\infty u(\tau)e^{-s\tau}\,d\tau\right]$$

or

$$\hat{y}(s) = \hat{g}(s)\hat{u}(s) \qquad (2\text{-}19)$$

This is an algebraic equation. We see that by applying the Laplace transform, a convolution integral is transformed into an algebraic equation. The function $\hat{g}(s)$ is called the *transfer function* of the system. It is, by definition, the Laplace transform of the impulse response. It can also be computed from Equation (2-19) as

$$\hat{g}(s) = \left.\frac{\mathscr{L}[y(t)]}{\mathscr{L}[u(t)]}\right|_{\text{relaxed at }t=0} = \left.\frac{\hat{y}(s)}{\hat{u}(s)}\right|_{\text{relaxed at }t=0} \qquad (2\text{-}20)$$

The concept of a transfer function is important not only in control systems but also in circuit theory. If u is a current source connected to a circuit and y is the voltage across the source, then the $\hat{g}(s)$ defined in Equation (2-20) is called the *impedance* of the circuit. If u is a voltage source connected to a circuit and y is the current flowing into the circuit, then the $\hat{g}(s)$ defined in (2-20) is called the

2-4 THE INPUT-OUTPUT DESCRIPTION OF SINGLE-VARIABLE SYSTEMS

admittance of the circuit. Since the transfer function is derived from the convolution integral (2-6), whenever a transfer function is used to describe a system, the system is always implicitly assumed to be linear, time-invariant, and relaxed at $t_0 = 0$.

The Laplace transform variable s in a transfer function $\hat{g}(s)$, as will be seen later, can be associated with frequency; hence the transfer-function description of a system is often referred to as in the *frequency domain*, whereas the impulse-response description is said to be in the *time domain*. For the class of systems studied in this book, it is generally easier to derive transfer functions by using (2-20) than to determine impulse responses. Furthermore most design techniques in control theory are developed for transfer functions rather than impulse responses. Hence in the remainder of this book, we deal exclusively with transfer functions. Certainly if the transfer function of a system is known, its impulse response can be obtained by just taking its inverse Laplace transform. (See Section 4-2.)

Example 1

Consider the resistive network shown in Figure 2-5. By loop analysis it is easy to show that $y(t) = \frac{1}{5}u(t)$, or $\hat{y}(s) = \frac{1}{5}\hat{u}(s)$. Hence the transfer function from u to y is $\frac{1}{5}$. Its impulse response is $\frac{1}{5}\delta(t)$, which is obtained by taking the inverse Laplace transform of $\frac{1}{5}$.

Example 2

Consider the system shown in Figure 2-6. Let k_1 be the viscous friction coefficient between the mass and the floor, and let k_2 be the spring constant. Then Newton's law yields

$$m\frac{d^2y}{dt^2} = u - k_1\frac{dy}{dt} - k_2 y$$

Taking the Laplace transform and assuming zero initial conditions, we obtain

$$ms^2\hat{y}(s) = \hat{u}(s) - k_1 s\hat{y}(s) - k_2\hat{y}(s)$$

Hence the transfer function from u to y is

$$\hat{g}(s) = \frac{1}{ms^2 + k_1 s + k_2} \tag{2-21}$$

If $m = 1$, $k_1 = 2$, $k_2 = 1$, then the impulse response of the system is

$$g(t) = \mathscr{L}^{-1}\left[\frac{1}{(s+1)^2}\right] = te^{-t} \quad \text{for } t \geq 0$$

Example 3

Consider the network shown in Figure 2-8. By loop analysis, we obtain, in the frequency domain,

$$\hat{\imath}_1(s)\left(1 + \frac{1}{s}\right) - \hat{\imath}_2(s)\frac{1}{s} = \hat{u}(s)$$

$$\hat{\imath}_1(s)\frac{1}{s} - \hat{\imath}_2(s)\frac{3}{s} = 0$$

Since

$$\hat{y}(s) = \frac{1}{s}\hat{\imath}_2(s)$$

it can be readily verified that the transfer function from u to y is

$$\hat{g}(s) = \frac{\hat{y}(s)}{\hat{u}(s)} = \frac{1}{3s + 2} \qquad (2\text{-}22)$$

■

We introduce the following definition to conclude this section.

Definition 2-5

A rational function is said to be *proper* if the degree of the numerator is equal to or smaller than that of the denominator. A rational function is said to be *strictly proper* if the degree of the numerator is smaller than that of the denominator. ■

The transfer function in Example 1 is proper because the degree of its numerator and denominator are both equal to zero. The transfer functions in Examples 2 and 3 are strictly proper. Most of the transfer functions encountered in this book are proper rational functions.

2-5 The State-Variable Description of Single-Variable Systems

The input-output description, in particular the transfer function, has two drawbacks. First, it is applicable only when a system is initially relaxed. Second, it describes only the input and output relation; no information concerning the internal structure of a system can be abstracted from this description. In order to overcome these drawbacks, we introduce in the following the state-variable description or internal description of systems.

The state **x** of a system is, by definition, the minimum amount of information that, together with the input u, determines uniquely the responses of the system. The set of equations that describes these unique relations among the input,

2-5 THE STATE-VARIABLE DESCRIPTION OF SINGLE-VARIABLE SYSTEMS

the output, and the state will be called a dynamical equation. In this book we study only dynamical equations of the form

$$\begin{cases} \dot{x}_1 = a_{11}x_1 + a_{12}x_2 + \cdots + a_{1n}x_n + b_1 u \\ \dot{x}_2 = a_{21}x_1 + a_{22}n_2 + \cdots + a_{2n}x_n + b_2 u \\ \vdots \\ \dot{x}_n = a_{n1}x_1 + a_{n2}x_2 + \cdots + a_{nn}x_n + b_n u \end{cases} \quad \text{(2-23a)}$$

$$y = c_1 x_1 + c_2 x_2 + \cdots + c_u x_n + du \quad \text{(2-23b)}$$

or in matrix form

$$\dot{\mathbf{x}}(t) = \mathbf{A}\mathbf{x}(t) + \mathbf{b}u(t) \quad \text{(state equation)} \quad \text{(2-24a)}$$

$$y(t) = \mathbf{c}\mathbf{x}(t) + du(t) \quad \text{(output equation)} \quad \text{(2-24b)}$$

with

$$\mathbf{x} = \begin{bmatrix} x_1 \\ x_2 \\ \vdots \\ x_n \end{bmatrix} \quad \mathbf{A} = \begin{bmatrix} a_{11} & a_{12} & \cdots & a_{1n} \\ a_{21} & a_{22} & \cdots & a_{2n} \\ \vdots & \vdots & & \vdots \\ a_{n1} & a_{n2} & \cdots & a_{nn} \end{bmatrix} \quad \mathbf{b} = \begin{bmatrix} b_1 \\ b_2 \\ \vdots \\ b_n \end{bmatrix}$$

$$\mathbf{c} = \begin{bmatrix} c_1 & c_2 & \cdots & c_n \end{bmatrix}$$

where $\mathbf{x}$, u, and y are, respectively, the state, the input, and the output of a system; $\mathbf{A}$ is an $n \times n$ real constant matrix; $\mathbf{b}$ is an $n \times 1$ real column vector; $\mathbf{c}$ is a $1 \times n$ real row vector; and d is a real constant. Equation (2-24a) governs the behavior of the state and is called the *state equation*. Equation (2-24b) gives the output and is called the *output equation*. As we see from (2-23a), the state equation in fact consists of a set of n first-order differential equations with constant coefficients. The set of two equations in (2-24) is often more informatively called an *n-dimensional linear time-invariant dynamical equation*. It can be verified that the input, the state, and the output governed by Equation (2-24) indeed satisfy the linearity and time-invariance properties. Conversely all lumped systems that are linear and time-invariant can be described by equations of the form of (2-24).

Example 1

Consider the resistive network shown in Figure 2-5. There is no state variable in the system. The system can be described by the zero-dimensional linear time-invariant dynamical equation $y(t) = \tfrac{1}{3}u(t)$.

Example 2

Consider the system shown in Figure 2-6. Let us choose the position and velocity of the mass as the state variables; that is, $x_1 = y$, $x_2 = \dot{y} = \dot{x}_1$. Newton's law yields

$$m\ddot{y} = u - k_1 \dot{y} - k_2 y$$

or

$$m\dot{x}_2 = u - k_1 x_2 - k_2 x_1$$

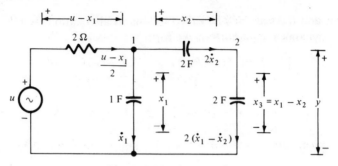

Figure 2-13 A network.

Hence we have

$$\begin{bmatrix} \dot{x}_1 \\ \dot{x}_2 \end{bmatrix} = \begin{bmatrix} 0 & 1 \\ -\dfrac{k_2}{m} & -\dfrac{k_1}{m} \end{bmatrix} \begin{bmatrix} x_1 \\ x_2 \end{bmatrix} + \begin{bmatrix} 0 \\ \dfrac{1}{m} \end{bmatrix} u$$

$$y = \begin{bmatrix} 1 & 0 \end{bmatrix} \begin{bmatrix} x_1 \\ x_2 \end{bmatrix}$$

This is a two-dimensional linear time-invariant dynamical equation.

Example 3

Consider the network shown in Figure 2-13. As discussed in Example 4 of Section 2-2, any two of the three capacitor voltages qualify as the state variables. Let us choose x_1 and x_2 as state variables. Now we express the characteristics of each branch in terms of x_1 and x_2 as shown. The application of Kirchhoff's current law to nodes 1 and 2 yields

$$\frac{u - x_1}{2} = \dot{x}_1 + 2\dot{x}_2$$

and

$$2\dot{x}_2 = 2(\dot{x}_1 - \dot{x}_2)$$

These two equations can be arranged, after manipulation, into matrix form as

$$\begin{bmatrix} \dot{x}_1 \\ \dot{x}_2 \end{bmatrix} = \begin{bmatrix} -\tfrac{1}{4} & 0 \\ -\tfrac{1}{8} & 0 \end{bmatrix} \begin{bmatrix} x_1 \\ x_2 \end{bmatrix} + \begin{bmatrix} \tfrac{1}{4} \\ \tfrac{1}{8} \end{bmatrix} u$$

$$y = \begin{bmatrix} 1 & -1 \end{bmatrix} \begin{bmatrix} x_1 \\ x_2 \end{bmatrix}$$

This is a two-dimensional linear time-invariant dynamical equation. If x_2 and x_3 are chosen as state variables, then we have (see Problem 2-7)

$$\begin{bmatrix} \dot{x}_2 \\ \dot{x}_3 \end{bmatrix} = \begin{bmatrix} -\frac{1}{8} & -\frac{1}{8} \\ -\frac{1}{8} & -\frac{1}{8} \end{bmatrix} \begin{bmatrix} x_2 \\ x_3 \end{bmatrix} + \begin{bmatrix} \frac{1}{8} \\ \frac{1}{8} \end{bmatrix} u \quad (2\text{-}25a)$$

$$y = \begin{bmatrix} 0 & 1 \end{bmatrix} \begin{bmatrix} x_2 \\ x_3 \end{bmatrix} \quad (2\text{-}25b)$$

This dynamical equation also describes the network in Figure 2-13. ∎

As shown in the foregoing example, in the state-variable description, depending on the state variables chosen, we may have different dynamical equations to describe the same system. This situation will however never occur in the input-output description; no matter which analysis method is used, we always obtain the same transfer function. As in the input-output description, we may apply the Laplace transform to a dynamical equation. The application of the Laplace transform to (2-24) yields

$$s\hat{\mathbf{x}}(s) - \mathbf{x}(0) = \mathbf{A}\hat{\mathbf{x}}(s) + \mathbf{b}\hat{u}(s) \quad (2\text{-}26a)$$

$$\hat{y}(s) = \mathbf{c}\hat{\mathbf{x}}(s) + d\hat{u}(s) \quad (2\text{-}26b)$$

or

$$\hat{\mathbf{x}}(s) = \underbrace{(s\mathbf{I} - \mathbf{A})^{-1}\mathbf{x}(0)}_{\text{zero-input response}} + \underbrace{(s\mathbf{I} - \mathbf{A})^{-1}\mathbf{b}\hat{u}(s)}_{\text{zero-state response}} \quad (2\text{-}27)$$

$$\hat{y}(s) = \mathbf{c}(s\mathbf{I} - \mathbf{A})^{-1}\mathbf{x}(0) + [\mathbf{c}(s\mathbf{I} - \mathbf{A})^{-1}\mathbf{b} + d]\hat{u}(s) \quad (2\text{-}28)$$

They are algebraic equations. We see that the responses can be decomposed into the zero-input response (the response due to a nonzero initial state) and the zero-state response (the response due to a nonzero input). This is not surprising, since the decomposition is a general property of every linear system, as indicated in Equation (2-5).

Unlike the input-output description where the frequency-domain description (transfer function) is mainly used, in the state-variable description the frequency-domain equations, (2-27) and (2-28), are hardly used. Most of the analysis and design are carried out by using the time-domain equation, (2-24).

Every linear time-invariant lumped system has both the input-output description and the state-variable description. It is natural to establish their relationship. Letting $\mathbf{x}(0) = \mathbf{0}$ and comparing (2-19) and (2-28), we obtain immediately

$$\hat{g}(s) = \mathbf{c}(s\mathbf{I} - \mathbf{A})^{-1}\mathbf{b} + d \quad (2\text{-}29)$$

This is the link between the two descriptions. If $\mathbf{A}$, $\mathbf{b}$, $\mathbf{c}$, and d are known, the transfer function $\hat{g}(s)$ can be computed directly from (2-29). Conversely, given a $\hat{g}(s)$, it is possible to find $\mathbf{A}$, $\mathbf{b}$, $\mathbf{c}$, and d to satisfy (2-29). This is called the *realization problem* and will be studied in Chapter 5.

2-6 Mathematical Descriptions of Multivariable Systems

The input-output description and the state-variable description of single-variable systems introduced in the preceding sections will be extended in this section to multivariable systems. We recall that a system is called *single variable* if there is only one input terminal and only one output terminal. By contrast, a system that has two or more input terminals and/or two or more output terminals is called a *multivariable system*. Consider the multivariable system with p input terminals and q output terminals shown in Figure 2-14. The inputs are denoted by $u_1, u_2, \ldots, u_p$, and the outputs by $y_1, y_2, \ldots, y_q$. For conciseness, we use the vectors

$$\mathbf{u} \triangleq \begin{bmatrix} u_1 \\ u_2 \\ \vdots \\ u_p \end{bmatrix} \quad \text{and} \quad \mathbf{y} \triangleq \begin{bmatrix} y_1 \\ y_2 \\ \vdots \\ y_q \end{bmatrix} \tag{2-30}$$

to denote the set of inputs and the set of outputs. If the multivariable system is linear, time-invariant, and relaxed at $t = 0$, then its input and output, as in Equation (2-19), can be related by

$$\hat{\mathbf{y}}(s) = \hat{\mathbf{G}}(s)\hat{\mathbf{u}}(s) \tag{2-31}$$

with

$$\hat{\mathbf{G}}(s) = \begin{bmatrix} \hat{g}_{11}(s) & \hat{g}_{12}(s) & \cdots & \hat{g}_{1p}(s) \\ \hat{g}_{21}(s) & \hat{g}_{22}(s) & \cdots & \hat{g}_{2p}(s) \\ \vdots & \vdots & & \vdots \\ \hat{g}_{q1}(s) & \hat{g}_{q2}(s) & \cdots & \hat{g}_{qp}(s) \end{bmatrix} \tag{2-32}$$

where $\hat{g}_{ij}(s)$ is the transfer function from the jth input terminal to the ith output terminal, and can be computed from

$$\hat{g}_{ij}(s) = \frac{\hat{y}_i(s)}{\hat{u}_j(s)} \bigg|_{\substack{\text{relaxed at } t=0 \\ \hat{u}_k(s) = 0 \text{ for } k \neq j}} \tag{2-33}$$

In computing $\hat{g}_{ij}(s)$, the output $\hat{y}_i(s)$ should be excited solely by $\hat{u}_j(s)$; hence it is required that $\hat{u}_k(s) = 0$ for $k \neq j$. Clearly $\hat{\mathbf{G}}(s)$ is a $q \times p$ matrix; it is called the *transfer-function matrix* of the system.

The extension of Equation (2-24) to the multivariable case takes the form

$$\dot{\mathbf{x}} = \mathbf{A}\mathbf{x} + \mathbf{B}\mathbf{u} \quad \text{(state equation)} \tag{2-34a}$$

$$\mathbf{y} = \mathbf{C}\mathbf{x} + \mathbf{D}\mathbf{u} \quad \text{(output equation)} \tag{2-34b}$$

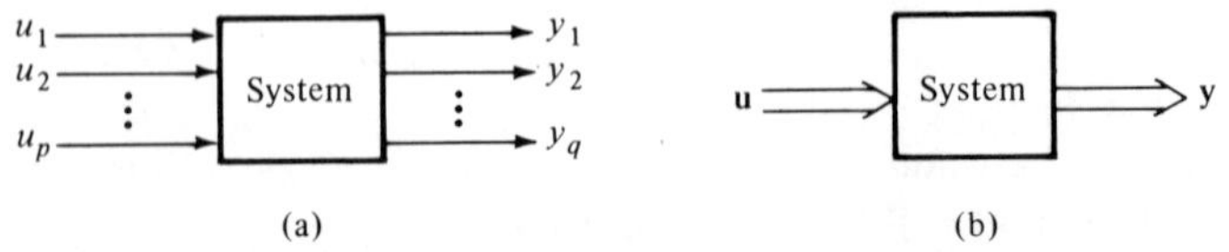

Figure 2-14 A multivariable system.

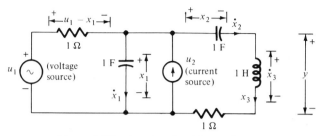

Figure 2-15 A network with two inputs.

where

$$\mathbf{A} = \begin{bmatrix} a_{11} & a_{12} & \cdots & a_{1n} \\ a_{21} & a_{22} & \cdots & a_{2n} \\ \vdots & \vdots & & \vdots \\ a_{n1} & a_{n2} & \cdots & a_{nn} \end{bmatrix} \quad \mathbf{B} = \begin{bmatrix} b_{11} & b_{12} & \cdots & b_{1p} \\ b_{21} & b_{22} & \cdots & b_{2p} \\ \vdots & \vdots & & \vdots \\ b_{n1} & b_{n2} & \cdots & b_{np} \end{bmatrix}$$

$$\mathbf{C} = \begin{bmatrix} c_{11} & c_{12} & \cdots & c_{1n} \\ c_{21} & c_{22} & \cdots & c_{2n} \\ \vdots & \vdots & & \vdots \\ c_{q1} & c_{q2} & \cdots & c_{qn} \end{bmatrix} \quad \mathbf{D} = \begin{bmatrix} d_{11} & d_{12} & \cdots & d_{1p} \\ d_{21} & d_{22} & \cdots & d_{2p} \\ \vdots & \vdots & & \vdots \\ d_{q1} & d_{q2} & \cdots & d_{qp} \end{bmatrix}$$

Clearly, $\mathbf{A}$, $\mathbf{B}$, $\mathbf{C}$, and $\mathbf{D}$ are, respectively, $n \times n$, $n \times p$, $q \times n$, and $q \times p$ real constant matrices. Equation (2-34) is again called an *n-dimensional linear time-invariant dynamical equation*. If a multivariable system has the transfer-function description (2-31) and the state-variable description (2-34), then they are related by, as in (2-29),

$$\hat{\mathbf{G}}(s) = \mathbf{C}(s\mathbf{I} - \mathbf{A})^{-1}\mathbf{B} + \mathbf{D} \tag{2-35}$$

Example

Consider the network shown in Figure 2-15. There are two inputs: one voltage source and one current source. The output is the voltage across the inductor. Hence its transfer-function matrix is a 1×2 matrix of the form $\mathbf{G}(s) = [\hat{g}_{11}(s) \; \hat{g}_{12}(s)]$. In computing $\hat{g}_{11}(s)$, we assume $u_2 \equiv 0$, and the network reduces to the one in Figure 2-16(a). By loop analysis, we obtain

$$\left(1 + \frac{1}{s}\right)\hat{i}_1(s) - \frac{1}{s}\hat{i}_2(s) = \hat{u}_1(s)$$

$$-\frac{1}{s}\hat{i}_1(s) + \left(1 + s + \frac{2}{s}\right)\hat{i}_2(s) = 0$$

Solving for $\hat{i}_2(s)$, and using $\hat{y}(s) = s\hat{i}_2(s)$, $\hat{g}_{11}(s)$ can be computed as

$$\hat{g}_{11}(s) = \frac{\hat{y}(s)}{\hat{u}_1(s)}\bigg|_{\substack{\hat{u}_2(s)=0 \\ \text{initially relaxed}}} = \frac{s\hat{i}_2(s)}{\hat{u}_1(s)} = \frac{s^2}{s^3 + 2s^2 + 3s + 1}$$

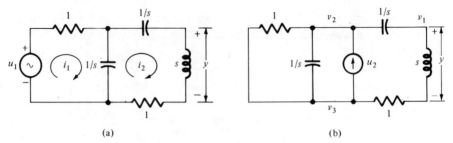

Figure 2-16 Two reduced networks.

In computing $\hat{g}_{12}(s)$, we assume $u_1 \equiv 0$, and the network reduces to the one in Figure 2-16(b). By node analysis, we obtain

$$\left(\frac{1}{s} + s\right)\hat{v}_1(s) - s\hat{v}_2(s) = 0$$

$$-s\hat{v}_1(s) + (1 + 2s)\hat{v}_2(s) - (1 + s)\hat{v}_3(s) = \hat{u}_2(s)$$

$$-(1 + s)\hat{v}_2(s) + (2 + s)\hat{v}_3(s) = -\hat{u}_2(s)$$

Solving for $\hat{v}_1(s)$ yields

$$\hat{g}_{12}(s) = \frac{\hat{y}(s)}{\hat{u}_2(s)}\bigg|_{\substack{\hat{u}_1(s)=0 \\ \text{initially relaxed}}} = \frac{s^2}{s^3 + 2s^2 + 3s + 1}$$

Hence the input-output description in the frequency domain of the network is

$$\hat{y}(s) = [\hat{g}_{11}(s) \quad \hat{g}_{12}(s)]\begin{bmatrix}\hat{u}_1(s) \\ \hat{u}_2(s)\end{bmatrix}$$

$$= \left[\frac{s^2}{s^3 + 2s^2 + 3s + 1} \quad \frac{s^2}{s^3 + 2s^2 + 3s + 1}\right]\begin{bmatrix}\hat{u}_1(s) \\ \hat{u}_2(s)\end{bmatrix} \quad (2\text{-}36)$$

We develop in the following the state-variable description of the network. Let us choose the capacitor voltages and the inductor current as state variables as shown in Figure 2-15. The application of Kirchhoff's current and voltage laws yields

$$\dot{x}_2 = x_3$$
$$(u_1 - x_1) - \dot{x}_1 + u_2 + \dot{x}_2 = 0$$
$$\dot{x}_3 + x_3 - x_1 + x_2 = 0$$
$$y = \dot{x}_3 = x_1 - x_2 - x_3$$

They can be arranged in matrix form as

$$\begin{bmatrix}\dot{x}_1 \\ \dot{x}_2 \\ \dot{x}_3\end{bmatrix} = \begin{bmatrix}-1 & 0 & -1 \\ 0 & 0 & 1 \\ 1 & -1 & -1\end{bmatrix}\begin{bmatrix}x_1 \\ x_2 \\ x_3\end{bmatrix} + \begin{bmatrix}1 & 1 \\ 0 & 0 \\ 0 & 0\end{bmatrix}\begin{bmatrix}u_1 \\ u_2\end{bmatrix} \quad (2\text{-}37\text{a})$$

$$y = [1 \quad -1 \quad -1]\mathbf{x} + [0 \quad 0]\mathbf{u} \quad (2\text{-}37\text{b})$$

This is the state-variable description of the network.

As mentioned earlier, the input-output description can also be obtained from the state-variable description by using Equation (2-35). We shall do so for this example. Let us compute

$$\mathbf{C}(s\mathbf{I} - \mathbf{A})^{-1}\mathbf{B} + \mathbf{D} = \begin{bmatrix} 1 & -1 & -1 \end{bmatrix} \begin{bmatrix} s+1 & 0 & 1 \\ 0 & s & -1 \\ -1 & 1 & s+1 \end{bmatrix}^{-1} \begin{bmatrix} 1 & 1 \\ 0 & 0 \\ 0 & 0 \end{bmatrix}$$

$$= \begin{bmatrix} 1 & -1 & -1 \end{bmatrix} \frac{1}{s^3 + 2s^2 + 3s + 1} \begin{bmatrix} s^2 + s + 1 & 1 & s \\ 1 & s^2 + 2s + 2 & s+1 \\ s & -(s+1) & s^2 + s \end{bmatrix}$$

$$\times \begin{bmatrix} 1 & 1 \\ 0 & 0 \\ 0 & 0 \end{bmatrix}$$

$$= \frac{1}{s^3 + 2s^2 + 3s + 1} \begin{bmatrix} s^2 & s^2 \end{bmatrix}$$

This is equal to (2-36), as expected. ∎

2-7 Complete Characterization

A linear time-invariant system can be described by a dynamical equation or by a transfer function. As discussed earlier, a dynamical equation describes not only the input-output relation but also the internal structure of a system and is applicable whether the system is initially relaxed or not. A transfer function, on the other hand, describes only the relationship between the input and output terminals of a system and is applicable only when the system is initially relaxed. Hence when a system is described by a transfer function, the transfer function will not reveal what will happen if the system is not initially relaxed, nor will it reveal the behavior inside the system.

Example 1

Consider the network shown in Figure 2-17. The input u is a current source; the output is the voltage y as shown. If the initial voltage across the capacitor is zero, the current will divide evenly in the two branches; that is, $i_1 = u/2$. Hence the transfer function from u to y is 1.

We study now the behavior of the network with nonzero initial conditions. Let x_1 be the voltage across the capacitor. From loop equations we obtain

$$i_1 = \frac{u + x_1}{2} \tag{2-38}$$

$$\dot{x}_1 = x_1 \tag{2-39}$$

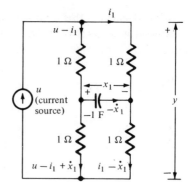

Figure 2-17
A network that is not completely characterized by its transfer function, 1.

It can be easily verified that the solution of Equation (2-39) is $x_1(t) = e^t x_1(0)$. Hence if $x_1(0) \neq 0$, the voltage across the capacitor will increase exponentially with time. Consequently so will the voltage across each resistor. This phenomenon however can never be detected from the transfer function. ∎

We see from this example that the transfer function is not a very good mathematical description of a system. The reason for this difficulty is that the capacitor cannot be excited by the input or detected from the output. In the following we shall develop a criterion to detect this kind of difficulty. Before proceeding we need some preliminary concepts.

Definition 2-6

A rational function is said to be *irreducible* if and only if there is no common factor (except a constant) between the numerator and the denominator. ∎

For example, the rational function $(s^2 + 2s + 1)/(s^2 + 3s + 2)$ is not irreducible because the numerator and the denominator have a common factor, $(s + 1)$. The rational function $(s + 1)/(s + 2)$ however is irreducible. Every rational function can be reduced to an irreducible one by canceling out the common factor; hence we shall assume throughout this book that every rational function is irreducible.

We recall that a rational function is said to be proper if the degree of the numerator is smaller than or equal to that of the denominator. We now introduce the concept of degree for proper rational functions. The *degree* of a proper (irreducible) rational function is defined as the degree of its denominator. Without the irreducibility assumption, the degree of a proper rational function is not well defined. For example, the degree of the rational function

$$\hat{g}(s) = \frac{s + 1}{s^2 + 3s + 2} = \frac{1}{s + 2}$$

is 1, not 2. The concept of degree will now be extended to the vector case. Consider the vector rational function

$$\hat{\mathbf{G}}(s) = \begin{bmatrix} \hat{g}_1(s) \\ \hat{g}_2(s) \\ \vdots \\ \hat{g}_q(s) \end{bmatrix} \quad \text{or} \quad \hat{\mathbf{G}}(s) = [\hat{g}_1(s) \quad \hat{g}_2(s) \quad \cdots \quad \hat{g}_p(s)]$$

The matrix $\hat{\mathbf{G}}(s)$ is called a *proper rational* matrix if *every* $\hat{g}_i(s)$ is a proper rational function; $\hat{\mathbf{G}}(s)$ is said to be *irreducible* if every $\hat{g}_i(s)$ is irreducible.

Definition 2-7

The *characteristic denominator* of a proper irreducible vector rational function $\hat{\mathbf{G}}(s)$ is defined as the *least common denominator* of all entries of $\hat{\mathbf{G}}(s)$. The degree of $\hat{\mathbf{G}}(s)$, denoted by $\delta \hat{\mathbf{G}}(s)$, is defined as the degree of the characteristic denominator of $\hat{\mathbf{G}}(s)$. ∎

Example 2

Consider the 1 × 3 rational matrix

$$\hat{\mathbf{G}}(s) = \begin{bmatrix} \dfrac{1}{(s+1)^2} & \dfrac{1}{s-1} & \dfrac{s+3}{(s+1)(s-1)^2} \end{bmatrix}$$

Its characteristic denominator is $(s+1)^2(s-1)^2$. Hence the degree of $\hat{\mathbf{G}}(s)$ is 4. ∎

We observe that if $\hat{\mathbf{G}}(s)$ is a scalar (1 × 1) rational function, then Definition 2-7 reduces to the usual definition of denominator. The concepts of characteristic denominator and degree can also be defined for general rational matrices. They are defined in terms of the minors of a rational matrix. The interested reader is referred to Reference [1]. With the concept of degree, we are ready to introduce the following definition.

Definition 2-8

A system is said to be *completely characterized* by its proper (irreducible) rational transfer function matrix $\hat{\mathbf{G}}(s)$ if the degree of $\hat{\mathbf{G}}(s)$ is equal to the number of the state variables of the system. ∎

The state is defined as the minimum set of initial conditions; hence the number of state variables in a system is unique. Therefore there is no ambiguity in this definition. According to this definition, the network in Figure 2-17 is not completely characterized by its transfer function, because the degree of the transfer function is 0, whereas the number of the state variable is 1. The system shown in Figure 2-15 has three state variables; its vector transfer function is computed in (2-36) and has degree 3. Hence the system is completely characterized by its vector transfer function. Similarly it can be verified that the systems in Figures 2-5 and 2-6 are completely characterized by their transfer functions. On the other hand, the system in Figure 2-8 is not completely characterized by its transfer function computed in (2-22).

As discussed in Example 1, a transfer function does not necessarily reveal the internal behavior of the system or the behavior due to nonzero initial conditions. However if a system is completely characterized by its transfer function, then it can be shown that the responses inside the system and the responses due to a nonzero initial state will appear as parts of the responses at the output terminals. Hence the behavior inside the system is, more or less, determinable from the input and output terminals. In other words, *if a system is completely characterized by its transfer function, then no essential information is lost in the transfer function. Hence in this case no difficulty or unexpected result will occur in the analysis and design of systems by the use of transfer functions.*

The concept of complete characterization is defined for transfer functions. We shall now introduce the equivalent concepts in dynamical equations. Consider the *n*-dimensional linear time-invariant dynamical equation

$$\dot{\mathbf{x}} = \mathbf{A}\mathbf{x} + \mathbf{B}\mathbf{u} \qquad (2\text{-}40\text{a})$$

$$\mathbf{y} = \mathbf{C}\mathbf{x} + \mathbf{D}\mathbf{u} \qquad (2\text{-}40\text{b})$$

where $\mathbf{x}$, $\mathbf{u}$, and $\mathbf{y}$ are, respectively, the state, the input, and the output; $\mathbf{A}$, $\mathbf{B}$, $\mathbf{C}$, and $\mathbf{D}$ are, respectively, $n \times n$, $n \times p$, $q \times n$, and $q \times p$ real constant matrices. Equation (2-40) is said to be *controllable*[3] if we can transfer any state to any other state in a finite time by the application of an input. The equation is said to be *observable* if we can determine the initial state from the knowledge of the input and the output over a finite time interval. For example, if the initial state of the system in Figure 2-17 is zero, no matter what input we apply, the state 0 cannot be transferred to another value. Hence the dynamical equation describing the system is not controllable. The output of the system in Figure 2-17 is always equal to the input, no matter what the initial state is; hence we are not able to determine its initial state from the knowledge of the input and the output. Therefore we may conclude that the dynamical equation describing the system is not observable.

The controllability and observability of Equation (2-40) are the properties of matrices $\mathbf{A}$, $\mathbf{B}$, and $\mathbf{C}$. The matrix $\mathbf{D}$ does not play any role here. Define the composite matrix

$$\mathbf{U} \triangleq [\mathbf{B} \;\vdots\; \mathbf{AB} \;\vdots\; \cdots \;\vdots\; \mathbf{A}^{n-1}\mathbf{B}] \qquad (2\text{-}41)$$

Note that it consists of n blocks of matrices of form $\mathbf{A}^i\mathbf{B}$, each of dimension $n \times p$. Hence the matrix $\mathbf{U}$ has dimension $n \times np$ and is called the *controllability*

---

[3] The concepts of controllability and observability are not really required in the understanding of the remainder of this text; hence they are introduced only briefly.

matrix. It can be shown that if the controllability matrix **U** has rank n, then Equation (2-40) is controllable; otherwise it is not controllable. See Reference [1]. Similarly we define

$$\mathbf{V} \triangleq \begin{bmatrix} \mathbf{C} \\ \mathbf{CA} \\ \vdots \\ \mathbf{CA}^{n-1} \end{bmatrix} \qquad (2\text{-}42)$$

The matrix **V** has dimension $qn \times n$, and is called the *observability matrix*. If the observability matrix **V** has rank n, then Equation (2-40) is observable; otherwise it is not observable. We note that the controllability property depends only on matrices **A** and **B**, whereas the observability property depends only on **A** and **C**.

Example 3

Consider the dynamical equation

$$\dot{\mathbf{x}} = \begin{bmatrix} 0 & 0 & 1 \\ 0 & 2 & 0 \\ 0 & 0 & 0 \end{bmatrix} \mathbf{x} + \begin{bmatrix} 1 & 2 \\ 1 & 1 \\ 0 & 0 \end{bmatrix} \mathbf{u}$$

$$y = \begin{bmatrix} 1 & 1 & 0 \end{bmatrix} \mathbf{x}$$

Its controllability matrix is

$$\mathbf{U} = \begin{bmatrix} 1 & 2 & 0 & 0 & 0 & 0 \\ 1 & 1 & 2 & 2 & 4 & 4 \\ 0 & 0 & 0 & 0 & 0 & 0 \end{bmatrix}$$

It has a rank smaller than 3. Hence the equation is not controllable. Its observability matrix is

$$\mathbf{V} = \begin{bmatrix} 1 & 1 & 0 \\ 0 & 2 & 1 \\ 0 & 4 & 0 \end{bmatrix}$$

which has a rank 3. Hence the equation is observable. ∎

With these two concepts, we are ready to discuss the implication of complete characterization in a dynamical equation. Suppose that a system has both the transfer-function description and the dynamical-equation description. If the system is completely characterized by its transfer function, then the dynamical equation must be controllable and observable. Conversely, if the dynamical equation is known to be controllable and observable, then the system must be completely characterized by its transfer function. Hence the concept of complete characterization in the input-output description is equivalent to the concepts of controllability and observability in the state-variable description.

2-8 Remarks and Review Questions

In analytical study, we deal mainly with models of physical systems. Hence modeling is a very important problem in engineering. The way to develop a proper model for a system often requires special knowledge and is best studied in a course dealing with the system.

How do you define linearity? Time-invariance?

Is it true that the response of any system can be decomposed into the zero-state response and the zero-input response?

Is the assignment of state variables to a system unique?

We have introduced in this chapter four mathematical equations to describe linear time-invariant lumped systems. These equations are listed in Table 2-1.

Table 2-1 Mathematical Description of Linear Time-Invariant Lumped Systems

	Input-output description	State-variable description
Time domain	$\mathbf{y}(t) = \int_0^t \mathbf{G}(t - \tau)\mathbf{u}(\tau)\,d\tau$	$\dot{\mathbf{x}} = \mathbf{A}\mathbf{x} + \mathbf{B}\mathbf{u}$ $\mathbf{y} = \mathbf{C}\mathbf{x} + \mathbf{D}\mathbf{u}$
Frequency domain	$\hat{\mathbf{y}}(s) = \hat{\mathbf{G}}(s)\hat{\mathbf{u}}(s)$ $\hat{\mathbf{G}}(s) = \mathscr{L}[\mathbf{G}(t)]$ $\hat{\mathbf{G}}(s) = \mathbf{C}(s\mathbf{I} - \mathbf{A})^{-1}\mathbf{B} + \mathbf{D}$	$\hat{\mathbf{x}}(s) = (s\mathbf{I} - \mathbf{A})^{-1}\mathbf{x}(0)$ $\quad + (s\mathbf{I} - \mathbf{A})^{-1}\mathbf{B}\hat{\mathbf{u}}(s)$ $\hat{\mathbf{y}}(s) = \mathbf{C}(s\mathbf{I} - \mathbf{A})^{-1}\mathbf{x}(0)$ $\quad + [\mathbf{C}(s\mathbf{I} - \mathbf{A})^{-1}\mathbf{B} + \mathbf{D}]\hat{\mathbf{u}}(s)$

Both the input-output description and the state-variable description are useful in practice. However in the input-output description the frequency-domain equation is mainly used, whereas in the state-variable description the time-domain equation is mainly used. Hence in the engineering literature the frequency-domain approach is often referred to transfer functions, and the time-domain approach is referred to state-variable equations.

A dynamical equation actually consists of a set of first-order differential equations. One may wonder why we employ a set of first-order differential equations rather than higher-order differential equations. The reasons are manifold. First, every high-order differential equation can be reduced to a set of first-order differential equations. Second, the notations used to describe first-order equations are compact and simple. Finally, first-order differential equations, as will be shown later, can be readily simulated on a digital or an analog computer.

What is the Laplace transform of $\dot{x} = ax$ if $a = 2$? If $a = 2t$? Do you know now why the Laplace transform is not used in the time-varying system?

A lumped system is defined as one with a finite number of state variables. A different but equivalent definition is that it has a rational transfer function.

Are the input-output description and the state-variable description of a linear-invariant lumped system always equivalent? Which description is more general? Why?

What problem may you encounter in using a transfer function to describe a system?

What is the degree of a proper rational function? What is the degree of a vector proper rational matrix? Is this concept useful in engineering?

References

[1] Chen, C. T., *Introduction to Linear Systems Theory*. New York: Holt, Rinehart and Winston, 1970.
[2] Cooper, G. R., and C. D. McGillem, *Methods of Signal and System Analysis*. New York: Holt, Rinehart and Winston, 1967.
[3] Desoer, C. A., and E. S. Kuh, *Basic Circuit Theory*. New York: McGraw-Hill, 1969.
[4] Lynch, W. A., and J. G. Truxal, *Signals and Systems in Electrical Engineering*. New York: McGraw-Hill, 1962.
[5] Shearer, J. L., A. T. Murphy, and H. H. Richardson, *Introduction to Systems Dynamics*. Reading, Mass.: Addison-Wesley, 1967.

Problems

2-1 Consider the memoryless systems with characteristics shown in Figure P2-1, in which u denotes the input and y the output. Which of them is a linear system?

2-2 Consider the system with characteristic shown in Figure P2-1(b). If you consider $\bar{y} \triangleq (y - y_0)$ as the output, is the system linear?

2-3 Consider the pendulum system shown in Figure P2-2, where the applying force u is the input and the angular displacement θ is the output. Choose the state variables for the system. Is the system linear? Is it time-invariant?

2-4 If the angular displacement θ in Problem 2-3 is very small, can you consider the system as linear? Find the transfer function and a dynamical equation to describe the linearized system.

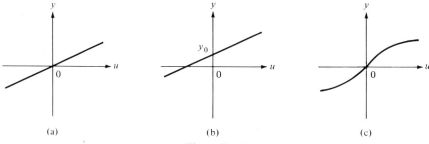

Figure P2-1

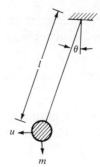

Figure P2-2

2-5 Consider a system that is initially relaxed. If we apply to it a unit step function—that is, $u(t) = 1$ for $t \geq 0$ and $u(t) = 0$ for $t < 0$—the response at the output terminal is measured as $1 - e^{-2t} + e^{-t}$. What is the transfer function of the system? What is the impulse response?

2-6 Is the system shown in Figure P2-3 a linear system? Is it true that if a system is linear, then all of its subsystems must be linear?

2-7 Consider the network shown in Figure 2-13. If x_2 and x_3 are chosen as the state variables, verify that its dynamical-equation description is given by Equation (2-25).

2-8 Consider a linear system with input u and output y. Three experiments are performed on this system using the inputs $u_1(t)$, $u_2(t)$, and $u_3(t)$, for $t \geq 0$. In each case, the initial state at $t = 0$, $\mathbf{x}(0)$, is the same. The corresponding observed outputs are $y_1(t)$, $y_2(t)$, and $y_3(t)$. Which of the following three predictions are true if $\mathbf{x}(0) \neq 0$?

a. If $u_3 = u_1 + u_2$, then $y_3 = y_1 + y_2$.
b. If $u_3 = \frac{1}{2}(u_1 + u_2)$, then $y_3 = \frac{1}{2}(y_1 + y_2)$.
c. If $u_3 = u_1 - u_2$, then $y_3 = y_1 - y_2$.

Which are true if $\mathbf{x}(0) = 0$? (*Answers:* No, yes, no, for $\mathbf{x}(0) \neq 0$; all yes, if $\mathbf{x}(0) = 0$).

2-9 Find the transfer function of the system whose input u and the output y are related by the differential equation

$$\frac{d^3y(t)}{dt^3} + 2\frac{d^2y(t)}{dt^2} + 3y(t) = 3\frac{du(t)}{dt} - 2u(t)$$

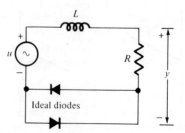

Figure P2-3

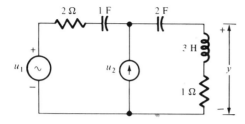

Figure P2-4

2-10 Find the transfer-function matrix and the state-variable description of the system shown in Figure P2-4.

2-11 What is the degree of each of the following proper rational matrices? Which of them are strictly proper?

a. $\hat{g}(s) = \dfrac{s^2 + 1}{2s^3 - 2s + 1}$

b. $\hat{\mathbf{G}}_1(s) = \begin{bmatrix} \dfrac{1}{s-1} & \dfrac{1}{s+1} & \dfrac{s-1}{(s+1)(s+2)} \end{bmatrix}$

c. $\hat{\mathbf{G}}_2(s) = \begin{bmatrix} \dfrac{s+1}{s-1} \\ \dfrac{s-1}{s+1} \\ \dfrac{1}{(s-1)(s+2)} \end{bmatrix}$

2-12 Are the systems in Problems 2-4 and 2-10 completely characterized by their transfer functions?

2-13 Let y_1 be the response of a linear time-invariant system due to the application of a unit step function input. Show that $g(t) = (d/dt)y_1(t)$, where g is the impulse response of the system.

2-14 Consider the network shown in Figure P2-5(a), in which T is a tunnel diode with the characteristic shown. Show that if v lies between a and b, then the circuit can be modeled as shown in Figure P2-5(b). Show also that if v lies between c and d, then the circuit can be modeled as shown in Figure P2-5(c), where $i_1' = i_1 - i_0$, $v' = v - v_0$. These linearized models are often called *linear incremental models*, or *small-signal models*.

2-15 Consider the simplified model of an aircraft shown in Figure P2-6. It is assumed that the aircraft is dynamically equivalent at the pitched angle θ_0, elevator angle u_0, altitude h_0, and cruising speed v_0. It is assumed that small deviations of θ and u from θ_0 and u_0 generate forces $f_1 = k_1\theta$ and $f_2 = k_2 u$, as shown in the figure. Let m be the mass of the aircraft, I the moment of inertia about the center of gravity P, $b\dot{\theta}$ the aerodynamic damping, and h the

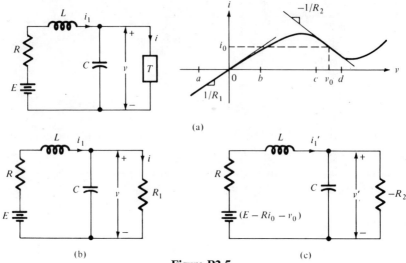

Figure P2-5

deviation of the altitude from h_0. Show that the transfer function from u to h is, by neglecting the effect of I,

$$\hat{g}(s) = \frac{\hat{h}(s)}{\hat{u}(s)} = \frac{k_1 k_2 l_2 - k_2 bs}{ms^2(bs + k_1 l_1)}$$

2-16 Consider a cart with a stick hinged on top of it, as shown in Figure P2-7. This could be a model of a space booster on takeoff. If the cart and the stick can move only on the plane of this paper, what are the state variables of the system? Is this system linear? If the angular displacement of the stick, θ, is very small, then the system can be described by

$$\ddot{\theta} = \theta + u$$
$$\ddot{y} = \beta\theta - u$$

where β is a constant, and u and β are expressed in appropriate units. Find the

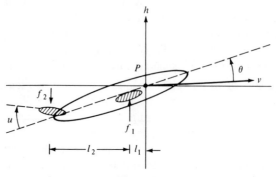

Figure P2-6

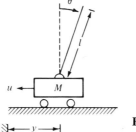

Figure P2-7

transfer function from u to θ. Is the system completely characterized by this transfer function? If we consider θ and y as outputs and u as the input, what is its transfer-function matrix? Is the system completely characterized by this transfer-function matrix?

2-17 The bridged-T network shown in Figure P2-8 can be used as a compensating network in control systems. Find its transfer function and state-variable description.

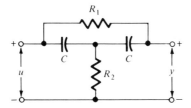

Figure P2-8

2-18 The soft landing phase of a lunar module descending on the moon can be modeled as shown in Figure P2-9. It is assumed that the thrust generated is proportional to $\dot{m}$, where m is the mass of the module. Then the system can be described by $m\ddot{y} = -k\dot{m} - mg$, where g is the gravity constant on the lunar surface. Define the state variables of the system as $x_1 = y$, $x_2 = \dot{y}$, $x_3 = m$, and $u = \dot{m}$. Find the dynamical equation description of the system.

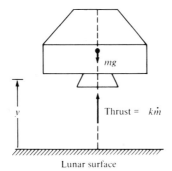

Figure P2-9

CHAPTER

3

Control Components and Systems

3-1 Introduction

Control systems are generally built by the interconnection of various components. These components can be mechanical, electrical, hydraulic, optical, or other devices. In this chapter, some of most commonly used devices and some typical control systems will be introduced.

The design of control systems can be divided into two categories. One concerns the design of better components for control systems; the other involves the design of overall systems by utilizing existing devices. The former category belongs to the domain of instrumentation engineers; the latter belongs to that of system engineers. As system engineers, we shall be mainly concerned with the design of overall systems, and hence the detailed structures of control components will not be studied. We shall instead study only the functions of the devices. In the process of studying them we shall first introduce models for the devices and then develop their mathematical descriptions. The limitation of the models will also be discussed.

The objective of this chapter is twofold: First, to show the reader how the mathematical equations that we shall be dealing with in the remainder of the book arise from physical systems. Second, to demonstrate the practical limitations of the linear models so that the reader will appreciate the importance of taking these limitations into consideration in the design. However we shall not

3-2 AC and DC Devices

A device is often designed to operate in a specified frequency range. For example, a direct-coupled amplifier is designed for very-low-frequency signals, from zero to few cycles per second (hertz, or Hz). An audio-frequency amplifier is designed for signals with frequencies ranging from 20 to 20,000 Hz. Some motors are designed to be driven by direct current, others by 60- or 400-Hz sinusoidal currents. Because of this operational frequency range, control components are classified as dc and ac devices. A dc device handles zero- or very-low-frequency signals, whereas an ac device handles higher-frequency signals. A control system may consist of exclusively ac devices, exclusively dc devices, or a combination of ac and dc devices.

The information-bearing signals in a control system are usually of low frequency, in the range of 0 to 20 Hz. Let $v(t)$ be a control signal defined over the time interval $[0, \infty)$, and let $\hat{v}(s)$ be its Laplace transform. The plot of the magnitude of $\hat{v}(j\omega)$ versus ω is called the *frequency spectrum* of v. As a control signal, the spectrum will be centered around $\omega = 0$, and will rarely go beyond 20 Hz, as shown in Figure 3-1. If this signal is connected to an ac device whose operational frequency range lies outside the signal frequency spectrum, then the device will not be operative, or the signal will not go through. This difficulty however can be overcome by the introduction of modulation. This is achieved by the multiplication of $v(t)$ by $\cos \omega_c t$; that is,

$$v_m(t) = v(t) \cos \omega_c t \qquad (3\text{-}1)$$

where ω_c is at least twice as great as the largest frequency component in the spectrum of v, and v_m is called a *modulated signal*. A plot of v_m is shown in Figure 3-2(a). Note that the envelope of v_m is identical to v. The polarity of v dictates the phase of v_m with respect to that of $\cos \omega_c t$. If v is positive, then v_m and $\cos \omega_c t$ have the same phase; otherwise they have a 180-degree phase

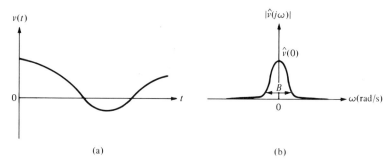

(a)   (b)

Figure 3-1 (a) A time function. (b) Its frequency spectrum.

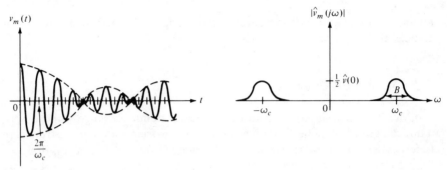

Figure 3-2 (a) The signal in Figure 3-1(a) after modulation. (b) Its frequency spectrum.

difference. Taking the Laplace transform of v_m yields

$$\hat{v}_m(s) = \tfrac{1}{2}[\hat{v}(s + j\omega_c) + \hat{v}(s - j\omega_c)] \tag{3-2}$$

The frequency spectrum of v_m is shown in Figure 3-2(b). It is now centered around ω_c. Hence by a proper choice of ω_c, a modulated control signal can be used to drive an ac device.

A device that transforms signal v into signal v_m is called a *modulator*. Conversely, a device that extracts signal v from v_m is called a *demodulator*. Many devices can function as modulators or demodulators by merely interchanging the input and output terminals. Two such devices are shown in Figure 3-3. The one in Figure 3-3(a) is an electronic device; the one in Figure 3-3(b) is an electromechanical device.

We discuss now the mathematical description of a modulator. If we consider v as the input and v_m in Equation (3-1) as the output, then the mathematical description of the modulator will be rather complicated. Recall that the reason for the introduction of v_m is to match the operational frequency ranges; therefore the information is not stored in v_m itself, but rather in its *envelope*. Hence we should consider v as the input and the envelope of v_m as the output of the modulator. Therefore the transfer function of a modulator is just equal to 1, for the envelope of v_m is identical to v. Similarly, the transfer function from the envelope of the input signal to the output signal of a demodulator is also equal to 1. Hence modulators and demodulators need not be considered in the process of designs of control systems. They merely serve as links between ac devices and dc devices.

3-3 Servomotors

Servomotors are the prime movers in control systems. They are used to move loads such as antennas, valves, the rudder of a ship, or the wing of a sweptwing aircraft. The servomotors that will be discussed in this section are the armature-controlled dc motor, two-phase ac servomotor, and hydraulic servomotor.

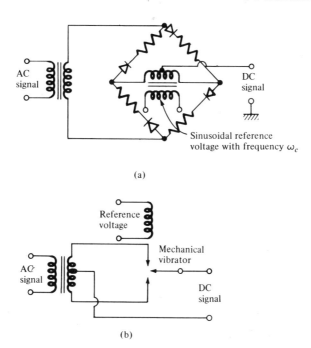

Figure 3-3 Two devices that perform modulation and demodulation.

Armature-controlled dc motor. Most of the dc motors used in control systems are of the separately excited type shown in Figure 3-4. If the armature current i_a is kept constant and the input signal is applied to the field circuit, then the motor is said to be field-controlled. Because of the difficulty in maintaining i_a constant, a field-controlled dc motor is less often used than an armature-controlled dc motor. In an armature-controlled dc motor the input voltage signal is applied to the armature circuit. The field current i_f is kept constant, or the field circuit may be replaced by a permanent magnetic field. We shall now develop the mathematical description of an armature-controlled dc motor, based on the following assumptions:

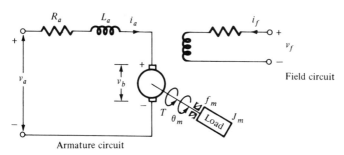

Figure 3-4 A dc motor.

1. The torque T generated on the motor shaft is linearly proportional to the armature current i_a; that is,

$$T = k_t i_a \qquad (3\text{-}3)$$

2. The back electromotive force voltage v_b developed on the armature circuit is linearly proportional to the angular velocity of the motor shaft;

$$v_b = k_b \frac{d\theta_m}{dt} \qquad (3\text{-}4)$$

where θ_m is the angular displacement of the motor shaft.

3. The Coulomb and static frictions are neglected (see Figure 2-10). The torque T is generated to drive the load and to overcome the viscous friction; that is,

$$T = J_m \frac{d^2\theta_m}{dt^2} + f_m \frac{d\theta_m}{dt} \qquad (3\text{-}5)$$

where J_m is the total moment of inertia on the motor shaft, including the inertias of load, rotor, and shaft, and f_m is the viscous friction coefficient.

4. No saturation occurs on the armature circuit if the magnitude of applied signal is smaller than the rated voltage. Hence we have

$$R_a i_a + L_a \frac{di_a}{dt} + v_b = v_a \qquad (3\text{-}6)$$

Under these assumptions the motor can be considered as a linear and time-invariant system. Hence it can be described by a transfer function and a dynamical equation. Taking the Laplace transform and assuming zero initial conditions, we obtain, from Equations (3-3) through (3-6),

$$R_a \hat{i}_a(s) + L_a s \hat{i}_a + k_b s \hat{\theta}_m(s) = \hat{v}_a(s)$$

$$k_t \hat{i}_a(s) = J_m s^2 \hat{\theta}_m(s) + f_m s \hat{\theta}_m(s)$$

The elimination of $\hat{i}_a$ from these equations yields

$$\hat{g}(s) = \frac{\hat{\theta}_m(s)}{\hat{v}_a(s)} = \frac{k_t}{s[(J_m s + f_m)(R_a + L_a s) + k_t k_b]} \qquad (3\text{-}7)$$

This is the transfer function from v_a to θ_m of the armature-controlled dc motor. The dc motor has three state variables, namely i_a, θ_m, and $\dot{\theta}_m$, and its transfer function has degree 3; hence the motor is completely characterized by its transfer function.

We derive now the state-variable description of the motor. Let us choose the state variables as $x_1 = \theta_m$, $x_2 = \dot{\theta}_m = \dot{x}_1$, $x_3 = i_a$. Then Equations (3-5) and (3-6) imply

$$L_a \dot{x}_3 = v_a - k_b x_2 - R_a x_3$$

$$J_m \dot{x}_2 = k_t x_3 - f_m x_2$$

$$\dot{x}_1 = x_2$$

which can be written in matrix form as

$$\begin{bmatrix} \dot{x}_1 \\ \dot{x}_2 \\ \dot{x}_3 \end{bmatrix} = \begin{bmatrix} 0 & 1 & 0 \\ 0 & \dfrac{-f_m}{J_m} & \dfrac{k_t}{J_m} \\ 0 & \dfrac{-k_b}{L_a} & \dfrac{-R_a}{L_a} \end{bmatrix} \begin{bmatrix} x_1 \\ x_2 \\ x_3 \end{bmatrix} + \begin{bmatrix} 0 \\ 0 \\ \dfrac{1}{L_a} \end{bmatrix} v_a \qquad \text{(3-8a)}$$

$$\theta_m = \begin{bmatrix} 1 & 0 & 0 \end{bmatrix} \begin{bmatrix} x_1 \\ x_2 \\ x_3 \end{bmatrix} \qquad \text{(3-8b)}$$

This is the state-variable description of the motor.

The inductance L_a in an armature-controlled dc motor is usually very small. Hence it is often neglected in the analysis. In this case, by assuming $L_a = 0$, Equation (3-7) reduces to

$$\hat{g}(s) = \frac{k_t}{s(J_m R_a s + k_t k_b + f_m R_a)} = \frac{k_m}{s(\tau_m s + 1)} \qquad \text{(3-9)}$$

where

$$k_m \triangleq \frac{k_t}{k_t k_b + f_m R_a} \triangleq \text{dc motor gain constant} \qquad \text{(3-10)}$$

and

$$\tau_m \triangleq \frac{J_m R_a}{k_t k_b + f_m R_a} \triangleq \text{dc motor time constant} \qquad \text{(3-11)}$$

If L_a is assumed to be zero, then there is no more energy-storage element in the armature circuit. Hence the armature current i_a is no longer a state variable. Consequently the state variables of the dc motor are $x_1 = \theta_m$ and $x_2 = \dot{\theta}_m = \dot{x}_1$. Using Equations (3-5) and (3-6), and assuming $L_a = 0$, the state-variable description of the motor with $L_a = 0$ can be obtained as

$$\begin{bmatrix} \dot{x}_1 \\ \dot{x}_2 \end{bmatrix} = \begin{bmatrix} 0 & 1 \\ 0 & \dfrac{-1}{J_m}\left(\dfrac{k_t k_b}{R_a} + f_m\right) \end{bmatrix} \begin{bmatrix} x_1 \\ x_2 \end{bmatrix} + \begin{bmatrix} 0 \\ \dfrac{k_t}{R_a J_m} \end{bmatrix} v_a \qquad \text{(3-12a)}$$

$$y = \begin{bmatrix} 1 & 0 \end{bmatrix} \begin{bmatrix} x_1 \\ x_2 \end{bmatrix} \qquad \text{(3-12b)}$$

AC servomotor. Most of the ac motors used in control systems are of the two-phase induction type shown in Figure 3-5. The reference phase is connected to an ac power supply, either of 60 Hz, 400 Hz, or 1000 Hz. The input signal v is, after modulation, applied to the control phase. Although the input signal is

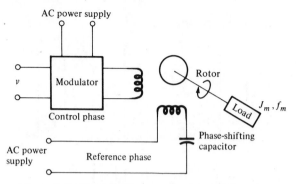

Figure 3-5 Model of a two-phase induction motor.

modulated, the torque generated on the motor shaft is a function of the envelope of the modulated input signal (that is, the original input signal v); hence an ac motor also serves as a demodulator. In order for an ac motor to be operative, the control phase and the reference phase should be 90 degrees out of phase; this can be achieved by the insertion of a capacitor, as shown in Figure 3-5.

A typical torque-speed relationship of a two-phase induction motor is shown in Figure 3-6 by use of dashed lines. In order to obtain a linear model, they are approximated by straight lines as shown and are assumed to be equally spaced. In other words, the generated torque is assumed to be a linear function of the input voltage v. The equation governing the torque-speed lines is clearly of the form

$$T = -k_n \frac{d\theta_m}{dt} + k_c v \tag{3-13}$$

where T is the torque generated on the motor shaft, $d\theta_m/dt$ is the motor shaft speed, v is the input signal, and k_n and k_c are constants to be determined. If we apply the rated voltage, E, of the motor to the control phase (that is, $v = E$), and prevent the motor shaft from rotating ($\theta_m \equiv 0$), then the measured torque

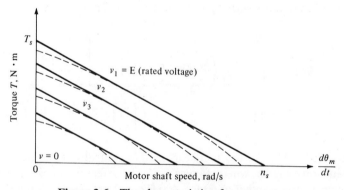

Figure 3-6 The characteristic of an ac motor.

T_s is called the *stall torque*. From the stall torque, k_c can be computed, by substituting $v = E$, $d\theta_m/dt = 0$, and $T = T_s$ into Equation (3-13), as

$$k_c = \frac{T_s}{E} \tag{3-14}$$

If we apply the rated voltage to the control phase (that is, $v = E$), and apply no load to the motor (this implies $T \doteq 0$, if motor inertia and friction are neglected), then the measured motor shaft speed n_s is called the *free speed*. From the free speed, Equations (3-13) and (3-14), k_n can be obtained as

$$k_n = \frac{k_c E}{n_s} = \frac{T_s}{n_s} \tag{3-15}$$

Note that $-k_n$ is just the slope of the straight lines in Figure 3-6. We also note that from the stall torque and free speed, the characteristic of an ac motor can be completely determined from Equations (3-13) through (3-15).

The generated torque is used to drive the load and overcome the viscous friction; hence

$$T = J_m \frac{d^2\theta_m}{dt^2} + f_m \frac{d\theta_m}{dt} \tag{3-16}$$

where J_m is the total moment of inertia on the motor shaft and f_m is the viscous friction coefficient. Equating (3-13) and (3-16) and taking the Laplace transform, we obtain

$$\hat{g}(s) = \frac{\hat{\theta}_m(s)}{\hat{v}(s)} = \frac{k_c}{s(J_m s + f_m + k_n)} \triangleq \frac{k_m}{s(\tau_m s + 1)} \tag{3-17}$$

where

$$k_m \triangleq \frac{k_c}{f_m + k_n} = \text{ac motor gain constant} \tag{3-18}$$

$$\tau_m \triangleq \frac{J_m}{f_m + k_n} = \text{ac motor time constant} \tag{3-19}$$

This is the transfer function of an ac motor. We note that although the input signal to the control phase of the ac motor is modulated, the transfer function is computed from v, the unmodulated signal, to θ_m.

In Equation (3-13), it is implicitly assumed that the torque depends instantaneously on the input signal; hence no state variable is associated with the control phase winding. The state variables are the motor-shaft angular displacement and velocity. The transfer function has degree 2, which is equal to the number of the state variables. Hence the ac motor is completely characterized by its transfer function.

By choosing the state variables as $x_1 = \theta_m$ and $x_2 = \dot{\theta}_m = \dot{x}_1$, and using Equations (3-13) and (3-16), we obtain

$$-k_n x_2 + k_c v = J_m \dot{x}_2 + f_m x_2$$

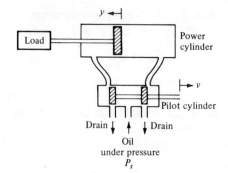

Figure 3-7
A hydraulic servomotor.

which, together with $\dot{x}_1 = x_2$, can be put into matrix form as

$$\begin{bmatrix} \dot{x}_1 \\ \dot{x}_2 \end{bmatrix} = \begin{bmatrix} 0 & 1 \\ 0 & -\dfrac{k_n + f_m}{J_m} \end{bmatrix} \begin{bmatrix} x_1 \\ x_2 \end{bmatrix} + \begin{bmatrix} 0 \\ \dfrac{k_c}{J_m} \end{bmatrix} v \qquad \text{(3-20a)}$$

$$y = \begin{bmatrix} 1 & 0 \end{bmatrix} \begin{bmatrix} x_1 \\ x_2 \end{bmatrix} \qquad \text{(3-20b)}$$

This is the state-variable description of the ac motor.

Hydraulic servomotor. A typical hydraulic servomotor is shown in Figure 3-7. The pilot cylinder is connected to an oil reservoir (not shown in the figure). The pressure P_s is maintained by a pump. The input signal v is used to move the piston of the pilot cylinder. This results in a pressure difference between P_1 and P_2 in the power cylinder, and hence a force to the load. A typical characteristic of a hydraulic servomotor is shown in Figure 3-8, in which F denotes the net force applied to the load and $\dot{y}$ is the velocity of the load. We see that if the curves in Figure 3-8 are approximated by straight and equidistant lines, then the resulting characteristic is similar to the one in Figure 3-6. By identifying the force F with the torque T, the linear speed $\dot{y}$ with the shaft angular speed $\dot{\theta}$,

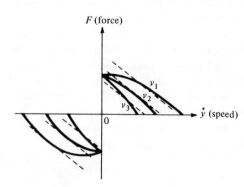

Figure 3-8
The characteristic of a hydraulic servomotor.

we conclude immediately that the transfer function of a hydraulic motor is of the form

$$\hat{g}(s) = \frac{\hat{y}(s)}{\hat{v}(s)} = \frac{k_m}{s(\tau_m s + 1)} \qquad (3\text{-}21)$$

where the output y is the load position and the input v is the piston position of the pilot cylinder. We note that Equation (3-21) is obtained under highly idealized conditions. In addition to the linearization, the oil leakage, compressibility of oil, and friction are not considered.

3-4 Gear Trains

In driving a load, the necessary minimum torque and the speed range are often specified. Certainly these are the factors used in deciding the size of a motor. However if motors alone are used, these specifications often cannot be met; it is then necessary to use gear trains to match the specifications. A gear-train system is shown in Figure 3-9, where T_i, θ_i, r_i, and N_i denote, respectively, the torque, angular displacement, radius, and number of teeth of gear i (where $i = 1$ or 2). It is clear that the number of teeth on a gear is linearly proportional to its radius; hence $N_1/r_1 = N_2/r_2$. That the linear distances traveled along the surfaces of both gears are the same implies that $\theta_1 r_1 = \theta_2 r_2$. The linear forces

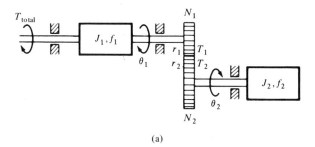

(a)

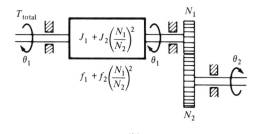

(b)

Figure 3-9 (a) A gear train. (b) Its equivalent.

developed at the contact point of both gears are equal; hence $T_1/r_1 = T_2/r_2$. By combining these equalities, we obtain

$$\frac{T_1}{T_2} = \frac{N_1}{N_2} = \frac{\theta_2}{\theta_1} \tag{3-22}$$

We see that (3-22) is a linear equation. It is again obtained under idealized conditions. In reality, there is always a certain amount of backlash between coupled gears; therefore the relationship between θ_1 and θ_2 should look like the one in Figure 3-10(b) rather than the one in Figure 3-10(a). Keeping the backlash small will inevitably increase the friction between the teeth and wear out the teeth faster. On the other hand, excessive amount of backlash will cause the so-called "jittering" problem in control systems.

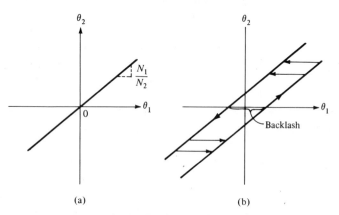

Figure 3-10 (a) Idealized characteristic of a gear train. (b) Actual characteristic of a gear train.

If gear 1 is connected to a motor shaft, in computing the transfer function of the motor, the inertia and friction on gear 2 must be transferred to gear 1. The torque T_2 on gear 2 is used to drive the inertia and overcome the viscous friction; hence

$$T_2 = J_2 \frac{d^2\theta_2}{dt^2} + f_2 \frac{d\theta_2}{dt} \tag{3-23}$$

where J_2 is the total moment of inertia on the shaft of gear 2, including the inertias of load, shaft, and gear, and f_2 is the viscous friction coefficient on gear 2. By use of (3-22), Equation (3-23) can be expressed as

$$T_1 = J_2 \left(\frac{N_1}{N_2}\right)^2 \frac{d^2\theta_1}{dt^2} + f_2 \left(\frac{N_1}{N_2}\right)^2 \frac{d\theta_1}{dt} \tag{3-24}$$

3-4 GEAR TRAINS

Hence the total torque required to drive the gear train is

$$T_{total} = J_1 \frac{d\theta_1^2}{dt^2} + f_1 \frac{d\theta_1}{dt} + T_1$$

$$= \left[J_1 + J_2 \left(\frac{N_1}{N_2}\right)^2 \right] \frac{d\theta_1^2}{dt^2} + \left[f_1 + f_2 \left(\frac{N_1}{N_2}\right)^2 \right] \frac{d\theta_1}{dt}$$

$$\triangleq J_{1_{eq}} \frac{d\theta_1^2}{dt^2} + f_{1_{eq}} \frac{d\theta_1}{dt} \qquad (3\text{-}25)$$

where

$$J_{1_{eq}} \triangleq J_1 + J_2 \left(\frac{N_1}{N_2}\right)^2 \qquad (3\text{-}26)$$

$$f_{1_{eq}} \triangleq f_1 + f_2 \left(\frac{N_1}{N_2}\right)^2 \qquad (3\text{-}27)$$

where J_1 and f_1 are the total moment of inertia and viscous friction on the shaft of gear 1. Consequently the gear train in Figure 3-9(a) is equivalent to the one in Figure 3-9(b). Similarly it can be shown (see Problem 2-2) that the gear trains in Figure 3-11 are equivalent.

Example

Consider the system shown in Figure 3-12(a). The characteristic of the ac motor is shown in Figure 3-12(b). Find the transfer function description of the system.

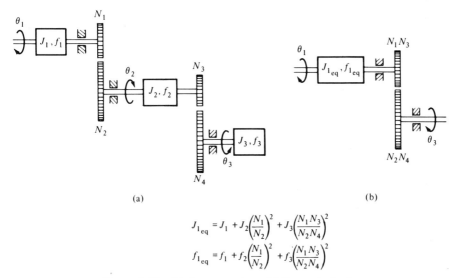

$$J_{1_{eq}} = J_1 + J_2 \left(\frac{N_1}{N_2}\right)^2 + J_3 \left(\frac{N_1 N_3}{N_2 N_4}\right)^2$$

$$f_{1_{eq}} = f_1 + f_2 \left(\frac{N_1}{N_2}\right)^2 + f_3 \left(\frac{N_1 N_3}{N_2 N_4}\right)^2$$

Figure 3-11 (a) A gear train. (b) Its equivalent.

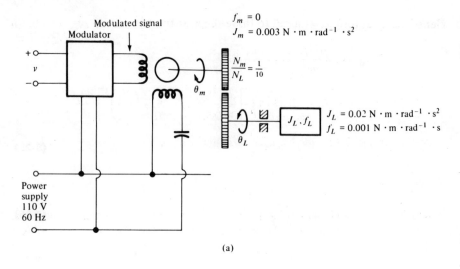

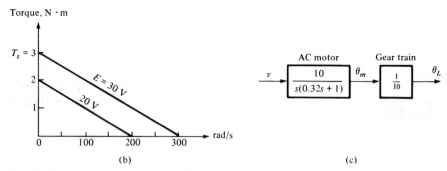

Figure 3-12 An ac motor driving a load. (a) The system. (b) The characteristic of the motor. (c) Block diagram.

From Figure 3-12(b) and Equations (3-14) and (3-15), the constants k_c and k_n can be computed as

$$k_c = \frac{T_s}{E} = \frac{3}{30} = 0.1$$

$$k_n = \frac{T_s}{n_s} = \frac{3}{300} = 0.01$$

The equivalent moment of inertia and equivalent friction coefficient on the motor shaft are, by use of (3-26) and (3-27),

$$J_{meq} = 0.003 + 0.02\left(\frac{1}{10}\right)^2 = 0.0032$$

$$f_{meq} = 0 + 0.001\left(\frac{1}{10}\right)^2 = 0.00001$$

Substituting J_{meq}, f_{meq}, k_c, and k_n into Equations (3-18) and (3-19), we obtain

$$k_m = \frac{k_c}{f_{meq} + k_n} = \frac{0.1}{0.00001 + 0.01} \doteq 10$$

$$\tau_m = \frac{J_{meq}}{f_{meq} + k_n} = \frac{0.0032}{0.00001 + 0.01} \doteq 0.32$$

Hence by using (3-17), the transfer function from v to θ_m is

$$\hat{g}(s) = \frac{10}{s(0.32s + 1)}$$

The transfer function from θ_m to θ_L is $N_m/N_L = \frac{1}{10}$. With these transfer functions, the system in Figure 3-12(a) can be represented by the block diagram[1] shown in Figure 3-12(c). ∎

3-5 Transducers

A transducer is a device which converts a signal from one form to another, for example, from a mechanical shaft position, temperature, or pressure to an electrical voltage. Two of the most often used transducers are potentiometers and tachometers. A potentiometer generates a voltage proportional to the displacement, whereas a tachometer generates a voltage proportional to the angular velocity. Both can be of the ac or dc type.

Potentiometers. A potentiometer is the simplest device that can be used to convert a linear or an angular displacement into a voltage. A wire-wound potentiometer with its characteristic is shown in Figure 3-13. The potentiometer consists of a finite number of turns of wiring; hence the contact point moves from turn to turn, and the generated voltage is stepwise, as shown. In analysis however the characteristic is always approximated by a straight line.

The applied voltage E of a potentiometer can be either a dc or an ac source. In the case of an ac source, the information is transmitted in the envelope of the output voltage signal. In either case, the transfer function of a potentiometer is equal to a constant; that is,

$$\frac{\hat{v}(s)}{\hat{\theta}(s)} = \frac{E}{\theta_{max}} \triangleq k \tag{3-28}$$

Its state-variable description is the zero-dimensional dynamical equation $v(t) = k\theta(t)$.

---

[1] If each system or device is represented by a block, the formed diagram is called a *block diagram*. It will be further discussed in Section 3-9.

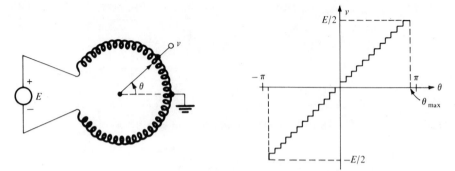

Figure 3-13 (a) A wire-wound potentiometer. (b) Its characteristic.

In employing a potentiometer, we should be cautious about the loading problem. Consider the circuit shown in Figure 3-14, in which R_p is the total resistance of the potentiometer and R_d is the input impedance of the device to which the output of the potentiometer is connected. If R_d is infinity, then the voltage v is linearly proportional to θ; that is, $v = (E/\theta_{max})\theta = k\theta$. If R_d is finite, then the voltage v can be computed as

$$v = \frac{E}{\theta_{max}\left[1 + \dfrac{R_p}{R_d}\dfrac{\theta}{\theta_{max}}\left(1 - \dfrac{\theta}{\theta_{max}}\right)\right]}\theta \tag{3-29}$$

which is no longer a linear equation. This kind of effect due to the introduction of a finite R_d is called the *loading problem*. It can be verified that if R_d is much larger than R_p (say $R_d \geq 10R_p$), then Equation (3-29) can be very well approximated by $v = k\theta$; otherwise the voltage v will not give a correct indication of the displacement θ. Hence in using a potentiometer we must make certain that the input impedance of the device to which the potentiometer is connected is much much larger than the resistance of the potentiometer.

Tachometers. A tachometer is a device used to measure angular velocity. It generates at its output a voltage proportional to its shaft velocity. Two of the

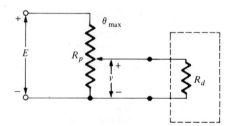

Figure 3-14
The loading problem in using a potentiometer.

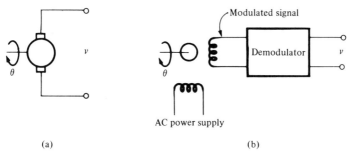

Figure 3-15 (a) A schematic diagram of a dc tachometer. (b) A schematic diagram of an ac tachometer.

most commonly used tachometers, a dc and an ac tachometer, are shown in Figure 3-15. The dc tachometer is essentially a small dc generator with a permanent magnetic field. The rotor of the tachometer is connected to the shaft to be measured. Its output generates a voltage proportional to the shaft angular velocity; that is,

$$v(t) = k \frac{d\theta(t)}{dt} \qquad (3\text{-}30)$$

where k is the sensitivity of the tachometer, in volts per radian per second. The ac tachometer is similar to the two-phase ac motor introduced in Figure 3-5. A sinusoidal voltage of rated value is applied to one winding; the envelope, $v(t)$, of the signal appearing across the second winding is proportional to the shaft angular velocity; that is, $v(t) = k \, d\theta(t)/dt$. Note that an ac tachometer also serves as a modulator. If we take the Laplace transform of (3-30), we obtain the transfer function from θ to v of an ac or a dc tachometer as

$$\hat{g}(s) = ks \qquad (3\text{-}31)$$

A differentiator having a transfer function that is the same as (3-31) cannot be used in the presence of noise, because it will amplify the noise. For example, consider the signal $\sin t$, which is contaminated by noise containing the term $0.01 \times \sin 10^3 t$. Note that most noises encountered in practice contain high frequencies. After differentiation, the signal becomes $\cos t$, whereas the noise becomes $10 \cos 10^3 t$, which now is the dominant term. Hence a differentiator is rarely used in practice.

A tachometer also has transfer function ks; therefore it is natural to ask why it can be used. A tachometer is usually attached to a shaft—for example, the shaft of an armature-controlled dc motor shown in Figure 3-16(a). The noise in the motor may arise from the input signal v_a, thermal noise in the resistor, and the commutators. In other words, the armature current is always

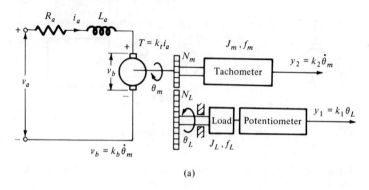

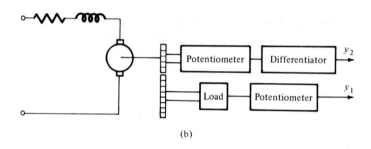

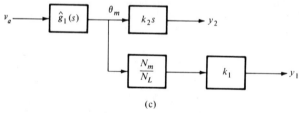

Figure 3-16 (a) A dc motor with a tachometer. (b) Measuring angular velocity by means of a potentiometer and a differentiator. (c) The block diagram of (a) and (b).

contaminated by noise. However if we use (3-3) and (3-5) and assume that $f_m = 0$, then the transfer function from i_a to θ_m is

$$\hat{\theta}_m(s) = \frac{k_t}{J_m s^2} \hat{i}_a(s) \qquad (3\text{-}32)$$

That is, the displacement θ_m is obtained by the integration of i_a twice. Therefore even if i_a is corrupted by noise, no differentiation of the noise ever occurs in the employment of the tachometer. This is the reason that tachometers can be and are widely used in practice. On the other hand, if we try to measure the angular velocity by use of the arrangement show in Figure 3-16(b), then because of the noise associated with the potentiometer and because of the differentiation of this noise by the differentiator, the resulting signal will be completely useless.

Example

Consider the system shown in Figure 3-16(a), with $R_a = 10\Omega,$[2] $L_a = 0.1$ H, $k_t = 1$ N·m·A^{-1}, $k_b = 1$ V·rad^{-1}·s, $N_m/N_L = \frac{1}{2}$, $k_1 = 1.5$ V·rad^{-1}, $k_2 = 0.8$ V·rad^{-1}·s, $J_L = 2$ N·m·rad^{-1}·s^2, $f_L = 0.02$ N·m·rad^{-1}·s, $J_m = 0.1$ N·m·rad^{-1}·s^2, and $f_m = 0.01$ N·m·rad^{-1}·s. The moment of inertia J_m includes the inertias of the tachometer, motor, and gear on the motor shaft. Find the transfer-function description and state-variable description of the system.

The equivalent total moment of inertia and viscous coefficient on the motor shaft can be computed from (3-26) and (3-27) as

$$J_{meq} = J_m + J_L \left(\frac{N_m}{N_L}\right)^2 = 0.1 + 2(\tfrac{1}{4}) = 0.6$$

$$f_{meq} = f_m + f_L \left(\frac{N_m}{N_L}\right)^2 = 0.01 + 0.02(\tfrac{1}{4}) = 0.015$$

Hence, from (3-7), the transfer function from v_a to θ_m is

$$\hat{g}_1(s) = \frac{\theta_m(s)}{\hat{v}_a(s)} = \frac{1}{s(0.6s + 0.015)(10 + 0.1s) + s}$$

$$= \frac{16.7}{s(s^2 + 100s + 19.2)}$$
(3-33)

With the aid of the block diagram in Figure 3-16(c), the transfer-function description of the system can be obtained as

$$\begin{bmatrix} \hat{y}_1(s) \\ \hat{y}_2(s) \end{bmatrix} = \begin{bmatrix} \dfrac{16.7 \times 0.5 \times 1.5}{s(s^2 + 100s + 19.2)} \\ \dfrac{16.7 \times 0.8}{s^2 + 100s + 19.2} \end{bmatrix} \hat{v}_a(s)$$
(3-34)

The state-variable description of an armature-controlled dc motor is computed in (3-8a) with $x_1 = \theta_m$, $x_2 = \dot{\theta}_m$, and $x_3 = i_a$. For this example, it becomes

$$\begin{bmatrix} \dot{x}_1 \\ \dot{x}_2 \\ \dot{x}_3 \end{bmatrix} = \begin{bmatrix} 0 & 1 & 0 \\ 0 & -\dfrac{0.015}{0.6} & \dfrac{1}{0.6} \\ 0 & -\dfrac{1}{0.1} & -\dfrac{10}{0.1} \end{bmatrix} \begin{bmatrix} x_1 \\ x_2 \\ x_3 \end{bmatrix} + \begin{bmatrix} 0 \\ 0 \\ \dfrac{1}{0.1} \end{bmatrix} v_a$$
(3-35a)

---

[2] We use the international (SI) system of abbreviations: A = ampere, V = volt, N = newton, rad = radian, H = henry, s = second, Ω = ohm, and m = meter.

By using $y_1 = k_1\theta_L = k_1(N_m/N_L)\theta_m$ and $y_2 = k_2\dot{\theta}_m$, the output equation is obtained as

$$\begin{bmatrix} y_1 \\ y_2 \end{bmatrix} = \begin{bmatrix} 0.75 & 0 & 0 \\ 0 & 0.8 & 0 \end{bmatrix} \begin{bmatrix} x_1 \\ x_2 \\ x_3 \end{bmatrix} \qquad (3\text{-}35a)$$

3-6 Error Detectors

An error detector generates an output signal that is proportional to the difference of two input signals. The error detectors to be introduced here are pairs of potentiometers, synchros, and operational amplifiers.

Pairs of potentiometers. A pair of potentiometers connected as shown in Figure 3-17 may perform the function of an error detector. The two potentiometers may be remotely located. For example, one may be located in a control tower and controlled by an operator, and the other may be attached to an antenna. The input signals θ_d and θ_o are mechanical positions, either linear or rotational; the output v is a voltage signal. They are related by

$$v(t) = k[\theta_d(t) - \theta_o(t)] \qquad (3\text{-}36)$$

or

$$\hat{v}(s) = k[\hat{\theta}_d(s) - \hat{\theta}_o(s)] \qquad (3\text{-}37)$$

where $k = E/\theta_{max}$. The pair of potentiometers can be represented schematically as shown in Figure 3-18. The circle at which the two signals meet is often called a *summing point*.

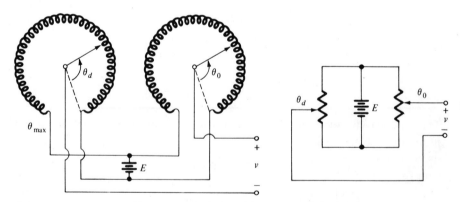

Figure 3-17 A pair of potentiometers.

Synchros. Another widely used error detector in control systems is a pair of synchros. Depending upon the manufacturers, synchros are also called *selsyns*, or *autosyns*. A pair of synchros is shown schematically in Figure 3-19; the unit CX is called a *synchro transmitter* and the unit CT is called a *synchro transformer*.

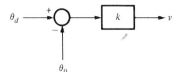

Figure 3-18
Schematic representation of an error detector.

The structures of these two units are similar except for the shapes of the rotors. The rotor of the synchro transmitter is geared or coupled to a knob or shaft whose position θ_d indicates the desired signal; the rotor of the synchro transformer is attached to the shaft to be controlled. Unlike the pairs of potentiometers, which can be excited by dc or ac sources, synchros can operate only under the excitations of ac sources. If a sinusoidal voltage source, usually of 60 or 400 Hz, is applied to the synchro transmitter, then the envelope, v, of the voltage signal developed on the synchro transformer can be shown to be

$$v(t) = k \sin\left[\theta_d(t) - \theta_0(t)\right] \quad \text{(3-38)}$$

where k, in volts per radian, is the sensitivity of the synchros. If the difference between θ_d and θ_0 is small, then Equation (3-38) can be approximated by

$$v(t) = k\left[\theta_d(t) - \theta_0(t)\right] \quad \text{(3-39)}$$

This is a linear equation, similar to (3-37). Hence a pair of synchros can also be represented, after linearization, by Figure 3-18.

Operational amplifiers. An operational amplifier is basically an amplifier with a very high gain and high input impedance. Its gain may range from 10^5 to 10^9, and its input impedance from 10^7 to 10^9 ohms. By using an operational amplifier, it is possible to add or subtract two or more signals.

Consider the operational amplifier circuit shown in Figure 3-20(a). The amplifier has negative gain $-A$ and input impedance Z_{in}. For an output voltage v_0 in the range of 10 volts or less and a very large A, the input voltage e_{in}

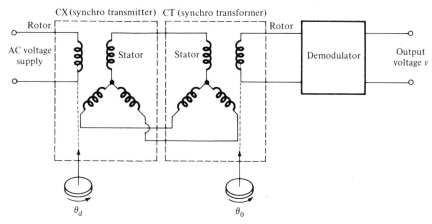

Figure 3-19 A pair of synchros.

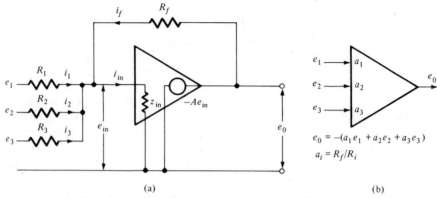

Figure 3-20 Operational amplifier with a feedback element. (a) The circuit. (b) Schematic representation.

to the amplifier is approximately equal to zero because $e_{in} = -e_0/A$. The current i_{in} flowing into the amplifier is again approximately equal to zero because of a very high input impedance. Hence we have

$$i_f = -(i_1 + i_2 + i_3)$$

That, in turn, together with $e_{in} \doteq 0$, implies

$$e_0 = -\left(\frac{R_f}{R_1} e_1 + \frac{R_f}{R_2} e_2 + \frac{R_f}{R_3} e_3\right) \tag{3-40}$$

We see from this equation that, by choosing R_f and R_i properly, some specified linear function of the input signals can be generated at the output of the operational amplifier circuit. For convenience, the circuit in Figure 3-20(a) is often represented schematically by the one shown in Figure 3-20(b). We note that Equation (3-40) can be extended to include more input signals or reduced to contain only one input signal. If there is only one input signal and if $R_f = R_1$, then the circuit changes only the sign of the signal. By connecting two operational amplifiers as shown in Figure 3-21, we can obtain an error detector.

An operational amplifier may have an operational frequency range from zero to 10^6 Hz; therefore it can operate on dc or ac (modulated) signals in control systems. Just as motors, gear trains, synchros, and potentiometers, operational amplifiers are available in packages from manufacturers.

Figure 3-21 An error detector.

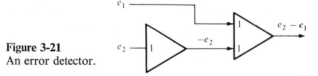

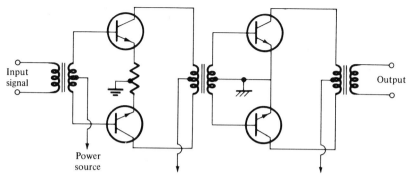

Figure 3-22 Two-stage transistor ac amplifier.

3-7 Amplifiers

The amplifiers used in control systems can be electronic, magnetic, hydraulic, or rotary. They are used to amplify error signals or to generate sufficient power to drive prime movers. Because of its high input impedance and low output impedance, an electronic amplifier is also used as an isolation amplifier between two circuits to reduce their loading effects (see Section 3-9). A two-stage push-pull transistor ac amplifier is shown in Figure 3-22. Control systems also employ dc electronic amplifiers. However unless the power supply is very steady, a dc amplifier may suffer the problem of drifting. The transfer function of a dc or an ac amplifier is rather unimpressive; it is just a constant k. This constant is obtained by approximation however. To be precise, the transfer function of an electronic amplifier is $k/(1 + \tau s)$. However for a very small τ, it can be approximated by k.

For driving a large load, an electronic amplifier may not be adequate. In this case, a rotating amplifier or other type of amplifier must be used.

Rotating power amplifiers. The single-stage rotating power amplifier shown in Figure 3-23(a) is essentially a dc generator. The rotor of the generator is driven by a motor with a constant angular velocity (not shown). The input voltage signal is applied to the field circuit. This input will generate a field current i_f according to

$$v = R_f i_f + L_f \frac{di_f}{dt} \tag{3-41}$$

The relationship between the field current i_f and the generated voltage e_g is shown in Figure 3.23(b). For small i_f it can be approximated by

$$e_g = k_g i_f \tag{3-42}$$

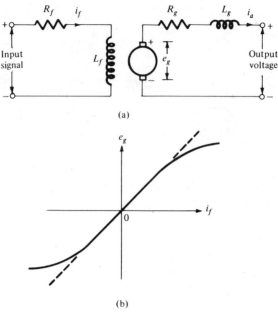

Figure 3-23 A dc generator.

where k_g is a real constant. Now if the armature current i_a is zero, then $e_g(t) = y(t)$. Hence the transfer function of the generator is

$$\hat{g}(s) = \frac{\hat{y}(s)}{\hat{v}(s)} = \frac{k_g}{L_f s + R_f} \quad (3\text{-}43)$$

Its state-variable description can also be readily obtained, by choosing the state variable $x = i_f$, as

$$\dot{x} = -\frac{R_f}{L_f} x + \frac{1}{L_f} v \quad (3\text{-}44\text{a})$$

$$y = k_g x \quad (3\text{-}44\text{b})$$

This is a one-dimensional linear time-invariant dynamical equation.

3-8 Compensation Networks

Networks are inexpensive and are effective in improving the performance of control systems. There are two types of networks: ac and dc networks. A dc network operates on the signal itself, whereas an ac network operates on the modulated signal. We show in Figure 3-24 some commonly used dc networks and their equivalent ac networks. By equivalent we mean that the effect of a dc network on the signal and the effect of the corresponding ac network on the

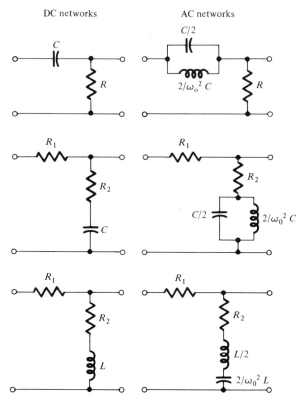

Figure 3-24 Schematic diagrams of equivalent dc and ac networks.

envelopes of the modulated signal are the same. The ac networks in Figure 3-24 are obtained from the dc one by applying the following rules:

1. A capacitor with capacitance C in a dc network is replaced by a parallel connection of a capacitor with capacitance $C/2$ and an inductor with inductance $2/C\omega_c^2$ in the equivalent ac network.

2. An inductor with inductance L in a dc network is replaced by the series connection of an inductor with inductance $L/2$ and a capacitor with capacitance $2/L\omega_c^2$ in the equivalent ac network.

3. The resistors remain unchanged.

We discuss briefly how these rules are developed. If the dc and ac networks shown in Figure 3-25 are equivalent, then the inputs and outputs of the networks

Figure 3-25 Block diagrams of equivalent dc and ac networks.

should be related as shown in the figure. If the transfer function of the dc network is $\hat{g}(s)$, then the transfer function of the ac network is

$$\hat{g}_c(s) \triangleq \frac{\mathscr{L}[y(t) \cos \omega_c t]}{\mathscr{L}[u(t) \cos \omega_c t]} = \frac{\hat{g}(s - j\omega_c)\hat{u}(s - j\omega_c) + \hat{g}(s + j\omega_c)\hat{u}(s + j\omega_c)}{\hat{u}(s - j\omega_c) + \hat{u}(s + j\omega_c)} \quad (3\text{-}45)$$

which is obtained by use of the formula

$$\mathscr{L}[f(t) \cos \omega t] = \mathscr{L}\left[f(t) \frac{e^{j\omega t} + e^{-j\omega t}}{2}\right] = \tfrac{1}{2}[\hat{f}(s - j\omega) + \hat{f}(s + j\omega)]$$

The transfer function $\hat{g}_c(s)$ in Equation (3-45) is a function of the input u. Hence for a different input u, we have a different equivalent ac network. Consequently it is not possible to obtain an exact equivalent ac network. Hence the best we can obtain is an approximated one. As discussed in Section 3-2, the frequency spectra of signals in control systems are narrow and centered around $\omega = 0$; hence we have

$$\hat{u}(j\omega \pm j\omega_c) \doteq 0$$

except in the neighborhood of $\omega \pm \omega_c = 0$. Consequently Equation (3-45) can be approximated by

$$\hat{g}_c(j\omega) \doteq \hat{g}(j\omega \mp j\omega_c) \quad (3\text{-}46)$$

for ω in the neighborhood of $\pm\omega_c$. This approximation asserts that the ac network should have the same frequency characteristics in the neighborhood of $\pm\omega_c$ as the dc network has in the neighborhood of $\omega = 0$. Hence given a dc network with $\hat{g}(j\omega)$, if we can find a network with frequency characteristics $\hat{g}(j\omega - j\omega_c)$, then the network is an equivalent ac network. This can be achieved by the following frequency transformation:

$$\omega \to p(\omega) \triangleq \frac{\omega}{2}\left(1 - \frac{\omega_c^2}{\omega^2}\right) = \frac{\omega}{2}\left(1 + \frac{\omega_c}{\omega}\right)\left(1 - \frac{\omega_c}{\omega}\right) \quad (3\text{-}47)$$

Note that $p(\omega)$ is approximately equal to $\omega \mp \omega_c$ in the neighborhood of $\pm\omega_c$. Hence if $\hat{g}(j\omega)$ is replaced by $\hat{g}(jp)$, an equivalent ac network can be obtained. Equation (3-47) transforms the network elements in a dc network into

$$j\omega L \to jpL = j\omega \frac{L}{2} + \frac{\omega_0^2 L}{j2\omega} \quad (3\text{-}48\text{a})$$

$$j\omega C \to jpC = j\omega \frac{C}{2} + \frac{\omega_0^2 C}{j2\omega} \quad (3\text{-}48\text{b})$$

$$R \to R$$

These are the rules introduced above.

Since the networks in Figure 3-24 consist of only passive elements (resistors, inductors, and capacitors), they are called *passive* networks. The passive networks shown in Figure 3-24 are the most commonly used compensation networks in control systems.

Operational amplifiers are now readily available, and the circuits built by using these amplifiers can also be used as compensation networks. It will be shown in Chapter 5 that every proper transfer function is synthesizable by using operational amplifier circuits. This is however not the case if we restrict ourself to the passive network elements. Hence operational amplifier circuits are more versatile and flexible than passive networks.

3-9 Examples of Control Systems

In this section some control systems and their mathematical descriptions will be introduced. Before proceeding, let us first discuss the problem of loading.

Loading problem and block diagram. If the transfer function of the tandem connection of two devices is different from the product of their individual transfer functions, then the devices are said to have a loading effect on each other. This concept is best illustrated by an example. Consider the networks shown in Figure 3-26(a). The transfer functions of networks M_1 and M_2 are, respectively, $\hat{g}_1 = s/(s + 1)$ and $\hat{g}_2 = 1$. After the connection of M_1 and M_2 in tandem, the overall transfer function can be easily shown to be $s/(2s + 1)$, which is different from the product of $\hat{g}_1(s)$ and $\hat{g}_2(s)$. Hence there is loading in this connection. In electrical networks, the effect of loading can be eliminated by the introduction of an isolating amplifier, as shown in Figure 3-26(b). Ideally the input impedance Z_{in} of the amplifier is infinity, and its output impedance Z_{out} is zero. Under this assumption, the overall transfer function is just equal to $\hat{g}_1(s)\hat{g}_2(s)$.

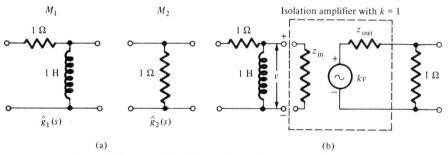

Figure 3-26 The loading of two networks.

The loading problem cannot always be eliminated in control systems. In that case, the transfer functions of connecting systems have to be properly modified. For example, in computing the transfer function of a motor driving a load, the inertia of the load must be included in the transfer function of the motor. It is not possible to separate the transfer function of the motor and that of the load.

Control systems are built by the interconnection of various subsystems, or

64 CONTROL COMPONENTS AND SYSTEMS

devices. If each subsystem is represented by a block, then a control system can be represented by a connection of blocks. This kind of diagram is called a *block diagram*. However because of the loading problem, each block in a block diagram may not represent a unit of a physical component. For example, consider the block diagram shown in Figure 3-12(c). Although the block denoted as the gear train represents the actual transfer function of the gear train, the block denoted as the motor has to include the inertia of the gear train and that of the load connected to the gear train. Therefore care must be taken in the interpretation of each block in a block diagram. The effect of loading on a block diagram will be further illustrated in the following examples.

Example 1

A heavy telephoto camera can be controlled by the system shown in Figure 3-27(a). Its detailed circuit diagram is shown in Figure 3-27(b). The camera is driven by an ac motor through a gear train and is designed to follow the movement of the spotting scope. A pair of synchros with sensitivity k_1 is used as an error detector to generate the error signal $e = k_1(\theta_d - \theta_L)$. An ac tachometer is connected to the motor shaft to generate $m = k_2\dot{\theta}_m$. The ac amplifier shown in Figure 3-22 is used to amplify the signal $(e - m)$. This system uses ac components exclusively. The block diagram of the system is shown in Figure 3-27(c). The transfer function of the block denoted as "motor and load" can be computed as in the Example in Section 3-4. From Figure 3-27(b) we can see that the gains k_1 and k_2 in Figure 3-27(c) can be easily adjusted.

Example 2

In a steel or paper mill, the products are moved by rollers, as shown in Figure 3-28(a). In order to maintain a prescribed uniform tension, the roller speeds are kept constant and equal to each other. This can be achieved by using the control system shown in Figure 3-28(b). Each roller is driven by an armature-controlled dc motor. A dc generator is used as a power amplifier. The desired roller speed is transformed into a voltage by means of a potentiometer. The loading problem associated with the potentiometer is eliminated by using an isolating amplifier. A dc tachometer picks up the roller speed. This control system uses dc devices exclusively.

The control system shown in Figure 3-28(b) can be represented by the block diagram shown in Figure 3-28(c). We note that the angular speed ω_1, rather than the angular displacement, is considered as the output of the motor; hence the transfer function of the tachometer is simply a constant, k_2. (If the input is displacement, then its transfer function is $k_2 s$.) The block denoted as "generator" is, to be precise, not correct, for the actual output of the generator is v_g, the voltage across the terminals denoted as E and F in Figure 3-28(b). But the transfer function from v to v_g is a function of the armature current i_a. In other words, there is a loading problem in the connection of the generator and the

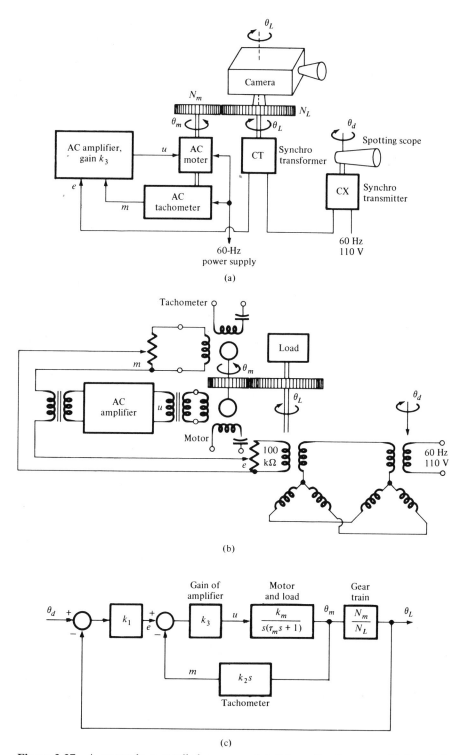

Figure 3-27 A remotely controlled camera. (a) Schematic diagram. (b) Detailed wiring diagram. (c) Block diagram.

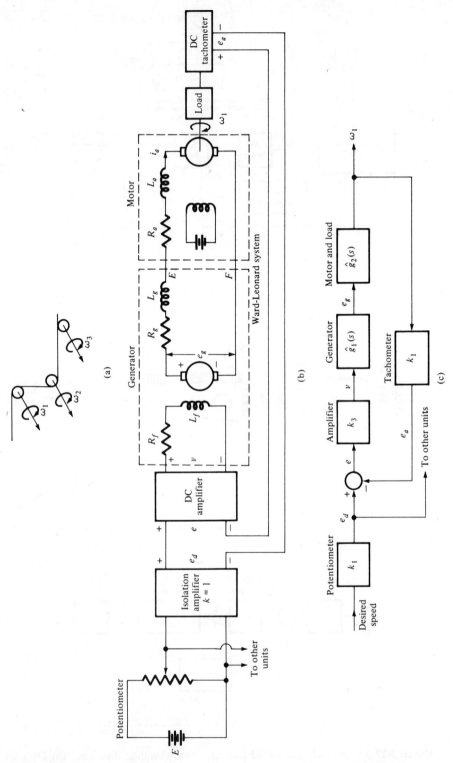

Figure 3-28 (a) Speed control of rollers. (b) Control system. (c) Block diagram.

motor. This loading problem can be eliminated however by considering e_g as the output. This is the reason we use e_g, rather than v_g, as the output of the block denoted as generator. However in computing the transfer function of the motor, the generator resistance R_g and inductance L_g have to be lumped with R_a and L_a. This is one way of solving the loading problem.

The transfer functions $\hat{g}_1(s)$ and $\hat{g}_2(s)$ can be obtained, with proper modification, from Equations (3-43) and (3-7) as

$$\hat{g}_1(s) = \frac{k_g}{L_f s + R_f} \tag{3-49}$$

$$\hat{g}_2(s) = \frac{k_t}{(J_m s + f_m)[(R_a + R_g) + (L_a + L_g)s] + k_t k_b} \tag{3-50}$$

In addition to the modification on R_a and L_a, Equation (3-50) is obtained from (3-7) by multiplying s, because the angular velocity, rather than displacement, is considered as the output of the motor in this example.

Example 3

The control system shown in Figure 3-29(a) uses both ac and dc devices. A demodulator (an ac device) is connected between the synchros and the dc amplifier. Its block diagram is shown in Figure 3-29(b). Since the transfer function of the demodulator is 1, the block denoted as "demodulator" can be eliminated. From this and the two previous examples, we see that there is no essential difference among the block diagrams of ac control systems, dc control systems, and hybrid control systems.

Example 4

In order to point the antenna toward the earth or the solar cells toward the sun, the attitude or orientation of a space vehicle has to be properly controlled. This can be achieved by the use of gas jets or reaction wheels. In the latter case, three sets of reaction wheels are needed to control the orientations of the vehicle in the three-dimensional space; however they are all identical.

A reaction wheel is actually a flywheel; it may be simply the rotor of a motor. It is assumed that the reaction wheel is driven by an armature control dc motor, as shown in Figure 3-30(a) and (b). The case of the motor is rigidly attached to the space vehicle. Because of the conservation of momentum, if the reaction wheel is driven in one direction, the space vehicle will rotate in the opposite direction. The orientation and its rate of change of the vehicle can be measured by use of a gyro and a rate gyro. The block diagram of the complete system is shown in Figure 3-30(c).

We derive in the following the mathematical description of the block denoted as "motor and vehicle." Let the angular displacements of the space vehicle and the reaction wheel with respect to the inertial coordinate be, respectively, θ

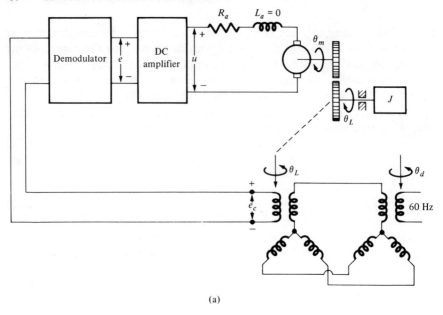

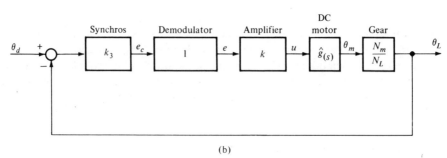

Figure 3-29 (a) A control system using both ac and dc devices. (b) Its block diagram.

and θ_m, as shown in Figure 3-30(a). They are chosen, for convenience, to be of opposite direction. Clearly the *relative* angular displacement of the reaction wheel (or the rotor of the motor) with respect to the space vehicle (or the stator or case of the motor) is $\theta + \theta_m$. Hence the armature circuit yields

$$v_a = R_a i_a + L_a \dot{i}_a + k_b (\dot{\theta} + \dot{\theta}_m) \tag{3-51}$$

The torque equation gives

$$k_t i_a = J_m \ddot{\theta}_m + f_m (\dot{\theta}_m + \dot{\theta}) \tag{3-52}$$

where J_m and f_m are, respectively, the moment of inertia and the viscous friction on the motor shaft. We note that the friction is generated between the motor shaft and the bearing that is attached to the space vehicle; hence the relative angular speed is used for the friction torque in Equation (3-52). Now the conservation of angular momentum yields

$$J_v \dot{\theta} = J_m \dot{\theta}_m \tag{3-53}$$

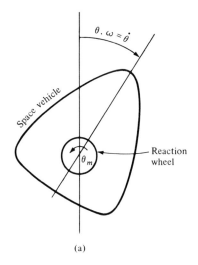

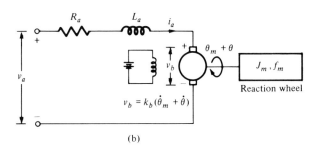

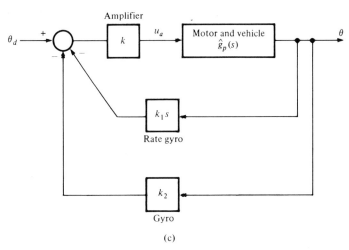

Figure 3-30 Control of the altitude of a space vehicle. (a) Space vehicle. (b) Dc motor driving a reaction wheel. (c) Block diagram.

where J_v is the moment of inertia of the vehicle. Using (3-51) through (3-53), it is straightforward to verify that

$$\hat{g}_p(s) = \frac{\hat{\theta}(s)}{\hat{v}_a(s)} = \frac{1}{s(R_a + L_a s)\left[\dfrac{J_v}{k_t}s + \dfrac{f_m}{k_t}\left(\dfrac{J_v + J_m}{J_m}\right)\right] + k_b\left(\dfrac{J_m + J_v}{J_m}\right)s}$$

(3-54)

By defining $x_1 = \theta$, $x_2 = \dot{\theta}$, and $x_3 = i_a$, the dynamical equation description can be obtained from Equations (3-51) through (3-53) as

$$\begin{bmatrix} \dot{x}_1 \\ \dot{x}_2 \\ \dot{x}_3 \end{bmatrix} = \begin{bmatrix} 0 & 1 & 0 \\ 0 & -\dfrac{f_m}{J_m}\left(1 + \dfrac{J_v}{J_m}\right) & \dfrac{k_t}{J_m} \\ 0 & -\dfrac{k_b}{L_a}\left(1 + \dfrac{J_v}{J_m}\right) & -\dfrac{R_a}{L_a} \end{bmatrix} \begin{bmatrix} x_1 \\ x_2 \\ x_3 \end{bmatrix} + \begin{bmatrix} 0 \\ 0 \\ \dfrac{1}{L_a} \end{bmatrix} v_a \quad \text{(3-55a)}$$

$$\theta = \begin{bmatrix} 1 & 0 & 0 \end{bmatrix} \mathbf{x} \quad \text{(3-55b)}$$

This is a three-dimensional dynamical equation.

Example 5

As a final example, consider the hydraulic control system shown in Figure 3-31(a). The antenna is driven by a hydraulic motor, which, in turn, is driven by an ac motor. The block diagram of the system is shown in Figure 3-31(b). The transfer functions of the ac and hydraulic motors are computed in Section 3-3. The angular displacement θ_L and the linear displacement x are related by $x = k_5 \theta_L$, where k_5 is the radius of the gear. The relationship between y and θ is however nonlinear. For small θ, it can be approximated by $\theta = k_6 y$. ∎

3-10 Remarks and Review Questions

Strictly speaking, control devices are all nonlinear. If so, why do we bother to develop linear models for them? The reasons are as follows: First, they can be very well approximated by linear models under their normal operating ranges. Second, and more important, if they are modeled as linear system, then a number of systematic design techniques can be applied. We shall however bear in mind that every analytical design is based on models; hence until real physical systems are tested, the design cannot be said to be completed. If models are chosen properly, theoretical design and practical result should be close.

What is the frequency spectrum of a signal?

3-10 REMARKS AND REVIEW QUESTIONS

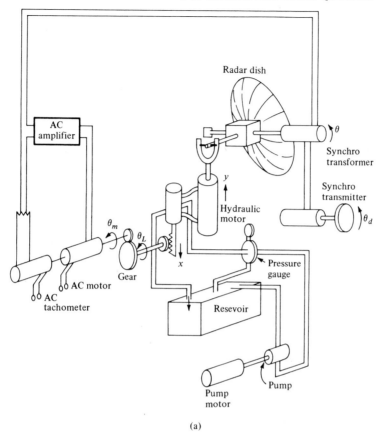

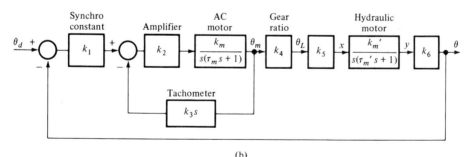

(b)

Figure 3-31 Hydraulic control system. (a) Schematic diagram. (b) Block diagram.

Why do we have to introduce modulation? Where is the information stored in a modulated signal?

What are the transfer functions of a modulator and a demodulator?

In the computation of the transfer function of an ac device, do we use the signal itself or its envelope as input and output?

What are the types of transfer functions of an armature-controlled dc motor (by assuming $L_a = 0$), a two-phase induction motor, and a hydraulic motor? Do they have the same form?

Differentiators are not used in practice because of the noise problem. Why are tachometers, which have transfer functions as do differentiators, used in practice?

What are the main properties of an operational amplifier?

How do you transfer a dc compensation network into an ac one?

Is it always true that the transfer function of the tandem connection of two systems is just the product of the transfer functions of the two systems?

Is there any way to eliminate the loading problem in electrical systems?

Is the output of the block denoted as generator in Figure 3-27 the voltage across the actual output terminals of a generator? If not, why?

Is it always possible to identify a block in a block diagram as exclusively a unit of a physical device?

A control system may consist of exclusively ac devices, exclusively dc devices, or a combination of dc and ac devices. However, in analytical design, no distinction between ac and dc devices is necessary, for the transfer functions of ac devices are computed for the envelopes of modulated signals. Therefore the design technique developed for dc devices are directly applicable to ac devices or hybrid systems.

The devices introduced in this chapter are far from complete; they will however provide sufficient physical background for the development of this text. For a more complete discussion of devices, the reader is referred to References [1], [4], [5], [9], and [11].

References

[1] Ahrendt, W. R., and C. J. Savant, Jr., *Servomechanism Practice*, 2d ed. New York: McGraw-Hill, 1960.
[2] Baeck, H. S., *Practical Servomechanical Design*. New York: McGraw-Hill, 1968.
[3] Cannon, R. H., *Dynamics of Physical Systems*. New York: McGraw-Hill, 1967.
[4] Charkey, E. S., *Electromechanical System Components*. New York: Wiley, 1972.
[5] Chestnut, H., and R. W. Mayer, *Servomechanisms and Regulating System Design*, vols. 1 and 2. New York: Wiley, 1951.
[6] Cooper, G. R., and C. D. McGillem, *Methods of Signal and System Analysis*. New York: Holt, Rinehart and Winston, 1967.
[7] Doebelin, E., *System Dynamics*. Columbus, Ohio: Merrill, 1972.
[8] Dorf, R. C., *Modern Control Systems*. Reading, Mass.: Addison-Wesley, 1967.

[9] Gibson, J. E., and F. B. Tuteur, *Control System Components*. New York: McGraw-Hill, 1958.
[10] Kuo, B. C., *Automatic Control Systems*. Englewood Cliffs, N.J.: Prentice-Hall, 1962.
[11] Norton, H. N., *Handbook of Transducers for Electronic Measuring Systems*. Englewood Cliffs, N.J.: Prentice-Hall, 1969.
[12] Raven, F. H., *Automatic Control Engineering*, 2d ed. New York: McGraw-Hill, 1968.
[13] Savant, C. J., Jr., *Control System Design*, 2d ed. New York: McGraw-Hill, 1964.
[14] Truxall, J. G., *Automatic Feedback Control System Synthesis*. New York: McGraw-Hill, 1955.

Problems

3-1 Find the frequency spectra of $v(t) = e^{-0.1t}$ and $v_m(t) = e^{-0.1t} \cos 1000t$.

3-2 Show that the constant k_t defined in Equation (3-3) is equal to the constant k_b defined in Equation (3-4). [*Hint:* The power flowing into the rotor is $v_b i_a$; the power delivered to the shaft is $T(d\theta_m/dt)$. They are equal under steady-state conditions.]

3-3 Consider the operational amplifier circuit shown in Figure P3-1. Show that the signal $e_0(t)$ is given by

$$e_0(t) = -\left[\frac{1}{R_1 C}\int_0^t e_1(\tau)\,d\tau + \frac{1}{R_2 C}\int_0^t e_2(\tau)\,d\tau\right] + e_0(0)$$

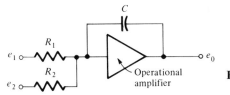

Figure P3-1

3-4 Show the equivalence of the gear trains in Figure 3-11.

3-5 Consider the control system shown in Figure P3-2. The tension of the tape is

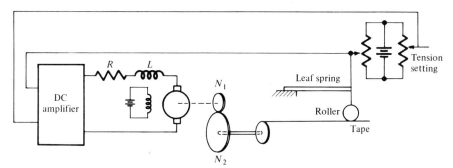

Figure P3-2

controlled by varying the motor shaft speed. The tension is measured by using a roller and a potentiometer as shown. Draw a block diagram for the system and indicate the type of transfer function for each block.

3-6 Find the block diagram and the transfer function of each block for the system shown in Figure P3-3.

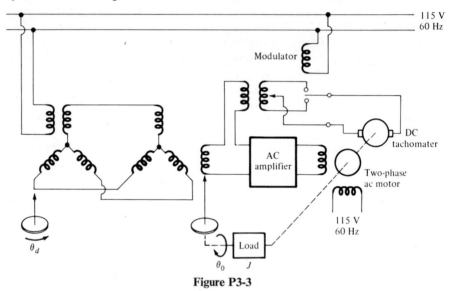

Figure P3-3

3-7 Find the block diagram and the transfer function of each block for the system shown in Figure P3-4. Find also the state-variable description of each block.

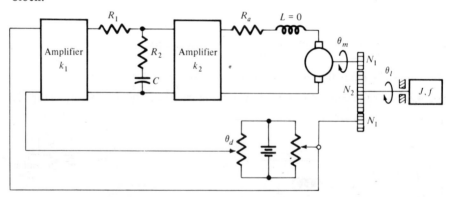

Figure P3-4

3-8 Consider the simplified voltage regulator shown in Figure P3-5. The generator is driven by a motor with a constant velocity (not shown). The generated voltage e_g is proportional to the field current i_f; that is, $e_g = k_g i_f$, with $k_g = 25$ V/A. The elements shown in Figure P3-5 have the following values:

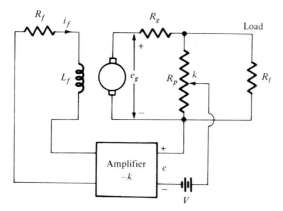

Figure P3-5

$R_g = 0.1\ \Omega$, $R_f = 10\ \Omega$, $R_l = 200\ \Omega$, $R_p = 10^4\ \Omega$, $k = 0.5$, $L_f = 1$ H, and $K = 1000$. Draw a block diagram and compute the transfer function of each block.

3-9 The control of temperature in a chemical process can be achieved as shown in Figure P3-6. The opening of the valve x is controlled by a solenoid; it is assumed that $x = k_s i_s$. The flow q of hot steam is proportional to the valve opening x (say, $q = k_q x$). The temperature θ and the steam flow q are related by the equation

$$\frac{d\theta}{dt} = -c\theta(t) + k_c q$$

Draw a block diagram for the system. Find the transfer-function description and state-variable description of each block.

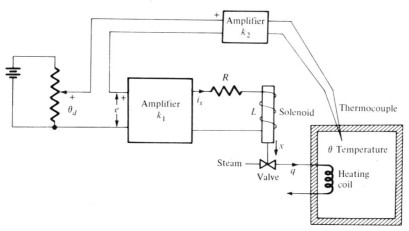

Figure P3-6

3-10 A garage door can be controlled by the system shown in Figure P3-7. Draw a block diagram for the system. If the speed-torque characteristics of the

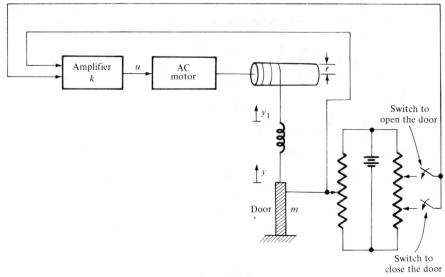

Figure P3-7

motor is as shown in Figure 3-12(b), the moment of inertia of the combination of the drum and rotor is 0.01 $N \cdot m \cdot rad^{-1} \cdot s$, the radius r of the drum is 0.1 m, the spring stiffness coefficient k is 1 N/m, and the mass of the door is 10 kg. What is the transfer function from u to y?

3-11 In addition to electromechanical transducers, optical transducers are also used in practice. The transducer shown in Figure P3-8 consists of a pulse generator and a pulse counter. By counting the number of pulses in a fixed interval of time, a signal proportional to the angular speed of the motor shaft can be generated. This signal is in digital form. By a proper conversion and by neglecting the so-called quantization problem, the transfer function from θ to v shown in Figure P3-8 can be approximated by ks. Draw a block diagram for the system. Indicate also the type of transfer function of each block.

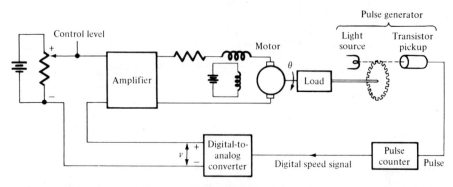

Figure P3-8

3-12 An amplidyne is often used in control systems where very large power amplification is required. An amplidyne is a two-stage rotary amplifier; it consists of two dc generators connected in series. A schematic diagram and a model of an amplidyne are shown in Figure P3-9. Find the transfer function of the amplidyne.

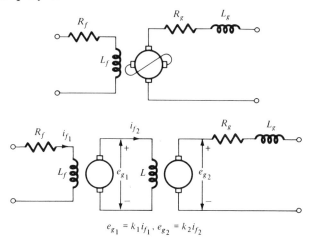

Figure P3-9

3-13 Machine tools can be controlled automatically by the instructions recorded on punched cards or tapes. This kind of control is called *numerical control* of machine tools. Numerical control can be classified as position control and continuous contour control. A schematic diagram of a position control system is shown in Figure P3-10. A feedback loop is introduced in the D/A (digital-to-

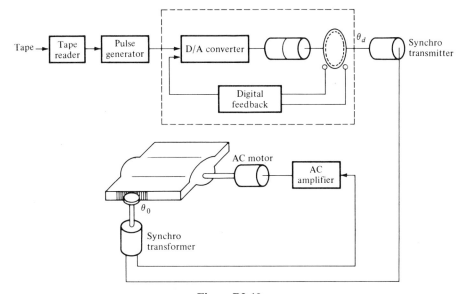

Figure P3-10

78 CONTROL COMPONENTS AND SYSTEMS

analog) converter to obtain a more accurate conversion. Draw a block diagram from θ_d to θ_o for the system. Indicate also the type of transfer function for each block.

3-14 In industry, a robot can be designed to replace a human operator to carry out repeated movements. Schematic diagrams of such a robot are shown in

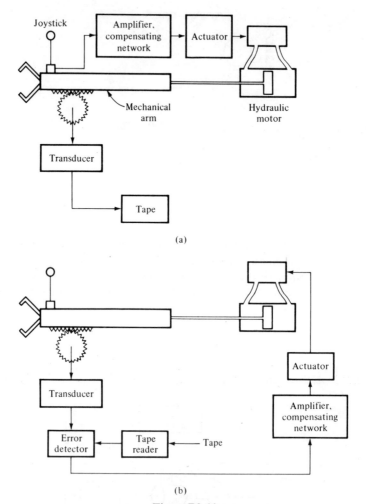

(a)

(b)

Figure P3-11

Figure P3-11. The desired movement is first applied by an operator to the joystick shown. The joystick activates the hydraulic motor and the mechanical arm. The movement of the arm is recorded on a tape, as shown in Figure P3-11(a).

The tape can then be used to control the mechanical arm, as shown in Figure P3-11(b). Draw a block diagram for the system in Figure P3-11(b). Indicate also the type of transfer function for each block.

3-15 Find the block diagram and the transfer function of each block for the system shown in Figure P3-12. Note that the compensation network is an ac network.

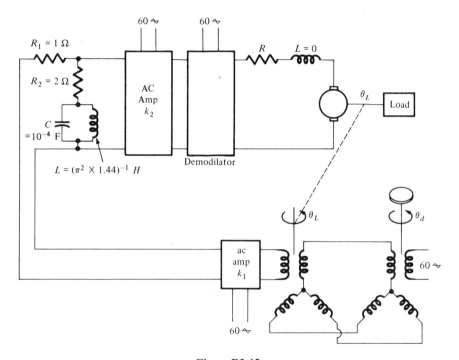

Figure P3-12

3-16 The voltage measuring device shown in Figure P3-13 is actually a control system. If the measured voltage e_2 is different from the actual voltage e_1, then the signal $(e_1 - e_2)$ will drive, after modulation and amplification, the ac motor to move the indicator until $e_1 = e_2$. Find the block diagram and the tranfers function of each block.

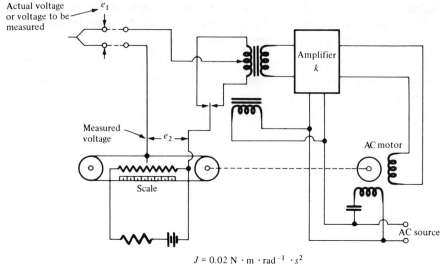

$$J = 0.02 \text{ N} \cdot \text{m} \cdot \text{rad}^{-1} \cdot s^2$$
$$f = 0.01 \text{ N} \cdot \text{m} \cdot \text{rad}^{-1} \cdot s$$

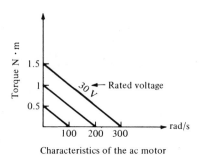

Characteristics of the ac motor

Figure P3-13

CHAPTER

4

Analysis: Quantitative and Qualitative

A great number of control systems can be described, as shown in the previous chapter, by both transfer functions and linear time-invariant dynamical equations. Once these mathematical descriptions are derived, the next step is naturally to analyze them. In this chapter the quantitative and qualitative analyses of these equations will be studied. In the quantitative analysis we are interested in the exact response of the equation due to some excitations. In the qualitative analysis we are interested in the general properties of the equations. For convenience of discussion, we study single-variable systems exclusively. Most of the results can however be easily extended to the multivariable case.

The analysis techniques developed in this chapter are applicable to any linear time-invariant system or any linear time-invariant composite system. By a composite system, we mean a system that consists of a number of subsystems. For example, the control systems introduced in Section 3-9 are all composite systems. In order to apply the analysis technique to composite systems, we must discuss first the mathematical descriptions of composite systems.

4-1 Mathematical Descriptions of Composite Systems

We use an example to illustrate the procedure of deriving mathematical equations for composite systems. Consider the feedback system shown in Figure 4-1.

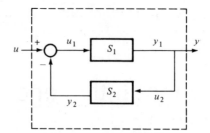

Figure 4-1
A feedback system.

It is assumed that the system S_i with input u_i and output y_i is described by the transfer function $\hat{g}_i(s)$ and the dynamical equation

$$\dot{x}_i = A_i x_i + b_i u_i \qquad (4\text{-}1a)$$

$$y_i = c_i x_i \qquad (4\text{-}1b)$$

Let u and y be, respectively, the input and output of the overall system. The problem is to find the transfer function and the state-variable descriptions of the overall system. The procedure is rather straightforward. First we shall establish the relationships between u, u_i, y, y_i from the configuration of the composite system, and then we use the description of S_i to eliminate u_i and y_i, for $i = 1$ and 2. Then the resulting equation will be in terms of u and y and will be the equation we are looking for. For example, for the composite system in Figure 4-1 we have

$$\hat{u}_1(s) = \hat{u}(s) - \hat{y}_2(s) \qquad (4\text{-}2)$$

$$\hat{u}_2(s) = \hat{y}_1(s) = \hat{y}(s) \qquad (4\text{-}3)$$

Substitution of $\hat{y}_1(s) = \hat{y}(s)$ into $\hat{y}_1(s) = \hat{g}_1(s)\hat{u}_1(s)$ yields $\hat{y}(s) = \hat{g}_1(s)\hat{u}_1(s)$, in which $\hat{u}_1(s)$ can be eliminated by using

$$\hat{u}_1(s) = \hat{u}(s) - \hat{y}_2(s) = \hat{u}(s) - \hat{g}_2(s)\hat{u}_2(s) = \hat{u}(s) - \hat{g}_2(s)\hat{y}(s)$$

This results in

$$\hat{y}(s) = \hat{g}_1(s)\hat{u}(s) - \hat{g}_1(s)\hat{g}_2(s)\hat{y}(s) \qquad (4\text{-}4)$$

or

$$[1 + \hat{g}_1(s)\hat{g}_2(s)]\hat{y}(s) = \hat{g}_1(s)\hat{u}(s) \qquad (4\text{-}5)$$

By definition, the overall transfer function $\hat{g}_f(s)$ is equal to $\hat{y}(s)/\hat{u}(s)$. Hence if $[1 + \hat{g}_1(s)\hat{g}_2(s)]^{-1}$ exists, then

$$\hat{g}_f(s) \triangleq \frac{\hat{y}(s)}{\hat{u}(s)} = \frac{\hat{g}_1(s)}{1 + \hat{g}_1(s)\hat{g}_2(s)} \qquad (4\text{-}6)$$

This is the overall transfer function of the feedback system shown in Figure 4-1. This equation will be constantly used and hence is worth remembering. Note that the feedback in Figure 4-1 is a negative feedback. If it were a positive feedback, then the overall transfer function would be $\hat{g}_1(s)/[1 - \hat{g}_1(s)\hat{g}_2(s)]$. (See Problem 4-1.)

4-1 MATHEMATICAL DESCRIPTIONS OF COMPOSITE SYSTEMS

In developing the state-variable description of a composite system we must first choose a state vector for the composite system. If $\mathbf{x}_i$ is the state vector of subsystem S_i, then the vector

$$\mathbf{x} = \begin{bmatrix} \mathbf{x}_1 \\ \mathbf{x}_2 \end{bmatrix} \qquad (4\text{-}7)$$

qualifies as the state for any connection of S_1 and S_2. For the feedback connection shown in Figure 4-1, we have $y_1 = y = u_2$, $u_1 = u - y_2$. Using Equation (4-1b), they can be written as

$$u_2 = y_1 = \mathbf{c}_1 \mathbf{x}_1$$

$$u_1 = u - y_2 = u - \mathbf{c}_2 \mathbf{x}_2$$

Substitution of these into Equation (4-1a) yields

$$\dot{\mathbf{x}}_1 = \mathbf{A}_1 \mathbf{x}_1 - \mathbf{b}_1 \mathbf{c}_2 \mathbf{x}_2 + \mathbf{b}_1 u$$

$$\dot{\mathbf{x}}_2 = \mathbf{A}_2 \mathbf{x}_2 + \mathbf{b}_2 \mathbf{c}_1 \mathbf{x}_1$$

Hence the overall state-variable description of Figure 4-1 is, using Equation (4-7),

$$\dot{\mathbf{x}} = \begin{bmatrix} \mathbf{A}_1 & -\mathbf{b}_1 \mathbf{c}_2 \\ \mathbf{b}_2 \mathbf{c}_1 & \mathbf{A}_2 \end{bmatrix} \mathbf{x} + \begin{bmatrix} \mathbf{b}_1 \\ 0 \end{bmatrix} u \qquad (4\text{-}8a)$$

$$y = \begin{bmatrix} \mathbf{c}_1 & 0 \end{bmatrix} \mathbf{x} \qquad (4\text{-}8b)$$

This example demonstrates the basic procedure of developing mathematical descriptions of composite systems. Although it is straightforward, it may become very tedious for complicated composite systems. Fortunately for the transfer-function description we have what is known as *Mason's formula*. The formula can be used to write down almost by inspection the overall transfer function of any composite system. For the state-variable description we have no similar formula and thus we have to use the procedure introduced above.

Mason's formula. We first introduce the concepts of *forward path* and *loop* of a composite system. Consider the composite system shown in Figure 4-2. Note that every branch is oriented—that is, unidirectional. A forward path from

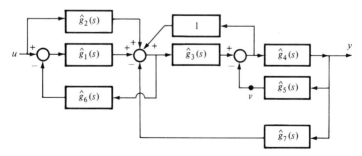

Figure 4-2 A composite system.

u to y is defined as any continuous connections of unidirectional branches from u to y along which no point is encountered more than once. For example, there are two forward paths from u to y in Figure 4-2: the paths consisting of the transfer functions $\{\hat{g}_1, \hat{g}_3, \hat{g}_4\}$ and $\{\hat{g}_2, \hat{g}_3, \hat{g}_4\}$. A loop is any unidirectional path that originates and terminates on the same point and along which no point is encountered more than once. The composite system in Figure 4-2 has four loops: $\{\hat{g}_4, -\hat{g}_5\}$, $\{\hat{g}_3, 1\}$, $\{\hat{g}_3, \hat{g}_4, -\hat{g}_7\}$, and $\{\hat{g}_1, -\hat{g}_6(s)\}$. It is important to remember that in designating transfer functions to a path or loop, the positive or negative sign at summing points should be included. Two loops are said to be *nontouching* if they have no point in common. For example, the loops $\{\hat{g}_4, -\hat{g}_5\}$ and $\{\hat{g}_1(s), -\hat{g}_6(s)\}$ are nontouching, but the loops $\{\hat{g}_1, -\hat{g}_6\}$ and $\{\hat{g}_3(s), 1\}$ are touching. Similarly a forward path and a loop are said to be nontouching if they have no point in common. A loop gain or path gain is defined as the product of all the transfer functions on the loop or path. With these concepts we are ready to introduce Mason's formula. The formula asserts that the transfer function from u to y is given by

$$\hat{g}(s) = \frac{\hat{y}(s)}{\hat{u}(s)} = \frac{\sum_i P_i \Delta_i}{\Delta} \quad (4\text{-}9)$$

where

$\Delta = 1 -$ (sum of all loop gains) $+$ (sum of the gain products of all possible combinations of two nontouching loops) $-$ (sum of the gain products of all possible combinations of three nontouching loops) $+ \cdots$
$P_i =$ the gain of the ith forward path from u to y
$\Delta_i =$ the Δ after replacing by zero the gains of those loops that touch the ith forward path

Note that the formation of Δ depends only on the loops of the system; it has nothing to do with the input u and output y. Hence Δ is an inherent characteristic of the system. Depending on different input and output, we have different P_i and consequently different Δ_i. We now give some examples of the application of Mason's formula.

Example 1

For the system shown in Figure 4-2, we have

$$\begin{aligned} \Delta &= 1 - (-\hat{g}_4\hat{g}_5 + \hat{g}_3 - \hat{g}_3\hat{g}_4\hat{g}_7 - \hat{g}_1\hat{g}_6) + (\hat{g}_1\hat{g}_6\hat{g}_4\hat{g}_5) \\ P_1 &= \hat{g}_1\hat{g}_3\hat{g}_4 \\ P_2 &= \hat{g}_2\hat{g}_3\hat{g}_4 \\ \Delta_1 &= 1 \\ \Delta_2 &= 1 \end{aligned} \quad (4\text{-}10)$$

Hence the transfer function from u to y of the system is

$$\hat{g}_a(s) = \frac{\hat{y}(s)}{\hat{u}(s)} = \frac{\hat{g}_1\hat{g}_3\hat{g}_4 + \hat{g}_2\hat{g}_3\hat{g}_4}{1 + \hat{g}_4\hat{g}_5 + \hat{g}_3\hat{g}_4\hat{g}_7 + \hat{g}_1\hat{g}_6\hat{g}_4\hat{g}_5 - \hat{g}_3 + \hat{g}_1\hat{g}_6} \quad (4\text{-}11)$$

Example 2

Find the transfer function from v to y of the system shown in Figure 4-2. The forward path from v to y is $\{-1, \hat{g}_4\}$. The gain -1 takes care of the negative feeding into the summing point. The formation of Δ is independent of forward paths; hence the Δ in this example is identical to the one in Example 1. From the definition we have

$$\Delta_1 = 1 - (-\hat{g}_1\hat{g}_6) = 1 + \hat{g}_1\hat{g}_6$$

Hence the transfer function from v to y is

$$\hat{g}_b(s) = \frac{\hat{y}(s)}{\hat{v}(s)} = \frac{-\hat{g}_4(1 + \hat{g}_1\hat{g}_6)}{\Delta}$$

where Δ is as given in Equation (4-10). Similarly the transfer function from v to u is zero, for there is no forward path from v to u. The transfer function from y to v is

$$\hat{g}_c(s) = \frac{\hat{v}(s)}{\hat{y}(s)} = \frac{\hat{g}_5(1 - \hat{g}_3 + \hat{g}_1\hat{g}_6)}{\Delta}$$

for $P_1 = \hat{g}_5$ and $\Delta_1 = 1 - (-\hat{g}_1\hat{g}_6 + \hat{g}_3 \cdot 1) = 1 - \hat{g}_3 + \hat{g}_1\hat{g}_6$. ∎

4-2 Transfer Function-Partial Fraction Expansion

In this section the response of a system will be studied from its transfer function. Since the transfer function describes only the zero-state response, the system will implicitly be assumed to be initially relaxed. The system may consist of only a single component or a large number of components. In either case the overall transfer function is assumed to be known. Before proceeding we first introduce the concepts of poles and zeros.

Definition 4-1

A number λ (real or complex) is said to be a *pole* of a rational function $\hat{g}(s)$ if $|\hat{g}(\lambda)| = \infty$. A number λ is said to be a *zero* of $\hat{g}(s)$ if $\hat{g}(\lambda) = 0$. ∎

If a rational function is irreducible—that is, if there is no common factor (except a constant) between its denominator and numerator (see Definition 2-6)—then the poles and zeros of the rational function are just, respectively, the roots of its denominator and numerator. Without the irreducibility assumption, the above statement is not correct. For example, $\lambda = -1$ is not a pole of $\hat{g}(s) = 2(s + 1)/(s^2 + 3s + 2)$, although it is a root of its denominator.

If a transfer function is factored in terms of poles and zeros, it will be of the form

$$\hat{g}(s) = k \frac{(s - z_1)(s - z_2) \cdots (s - z_m)}{(s - p_1)(s - p_2) \cdots (s - p_n)}$$

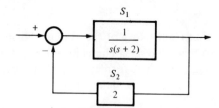

Figure 4-3
A feedback system.

where $p_1, p_2, \ldots$ are not necessarily distinct, nor are $z_1, z_2, \ldots$. Note that the knowledge of poles and zeros is not sufficient to determine the transfer function uniquely. In order to determine it uniquely, we need, in addition, the knowledge of the multiplication factor k.

Example 1

Consider the system shown in Figure 4-3. The transfer function of the overall system is, using Equation (4-6),

$$\hat{g}(s) = \frac{\dfrac{1}{s(s+2)}}{1 + \dfrac{2}{s(s+2)}} = \frac{1}{s^2 + 2s + 2} = \frac{1}{(s+1-j)(s+1+j)} \tag{4-12}$$

Hence the poles of the system are $-1 \pm j$. There is no zero in the system. We note that the poles of the overall system are different from those of systems S_1 and S_2. ∎

The input and output of every linear time-invariant lumped system that is initially relaxed at $t = 0$ can be described, in the frequency domain, by

$$\hat{y}(s) = \hat{g}(s)\hat{u}(s) \tag{4-13}$$

where $\hat{g}(s)$ is its transfer function; or in the time domain,

$$y(t) = \int_0^t g(t-\tau)u(\tau)\,d\tau \tag{4-14}$$

where $g(t)$ is its impulse response and is the inverse Laplace transform of $\hat{g}(s)$. In computing the response of the system, if the Laplace transform of $u(t)$ is not a rational function, it may be easier to compute y directly from Equation (4-14) than from (4-13). Equation (4-14) can be computed by either the graphical or numerical method. Since its computation is not needed in this text, its discussion is omitted. The interested reader is referred to Reference [2]. If the Laplace transform of u is a rational function, then it is very convenient to compute the output $y(t)$ from (4-13) by using the Laplace transform table (see Appendix A). The procedure is listed in the following.

4-2 TRANSFER FUNCTION-PARTIAL FRACTION EXPANSION

Computation of $y(t)$ from $\hat{y}(s) = \hat{g}(s)\hat{u}(s)$. The required computation is listed in the following:

1. Find the Laplace transform of $u(t)$ by using the Laplace transform table.
2. Compute the poles of $\hat{g}(s)\hat{u}(s)$; that is,

$$\hat{y}(s) = \hat{g}(s)\hat{u}(s) = \frac{N(s)}{\prod_{i=1}^{m}(s - \lambda_i)^{n_i}}$$

where λ_i, real or complex, for $i = 1, 2, \ldots, m$, are the distinct poles of $\hat{g}(s)\hat{u}(s)$. Since $\hat{g}(s)$ and $\hat{u}(s)$ are the Laplace transforms of real-valued time functions, their coefficients are real. Hence the complex conjugate poles must appear in pairs and have the same power.

3. By using partial fraction expansion, expand $\hat{y}(s)$ into

$$\hat{y}(s) = \sum_{i=1}^{m} \left[\frac{k_{i1}}{(s - \lambda_i)} + \frac{k_{i2}}{(s - \lambda_i)^2} + \cdots + \frac{k_{in_i}}{(s - \lambda_i)^{n_i}} \right] \quad \text{(4-15)}$$

where

$$k_{in_i} = \hat{g}(s)\hat{u}(s)(s - \lambda_i)^{n_i}\big|_{s=\lambda_i}$$

$$k_{i(n_i-1)} = \frac{d}{ds}\left[\hat{g}(s)\hat{u}(s)(s - \lambda_i)^{n_i}\right]\bigg|_{s=\lambda_i}$$

$$\vdots \quad \text{(4-16)}$$

$$k_{i1} = \frac{1}{(n_i - 1)!} \frac{d^{n_i-1}}{ds^{n_i-1}} \left[\hat{g}(s)\hat{u}(s)(s - \lambda_i)^{n_i}\right]\bigg|_{s=\lambda_i}$$

for $i = 1, 2, \ldots, m$. Note that the constants associated with complex conjugate poles are mutually complex conjugate. To be more specific, if $\lambda_j = \bar{\lambda}_l$, where the overbar denotes the complex conjugate, then $k_{j\alpha} = \bar{k}_{l\alpha}$, for $\alpha = 1, 2, \ldots, n_j$.

4. The output $y(t)$ is obtained, by taking the inverse Laplace transform, as

$$y(t) = \sum_{i=1}^{m}\left[k_{i1}e^{\lambda_i t} + k_{i2}te^{\lambda_i t} + \cdots + \frac{k_{in_i}}{(n_i - 1)!}t^{n_i-1}e^{\lambda_i t}\right]$$

We see that the output is a linear combination of terms of form $t^k e^{\lambda t}$, for $k = 0, 1, 2, \ldots$. These terms are dictated by the poles of $\hat{g}(s)$ and those of $\hat{u}(s)$. This shows the importance of the concept of poles.

Example 2

Find the output of the system shown in Figure 4-3 due to the application of a unit step function.

The transfer function of the system is computed in Equation (4-12). The Laplace transform of a unit step function [that is, $u(t) = 1$ for $t \geq 0$, and $u(t) = 0$ for $t < 0$] is $1/s$. Hence we have

$$\hat{y}(s) = \hat{g}(s)\hat{u}(s) = \frac{1}{(s + 1 + j)(s + 1 - j)s}$$

$$= \frac{1/[2j(j + 1)]}{s + 1 + j} + \frac{1/[2j(j - 1)]}{s + 1 - j} + \frac{\frac{1}{2}}{s}$$

88 ANALYSIS: QUANTITATIVE AND QUALITATIVE

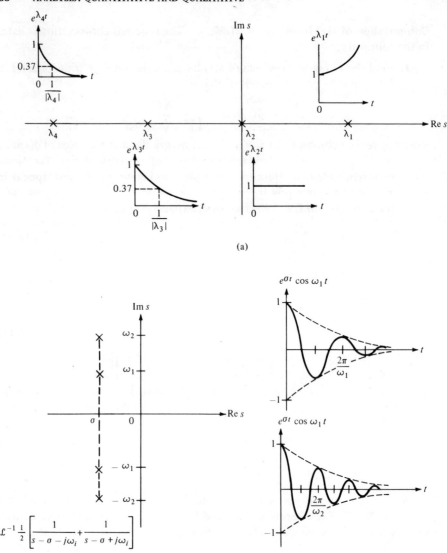

Figure 4-4 (a) The responses of real poles. (b) The responses of pairs of complex conjugate poles.

which implies

$$y(t) = \frac{1}{2} + \frac{1}{2j(j+1)} e^{-(1+j)t} + \frac{1}{2j(j-1)} e^{-(1-j)t}$$

$$= \frac{1}{2} + \frac{1}{2\sqrt{2}} e^{-t} \cos\left(t + \frac{3\pi}{4}\right) \qquad \text{for } t \geq 0$$

The time function of some real and complex poles are shown in Figure 4-4. For the complex poles we see that the envelope of the response is governed by the real part of the pole and the frequency of oscillation is governed by the imaginary part of the pole. If a pole lies in the open right-half s-plane,[1] then the response will increase exponentially to infinity; if a pole lies in the open left s-plane, then the response will decrease exponentially to zero. If a pole lies on the imaginary axis, then the response may either go to infinity or stay in oscillation, depending on whether it is a repeated or a simple pole. (Why?)

For a real or complex pole lying in the open left-half s-plane, the positive real number defined as $\tau \triangleq 1/|\lambda|$ if the pole is real, or $\tau \triangleq 1/|\alpha|$ if the pole is complex and α is its real part, is called the *time constant*. Its physical meaning is that the response or its envelope will reduce to $e^{-1} \times 100 = 36.8$ percent of its original value in each time constant. Hence the smaller the time constant, the faster the response decreases.

In the following we introduce a geometrical method of computing the constants k_i of a partial fraction expansion from the pole-zero configuration. We use an example to illustrate the procedure. Consider

$$\hat{y}(s) = \hat{g}(s)\hat{u}(s) = k \frac{(s - z_1)}{(s - \lambda_1)(s - \lambda_2)(s - \bar{\lambda}_2)}$$

It can be expanded as

$$y(s) = k \left(\frac{k_1}{s - \lambda_1} + \frac{k_2}{s - \lambda_2} + \frac{\bar{k}_2}{s - \bar{\lambda}_2} \right)$$

Here we have used the fact that the constants associated with complex conjugate poles are mutually conjugate. The constants k_1 and k_2 are, from Equation (4-16),

$$k_1 = \frac{(\lambda_1 - z_1)}{(\lambda_1 - \lambda_2)(\lambda_1 - \bar{\lambda}_2)}$$

$$k_2 = \frac{(\lambda_2 - z_1)}{(\lambda_2 - \lambda_1)(\lambda_2 - \bar{\lambda}_2)}$$

The poles and zero of $\hat{y}(s)$ are plotted in Figure 4-5. Note that $(\lambda_i - \lambda_j)$ is a vector from λ_j to λ_i. If every vector in k_1 and k_2 is expressed in terms of magnitude and phase, then k_1 and k_2 become

$$k_1 = \frac{l_2}{l_1 e^{j\theta_1} l_1 e^{j\theta_2}} = \frac{l_2}{l_1^2} e^{-j(\theta_1 + \theta_2)} = \frac{l_2}{l_1^2} e^{-j2\pi} = \frac{l_2}{l_1^2}$$

$$k_2 = \frac{l_3 e^{j\theta_4}}{l_1 e^{j\theta_3} l_4 e^{j\theta_5}} = \frac{l_3}{l_1 l_4} e^{j(\theta_4 - \theta_3 - \theta_5)}$$

---

[1] By *open* right-half s-plane we mean the right-half s-plane excluding the imaginary axis; the *closed* right s-plane includes the imaginary axis.

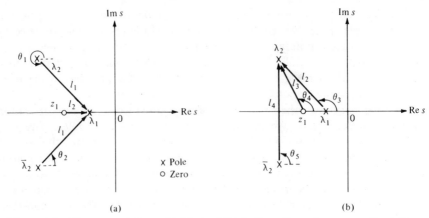

Figure 4-5 The computation of (a) k_1 and (b) k_2 from the pole-zero configuration.

where l_i denotes the magnitude and θ_i the phase of each vector as shown in Figure 4-5. Once l_i and θ_i are *measured* directly from the pole-zero configuration, the constants k_1 and k_2 can be readily obtained.

From this graphical method of computing k_i we can see that if a zero is very close to a pole, then the constant associated with the pole is very small. If there are two poles very close, then the constants associated with these two poles are very large. Hence it is possible to estimate the constant associated with a pole from the locations of the pole and zero. The output $\hat{y}(s)$ in this example has only simple poles. If $\hat{y}(s)$ has repeated poles, it is still possible to estimate the constants associated with simple poles from the pole-zero locations. However for the constants associated with repeated poles, some additional computation is required. See Reference [6].

4-3 Solutions of Dynamical Equations

In this section the solution of the linear time-invariant dynamical equation

$$\dot{x} = Ax + bu \qquad (4\text{-}17a)$$

$$y = cx + du \qquad (4\text{-}17b)$$

will be studied. We examine it first in the frequency domain and then in the time domain. By applying the Laplace transform, Equation (4-17) becomes, as derived in (2-27) and (2-29),

$$\hat{x}(s) = (sI - A)^{-1} x(0) + (sI - A)^{-1} b\hat{u}(s) \qquad (4\text{-}18a)$$

$$\hat{y}(s) = c\hat{x}(s) + d\hat{u}(s) \qquad (4\text{-}18b)$$

These are algebraic equations. After $\hat{x}(s)$ is computed algebraically, its time response can then be obtained by taking its inverse Laplace transform. This is best illustrated by an example.

Example 1

Find the response of

$$\dot{x} = \begin{bmatrix} 0 & 0 & -2 \\ 0 & 1 & 0 \\ 1 & 0 & 3 \end{bmatrix} x + \begin{bmatrix} 0 \\ 0 \\ 1 \end{bmatrix} u$$

$$y = \begin{bmatrix} 1 & 1 & 0 \end{bmatrix} x$$

due to the initial state $x(0) = \begin{bmatrix} 0 & 1 & 0 \end{bmatrix}'$ and a unit step function input, where the prime denotes the transpose of a vector.

By direct computation, we have

$$(sI - A)^{-1} = \begin{bmatrix} s & 0 & 2 \\ 0 & s-1 & 0 \\ -1 & 0 & s-3 \end{bmatrix}^{-1}$$

$$= \frac{1}{(s-1)^2(s-2)} \quad (4\text{-}19)$$

$$\times \begin{bmatrix} (s-1)(s-3) & 0 & -2(s-1) \\ 0 & (s-1)(s-2) & 0 \\ (s-1) & 0 & s(s-1) \end{bmatrix}$$

and, from (4-18a),

$$\hat{x}(s) = \begin{bmatrix} 0 \\ \frac{1}{s-1} \\ 0 \end{bmatrix} + \frac{1}{(s-1)^2(s-2)} \begin{bmatrix} -2(s-1) \\ 0 \\ s(s-1) \end{bmatrix} \frac{1}{s} = \begin{bmatrix} \frac{-2}{s(s-1)(s-2)} \\ \frac{1}{s-1} \\ \frac{1}{(s-1)(s-2)} \end{bmatrix}$$

$$= \begin{bmatrix} \frac{-1}{s} + \frac{2}{s-1} - \frac{1}{s-2} \\ \frac{1}{s-1} \\ \frac{-1}{s-1} + \frac{1}{s-2} \end{bmatrix}$$

Consequently the response of the system is

$$x(t) = \begin{bmatrix} -1 + 2e^t - e^{2t} \\ e^t \\ -e^t + e^{2t} \end{bmatrix} \quad (4\text{-}20)$$

and

$$y(t) = \begin{bmatrix} 1 & 1 & 0 \end{bmatrix} x(t) = -1 + 3e^t - e^{2t} \qquad \blacksquare$$

We discuss now the solution of (4-17) directly in the time domain. The solution hinges on the matrix function e^{At}. Following the infinite series

$$e^{at} = 1 + at + \frac{1}{2!} a^2 t^2 + \cdots + \frac{1}{n!} a^n t^n + \cdots$$

we *define*

$$e^{At} \triangleq I + At + \frac{1}{2!} A^2 t^2 + \cdots + \frac{1}{n!} A^n t^n + \cdots \qquad (4\text{-}21)$$

where I is a unit matrix and $A^n \triangleq AA \cdots A$ (n times). This is an infinite power series. It can be shown that the series converges for any finite t; hence the series is well defined. We derive from this definition some properties of e^{At}. At $t = 0$, (4-21) reduces to

$$e^0 = I \qquad (4\text{-}22)$$

Direct verification yields

$$e^{A(t-\tau)} = e^{At} e^{-A\tau} \qquad (4\text{-}23)$$

which implies, by choosing $t = \tau$, that

$$e^{At} e^{-At} = e^0 = I$$

or

$$[e^{At}]^{-1} = e^{-At} \qquad (4\text{-}24)$$

Taking the derivative of e^{At} term by term, we have

$$\frac{d}{dt} e^{At} = A + A^2 t + \frac{1}{2!} A^3 t^2 + \cdots + \frac{1}{(n-1)!} A^n t^{n-1} + \cdots$$

$$= A \left(I + At + \frac{1}{2!} A^2 t^2 + \cdots + \frac{1}{(n-1)!} A^{n-1} t^{n-1} + \cdots \right)$$

$$= \left(I + At + \frac{1}{2!} A^2 t^2 + \cdots + \frac{1}{(n-1)!} A^{n-1} t^{n-1} + \cdots \right) A$$

$$= A e^{At} = e^{At} A \qquad (4\text{-}25)$$

With these properties, the solution of (4-17) can now be found. We rewrite (4-17) as

$$\dot{x} - Ax = bu(t)$$

which implies that

$$\frac{d}{d\tau} (e^{-A\tau} x) = \left(\frac{d}{d\tau} e^{-A\tau} \right) x + e^{-A\tau} \left(\frac{d}{d\tau} x \right) = e^{-A\tau} (-Ax + \dot{x}) = e^{-A\tau} bu(\tau)$$

Hence we have

$$\int_0^t \frac{d}{d\tau} (e^{-A\tau} x) \, d\tau = e^{-A\tau} x(\tau) \Big|_0^t = \int_0^t e^{-A\tau} bu(\tau) \, d\tau$$

or

$$e^{-\mathbf{A}t}\mathbf{x}(t) - \mathbf{I}\mathbf{x}(0) = \int_0^t e^{-\mathbf{A}\tau}\mathbf{b}u(\tau)\,d\tau$$

or

$$\mathbf{x}(t) = e^{\mathbf{A}t}\left[\mathbf{x}(0) + \int_0^t e^{-\mathbf{A}\tau}\mathbf{b}u(\tau)\,d\tau\right] = e^{\mathbf{A}t}\mathbf{x}(0) + e^{\mathbf{A}t}\int_0^t e^{-\mathbf{A}\tau}\mathbf{b}u(\tau)\,d\tau$$

(4-26)

Hence if $e^{\mathbf{A}t}$ is known, the response of Equation (4-17) can be computed from (4-26) and (4-17a). If $e^{\mathbf{A}t}$ is computed from the infinite series (4-21), a closed-form solution generally cannot be obtained. In the following we introduce a different procedure for computing $e^{\mathbf{A}t}$, which will result in a closed-form solution. See Reference [1].

Problem. Given an $n \times n$ real constant matrix $\mathbf{A}$. Compute $e^{\mathbf{A}t}$.

Procedure

1. Compute $\Delta(\lambda) \triangleq \det(\lambda\mathbf{I} - \mathbf{A})$, where det stands for the determinant. $\Delta(\lambda)$ is called the *characteristic polynomial* of $\mathbf{A}$.
2. Find the roots of $\Delta(\lambda) = 0$, say

$$\Delta(\lambda) = (\lambda - \lambda_1)^{n_1}(\lambda - \lambda_2)^{n_2} \cdots (\lambda - \lambda_m)^{n_m}$$

where $n_1 + n_2 + \cdots + n_m = n$. In other words, $\Delta(\lambda)$ has root λ_i with multiplicity n_i. If λ_i is a complex number, then its complex conjugate is also a root of $\Delta(\lambda)$. The roots $\lambda_1, \lambda_2, \ldots, \lambda_m$ are called the *eigenvalues* of $\mathbf{A}$.
3. Form a polynomial $g(\lambda)$ of degree $n - 1$; that is,

$$g(\lambda) = \alpha_0 + \alpha_1\lambda + \alpha_2\lambda^2 + \cdots + \alpha_{n-1}\lambda^{n-1}$$

where the unknown parameters $\alpha_0, \alpha_1, \ldots, \alpha_{n-1}$ are to be solved in step 5.

4. Form the following n equations (for $i = 1, 2, \ldots, m$):

$$e^{\lambda_i t} = g(\lambda_i)$$
$$te^{\lambda_i t} = g'(\lambda_i)$$
$$\vdots$$
$$t^{n_i-1}e^{\lambda_i t} = g^{n_i-1}(\lambda_i)$$

5. Solve for the n unknown $\alpha_0, \alpha_1, \ldots, \alpha_{n-1}$ from the n equations in step 4. Then

$$e^{\mathbf{A}t} \triangleq g(\mathbf{A}) = \alpha_0\mathbf{I} + \alpha_1\mathbf{A} + \cdots + \alpha_{n-1}\mathbf{A}^{n-1}$$

Once $e^{\mathbf{A}t}$ is computed, the response of (4-17) can be obtained from (4-26) by direct integration.

Example 2

Find the response of

$$\dot{\mathbf{x}} = \begin{bmatrix} 0 & 0 & -2 \\ 0 & 1 & 0 \\ 1 & 0 & 3 \end{bmatrix} \mathbf{x} + \begin{bmatrix} 0 \\ 0 \\ 1 \end{bmatrix} u$$

$$y = \begin{bmatrix} 1 & 1 & 0 \end{bmatrix} \mathbf{x}$$

due to the initial state, $\mathbf{x}(0) = \begin{bmatrix} 0 & 1 & 0 \end{bmatrix}'$, and a unit step function input.

We first compute $e^{\mathbf{A}t}$. The characteristic polynomial of $\mathbf{A}$ is

$$\Delta(\lambda) = \det [\lambda \mathbf{I} - \mathbf{A}] = \det \begin{bmatrix} \lambda & 0 & 2 \\ 0 & \lambda - 1 & 0 \\ -1 & 0 & \lambda - 3 \end{bmatrix} = (\lambda - 1)^2 (\lambda - 2)$$

Let $g(\lambda) = \alpha_0 + \alpha_1 \lambda + \alpha_2 \lambda^2$. Then

$$e^{\lambda t} = g(\lambda) \text{ at } \lambda = 1 \Rightarrow e^t = \alpha_0 + \alpha_1 + \alpha_2$$

$$te^{\lambda t} = g'(\lambda) \text{ at } \lambda = 1 \Rightarrow te^t = \alpha_1 + 2\alpha_2$$

$$e^{\lambda t} = g(\lambda) \text{ at } \lambda = 2 \Rightarrow e^{2t} = \alpha_0 + 2\alpha_1 + 4\alpha_2$$

Solving these equations, we obtain $\alpha_0 = -2te^t + e^{2t}$, $\alpha_1 = 3te^t + 2e^t - 2e^{2t}$, and $\alpha_2 = e^{2t} - e^t - te^t$. Hence we have

$$e^{\mathbf{A}t} = g(\mathbf{A}) = (-2te^t + e^{2t})\mathbf{I} + (3te^t + 2e^t - 2e^{2t})\mathbf{A} + (e^{2t} - e^t - te^t)\mathbf{A}^2$$

$$= \begin{bmatrix} 2e^t - e^{2t} & 0 & 2e^t - 2e^{2t} \\ 0 & e^t & 0 \\ -e^t + e^{2t} & 0 & 2e^{2t} - e^t \end{bmatrix}$$

The solution $x(t)$ can now be computed from (4-26) as

$$x(t) = \begin{bmatrix} 2e^t - e^{2t} & 0 & 2e^t - 2e^{2t} \\ 0 & e^t & 0 \\ -e^t + e^{2t} & 0 & 2e^{2t} - e^t \end{bmatrix}$$

$$\times \left(\begin{bmatrix} 0 \\ 1 \\ 0 \end{bmatrix} + \int_0^t \begin{bmatrix} 2e^{-\tau} - e^{-2\tau} & 0 & 2e^{-\tau} - 2e^{-2\tau} \\ 0 & e^{-\tau} & 0 \\ -e^{-\tau} + e^{-2\tau} & 0 & 2e^{-2\tau} - e^{-\tau} \end{bmatrix} \begin{bmatrix} 0 \\ 0 \\ 1 \end{bmatrix} u(\tau) \, d\tau \right)$$

$$= \begin{bmatrix} 2e^t - e^{2t} & 0 & 2e^t - 2e^{2t} \\ 0 & e^t & 0 \\ -e^t + e^{2t} & 0 & 2e^{2t} - e^t \end{bmatrix} \times \left(\begin{bmatrix} 0 \\ 1 \\ 0 \end{bmatrix} + \begin{bmatrix} \int_0^t (2e^{-\tau} - 2e^{-2\tau}) \, d\tau \\ 0 \\ \int_0^t (2e^{-2\tau} - e^{-\tau}) \, d\tau \end{bmatrix} \right)$$

$$= \begin{bmatrix} -1 + 2e^t - e^{2t} \\ e^t \\ e^{2t} - e^{-t} \end{bmatrix}$$

The output $y(t)$ is given by

$$y(t) = \begin{bmatrix} 1 & 1 & 0 \end{bmatrix} \mathbf{x}(t) = -1 + 3e^t - e^{2t}$$

This result is identical to the one in Example 1. ∎

The solution of a linear time-invariant dynamical equation can be obtained either by using the Laplace transform or directly in the time domain. For the example presented in this section, it seems that the Laplace-transform method is slightly simpler than the time-domain method. It is not clear however that this is always the case. If a dynamical equation is solved in a digital computer (see Chapter 5), neither method is used. The solution is obtained by direct integration or by using the infinite series (4-21). There are many existing subroutines available, for example, in IBM/360 Scientific Subroutine Packages. These subroutines are well tested and can be easily employed.

4-4 BIBO, Asymptotic, and Total Stability

In this section a qualitative property of linear systems—namely, stability—will be introduced. The concept of stability is very important because almost every working system is designed to be stable. If a system is not stable, it is usually of no use in practice. Since the response of a linear system can always be decomposed into the zero-state response and the zero-input response, we shall first discuss the stability of the zero-state response and that of the zero-input response, and then discuss the stability of the total response.

Definition 4-2

The zero-state response of a system is said to be *BIBO (bounded-input bounded-output) stable* if for every bounded input, the excited output is bounded. ∎

A function $h(t)$ is said to be bounded if its magnitude does not go to infinity in the time interval $[0, \infty)$ or, equivalently, if there exists a real constant k such that $|h(t)| \leq k < \infty$ for all t in $[0, \infty)$.

Example 1

Consider the network shown in Figure 4-6. The input u is a current source, and the output y is the voltage across the inductor. The transfer function from u to

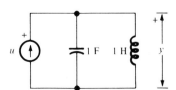

Figure 4-6
A network that is not BIBO stable.

y can be easily shown to be $s/(s^2 + 1)$. If we apply input $u(t) = \sin t$, and if the initial condition is zero, then the output is given by

$$y(t) = \mathscr{L}^{-1}\left(\frac{1}{s^2+1} \cdot \frac{s}{s^2+1}\right) = \frac{-1}{2} t \sin t$$

The input $u(t) = \sin t$ is bounded; however the output is not. Hence the system is not BIBO stable. ∎

To show that a system is not BIBO stable it is sufficient to find one bounded input that excites an unbounded output, as we did in this example. However even if we find a thousand bounded inputs that excite bounded outputs, we still cannot conclude that the system is BIBO stable.

The zero-state response of a system is characterized fully by its transfer functions; hence the BIBO stability is a property of the transfer function. If a system is described by a dynamical equation, say,

$$\dot{\mathbf{x}} = \mathbf{A}\mathbf{x} + \mathbf{b}u$$
$$y = \mathbf{c}\mathbf{x} + du$$

in order to check its BIBO stability, the transfer function

$$\hat{g}(s) = \mathbf{c}(s\mathbf{I} - \mathbf{A})^{-1}\mathbf{b} + d$$

must be first computed.

Theorem 4-1

A linear time-invariant system that is described by a proper rational transfer function $\hat{g}(s)$ is BIBO stable if and only if *all* the poles of $\hat{g}(s)$ have negative real parts or, equivalently, if all the poles of $\hat{g}(s)$ lie inside the open left-half s-plane. ∎

The open left-half s-plane is the left-half s-plane excluding the $j\omega$-axis. This theorem asserts that if none of the poles of $\hat{g}(s)$ have positive or zero real parts, then the system is BIBO stable; otherwise the system is not BIBO stable. We note that the zeros of $\hat{g}(s)$ play no role in the stability. This theorem is not proved here; the interested reader should see Reference [1], page 321. If we apply this theorem to Example 1, we conclude immediately that the system is not BIBO stable, for its transfer function has poles $+j$ and $-j$, which have zero real parts.

We study now the zero-input responses of a system. Let the system be described by dynamical equation

$$\dot{\mathbf{x}} = \mathbf{A}\mathbf{x} + \mathbf{b}u \qquad (4\text{-}27\text{a})$$
$$y = \mathbf{c}\mathbf{x} + du \qquad (4\text{-}27\text{b})$$

In the study of the zero-input response, the input u is assumed to be identically zero; hence (4-27a) reduces to

$$\dot{\mathbf{x}} = \mathbf{A}\mathbf{x} \qquad (4\text{-}28)$$

Note that in this equation the response is excited exclusively by a nonzero initial state. Clearly the transfer function and the concept of BIBO stability are not applicable here, for there is no input involved in (4-28). The stability concept we shall introduce for (4-28) is called *asymptotic stability*.

Definition 4-3

The zero-input response of (4-27a) or, equivalently, the response of $\dot{\mathbf{x}} = \mathbf{A}\mathbf{x}$ is said to be *asymptotically stable* if, for *any* initial state $\mathbf{x}_0$, the response due to $\mathbf{x}_0$ approaches zero eventually; that is, $\mathbf{x}(t) = e^{\mathbf{A}t}\mathbf{x}_0 \to \mathbf{0}$ as $t \to \infty$. ∎

Note that this definition is defined for the state. However if the state $\mathbf{x}(t)$ approaches zero, so does the output.

Example 2

Consider the network shown in Figure 4-6. If the input u is zero, the network reduces to the one in Figure 4-7(a). The state variables of the system are the capacitor voltage x_1 and inductor current x_2. Because there is no energy dissipative element (that is, resistor) in Figure 4-7(a), if the initial state is different from zero, the energy will transfer back and forth between the inductor and capacitor, and the state will never go to zero. Hence the network in Figure 4-7(a) is not asymptotically stable. For the network in Figure 4-7(b), because the energy will dissipate in the resistor, x_1 and x_2 will eventually approach zero. Hence the network in Figure 4-7(b) is asymptotically stable. ∎

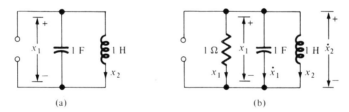

Figure 4-7 (a) A network that is not asymptotically stable. (b) An asymptotically stable network.

The condition for asymptotic stability is given in the following. Its proof can be found in Reference [1], page 336.

Theorem 4-2

The response of $\dot{\mathbf{x}} = \mathbf{A}\mathbf{x}$ is asymptotically stable if and only if all the eigenvalues of $\mathbf{A}$ [that is, all the roots of the characteristic polynomial of $\mathbf{A}$, $\Delta(\lambda) \triangleq \det(\lambda\mathbf{I} - \mathbf{A})$] have negative real parts. ∎

Example 3

The state-variable description of the network in Figure 4-7(b) can be found as

$$\begin{bmatrix} \dot{x}_1 \\ \dot{x}_2 \end{bmatrix} = \begin{bmatrix} -1 & -1 \\ 1 & 0 \end{bmatrix} \begin{bmatrix} x_1 \\ x_2 \end{bmatrix}$$

Its characteristic polynomial is

$$\Delta(\lambda) = \det(\lambda \mathbf{I} - \mathbf{A}) = \det \begin{bmatrix} \lambda + 1 & 1 \\ -1 & \lambda \end{bmatrix} = \lambda(\lambda + 1) + 1$$

Its roots are $(-1 \pm \sqrt{3})/2$ which have negative real parts. Hence the network is asymptotically stable. ∎

At this point it is natural to discuss the relationship between BIBO stability and asymptotic stability. The BIBO stability is determined by the poles of a transfer function, whereas the asymptotic stability is determined by the eigenvalues of a matrix $\mathbf{A}$. Therefore if the relationship between the poles and the eigenvalues can be established, then the posed problem can be resolved. The transfer function and the matrix $\mathbf{A}$ are related by

$$\hat{g}(s) = \mathbf{c}(s\mathbf{I} - \mathbf{A})^{-1}\mathbf{b} + d = \frac{1}{\det(s\mathbf{I} - \mathbf{A})} \mathbf{c} \operatorname{Adj}(s\mathbf{I} - \mathbf{A})\mathbf{b} + d$$

(4-29)

where Adj( ) stands for the adjoint of a matrix, and $\mathbf{c}$ Adj $(s\mathbf{I} - \mathbf{A})\mathbf{b}$ is a polynomial. The eigenvalues of $\mathbf{A}$ are the roots of det $(s\mathbf{I} - \mathbf{A})$. The poles of $\hat{g}(s)$ are part of the roots of det $(s\mathbf{I} - \mathbf{A})$ because of the possibility of having common factor between det $(s\mathbf{I} - \mathbf{A})$ and $\mathbf{c}$ Adj $(s\mathbf{I} - \mathbf{A})\mathbf{b}$. Hence we have

$$\{\text{poles of } \hat{g}(s)\} \subset \{\text{eigenvalues of } A\} \qquad (4\text{-}30)$$

where $\subset$ denotes "a subset of," or "included in." This fact implies that if all the eigenvalues of $\mathbf{A}$ have negative real parts, so have the poles of $\hat{g}(s)$. Hence *the asymptotic stability of the zero-input response implies the BIBO stability of the zero-state response.* Conversely, the fact that all the poles have negative real parts does not necessarily imply that all the eigenvalues have negative real parts. Hence BIBO stability may not imply asymptotic stability.

The BIBO stability is defined for the zero-state response, whereas the asymptotic stability is defined for the zero-input response. We shall now introduce a stability concept for the entire response.

Definition 4-4[2]

A system is said to be *totally stable* if its zero-input response is asymptotically stable, and for any initial state and for any bounded input, the output as well as all the state variables are bounded. ∎

---

[2] This definition is different from the total stability defined in Reference [1]. This is done to avoid the subtle difference between total stability and asymptotic stability and to simplify the subsequent development.

We see that total stability is more stringent than BIBO stability. It requires not only the boundedness of the output but also of all state variables; the boundedness must hold not only for the zero state but also for any initial state. A system that is BIBO stable may not function properly, because some of the state variables might increase with time, and the system will burn out, or at least saturate. Therefore in practice, every system is required to be totally stable.

As can be seen from Definition 4-4, total stability implies asymptotic stability. It turns out that asymptotic stability also implies total stability. This can be seen by writing

$$\dot{\mathbf{x}}(s) = (s\mathbf{I} - \mathbf{A})^{-1}\mathbf{x}(0) + (s\mathbf{I} - \mathbf{A})^{-1}\mathbf{b}\hat{u}(s)$$

$$= \frac{1}{\det(s\mathbf{I} - \mathbf{A})} \text{Adj}(s\mathbf{I} - \mathbf{A})[\mathbf{x}(0) + \mathbf{b}\hat{u}(s)]$$

and observing that the poles of the transfer functions from u to each component of $\mathbf{x}$ are just the roots of $\det(s\mathbf{I} - \mathbf{A})$. Hence if the zero-input response of the system is asymptotically stable, then all the poles have negative real parts. Hence, following Theorem 4-1, for any bounded u and any $\mathbf{x}(0)$, all state variables, and consequently the output, will be bounded. Therefore we have the following theorem.

Theorem 4-3

The system described by (4-27) is totally stable if and only if all eigenvalues of $\mathbf{A}$ have negative real parts. ∎

Example 4

Study the stability of the system described by

$$\dot{\mathbf{x}} = \begin{bmatrix} -1 & 1 \\ 0 & -2 \end{bmatrix} \mathbf{x} + \begin{bmatrix} 1 \\ 0 \end{bmatrix} u$$

$$y = \begin{bmatrix} 2 & 1 \end{bmatrix} \mathbf{x}$$

The eigenvalues of $\mathbf{A}$ are -1 and -2. Hence the system is asymptotically stable, totally stable, and, consequently, BIBO stable.

Example 5

Study BIBO, total, and asymptotic stability of the system described by

$$\begin{bmatrix} \dot{x}_1 \\ \dot{x}_2 \end{bmatrix} = \begin{bmatrix} -1 & 1 \\ 0 & 0 \end{bmatrix} \begin{bmatrix} x_1 \\ x_2 \end{bmatrix} + \begin{bmatrix} 1 \\ 0 \end{bmatrix} u \qquad (4\text{-}31)$$

$$y = \begin{bmatrix} 2 & 1 \end{bmatrix} \mathbf{x}$$

The eigenvalues of **A** are -1 and 0. One eigenvalue has a zero real part; hence the zero-input response of the system is not asymptotically stable, nor is the system totally stable.

To check BIBO stability of the system, we must compute the transfer function. The transfer function of the system is

$$\hat{g}(s) = \begin{bmatrix} 2 & 1 \end{bmatrix} \begin{bmatrix} s+1 & -1 \\ 0 & s \end{bmatrix}^{-1} \begin{bmatrix} 1 \\ 0 \end{bmatrix} = \frac{2}{s+1}$$

It has pole -1; hence the zero-state response of the system is BIBO stable. ∎

4-5 Stability and Complete Characterization

As mentioned in the previous section, control systems are often required to be totally stable; otherwise the systems may burn out or saturate. If the dynamical equation description of a system is available, its total stability can be checked from the matrix **A**. In the design of control systems, dynamical equations are however not always available; the only available description may be transfer functions. In this case it is necessary to check total stability from transfer functions. This can be done with the aid of the concept of complete characterization.

Recall from Section 2-7 that a system is defined to be completely characterized by its transfer function if the degree of the transfer function is equal to the number of the state variables. As indicated in Equation (4-30), the set of the poles of $\hat{g}(s)$ may be equal to or included in the set of the eigenvalues of **A**. However if a system is completely characterized, then the set of the poles of $\hat{g}(s)$ is equal to the set of the eigenvalues of **A**; that is,

$$\{\text{poles of } \hat{g}(s)\} = \{\text{eigenvalues of } \mathbf{A}\}$$

for the degree of $\hat{g}(s)$ is equal to the number of the poles, and the number of state variables is equal to the number of the eigenvalues. Hence if a system is completely characterized by its transfer function or, equivalently, its dynamical equation description is controllable and observable, then BIBO stability implies and is implied by total stability. We state this as a theorem.

Theorem 4-4

If a system is completely characterized by its transfer function $\hat{g}(s)$, then the system is BIBO stable, asymptotically stable, and totally stable if and only if all the poles of $\hat{g}(s)$ have negative real parts. ∎

Control systems are mostly built by interconnection of various components or devices. It is not unrealistic to assume that every component or device is completely characterized by its transfer function. Although every component is completely characterized, it does not follow that the composite system is completely characterized by its overall transfer function. Therefore Theorem 4-4 cannot always be employed. Before extending it to a more general case, we must study the complete characterization of composite systems.

There are many kinds of connections in composite systems; however they are basically built from the following three connections: the tandem, the parallel, and the feedback connections of two systems. Consider system S_i, $i = 1, 2$, which is completely characterized by $\hat{g}_i(s)$. If $\hat{g}_i(s)$ has degree n_i, then system S_i has, following Definition 2-8, n_i state variables. After the connection of S_1 and S_2, the number of the state variables of the composite system is clearly equal to $n_1 + n_2$. Hence in order for the overall transfer function $\hat{g}(s)$ to characterize completely the composite system, the degree of $\hat{g}(s)$ must be equal to $n_1 + n_2$. If the degree of $\hat{g}(s)$ is equal to a value of n that is smaller than $n_1 + n_2$, then the composite system is not completely characterized by $\hat{g}(s)$, and $n_1 + n_2 - n$ poles of the overall system will not appear as poles of $\hat{g}(s)$. Therefore in the design, not only should the poles of $\hat{g}(s)$ but also the missing poles be taken into consideration.

Theorem 4-5

Consider a composite system that consists of m subsystems. Let each subsystem be completely characterized by its proper transfer function $\hat{g}_i(s)$ with degree n_i. Let the degree of the overall transfer function $\hat{g}(s)$ of the composite system be n. If

$$n = \sum_{i=1}^{m} n_i$$

then the composite system is completely characterized by the overall transfer function $\hat{g}(s)$. Otherwise there are

$$\left(\sum_{i=1}^{m} n_1\right) - n$$

poles missing from $\hat{g}(s)$. ∎

The following theorem provides a way of detecting missing poles.

Theorem 4-6

Consider two single-variable systems, S_1 and S_2, that are completely characterized by their proper rational transfer functions $\hat{g}_1(s)$ and $\hat{g}_2(s)$.

1. The parallel connection of S_1 and S_2 is completely characterized by $\hat{g}(s) = \hat{g}_1(s) + \hat{g}_2(s)$ if and only if $\hat{g}_1(s)$ and $\hat{g}_2(s)$ do not have any pole in common.

2. The tandem connection of S_1 and S_2 is completely characterized by

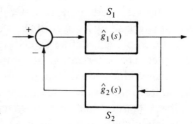

Figure 4-8
Feedback connection of S_1 and S_2.

$\hat{g}(s) = \hat{g}_1(s)\hat{g}_2(s)$ if and only if there is no pole-zero cancellation between $\hat{g}_1(s)$ and $\hat{g}_2(s)$.

3. The feedback connection of S_1 and S_2 shown in Figure 4-8 is completely characterized by $\hat{g}(s) = (1 + \hat{g}_1\hat{g}_2)^{-1}\hat{g}_1$ if and only if no pole of $\hat{g}_2(s)$ is canceled by any zero of $\hat{g}_1(s)$. ∎

A proof of this theorem can be found in Reference [4], page 359. This theorem implies that the systems in Figure 4-9(a), (b), and (c) are not completely characterized by their transfer functions. The missing pole in these systems is $(s - 1)$. Note that although there is a pole-zero cancellation in Figure 4-9(d), the system, according to part 3 of Theorem 4-6, is completely characterized by its overall transfer function.

With the concept of missing poles, we can now extend Theorem 4-4 to a more general case.

Theorem 4-7

A system is BIBO stable, asymptotically stable, and totally stable if and only if the poles of its transfer function and its missing poles all have negative real parts. ∎

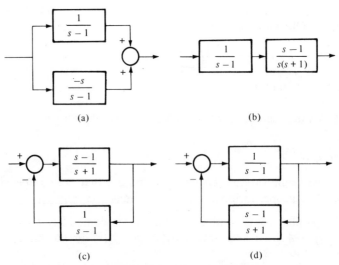

Figure 4-9 Pole-zero cancellation.

Example 1

Consider the feedback systems shown in Figure 4-10. It is assumed that systems S_1 and S_2 are, respectively, completely characterized by their transfer functions. This assumption implies that each system has one state variable. (See Definition 2-8.) Consequently the feedback system has two state variables. The transfer function of Figure 4-10(a) is

$$\hat{g}_{f1}(s) = \frac{\dfrac{s+2}{s-1}}{1 + \dfrac{2}{s-2} \cdot \dfrac{s-2}{s-1}} = \frac{s-2}{s+1}$$

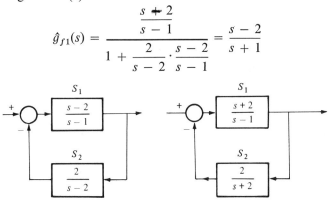

Figure 4-10 Two feedback systems.

which has degree 1. Hence $\hat{g}_{f1}(s)$ does not characterize completely the feedback connection in Figure 4-10(a); there is one pole missing from $\hat{g}_{f1}(s)$. The missing pole is $(s-2)$. Hence although the feedback system in Figure 4-10(a) is BIBO stable [since the pole of $\hat{g}_{f1}(s)$ has a negative real part], it is not totally stable.

On the other hand, the transfer function of Figure 4-10(b) is

$$\hat{g}_{f2} = \frac{s+2}{s+1}$$

which does not characterize the system completely. However the missing pole is $(s+2)$; hence the system is both BIBO stable and totally stable.

Example 2

Consider the feedback system shown in Figure 4-11. System S_i, $i = 1, 2, 3$, is assumed to be completely characterized by its transfer function. Study the stability of the system.

By employing part 3 of Theorem 4-6, we see that there is a missing pole, namely $s - 1$, in the feedback connection of S_1 and S_2. Indeed, the transfer function of S_{12} is

$$\hat{g}_{12}(s) = \frac{(s+1)(s-1)}{(s+3)(s^2+3s+1)}$$

which has a degree that is one smaller than the sum of the degrees of S_1 and S_2. The transfer function of the tandem connection of S_3 and S_{12} is $(s-1)/2(s^2+3s+1)$, which has a degree that is two smaller than the sum of the

104 ANALYSIS: QUANTITATIVE AND QUALITATIVE

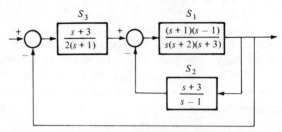

Figure 4-11 A feedback system.

degrees of S_3 and S_{12}. The missing poles are, from part 2 of Theorem 4.6, $(s + 1)$ and $(s + 3)$. The overall transfer function of the system is

$$\hat{g}(s) = \frac{s - 1}{2s^2 + 7s + 1}$$

Hence the system is BIBO stable. Since one of the missing poles has a positive real part, the system is not totally stable.   ∎

In the search of missing poles of a system, it is suggested that the configuration first be decomposed into the three basic connections. We then apply Theorem 4-6 and compute the transfer function of each connection as we did in this example. If we use Mason's formula to compute the overall transfer function, the missing poles cannot be so easily detected.

To conclude this section we stress again that stability of a system is determined by its overall mathematical description. It is possible that an overall system is BIBO stable or totally stable, although its subsystems are not. It is also possible that an overall system is not BIBO stable, although all its subsystems are BIBO stable.

4-6 Routh-Hurwitz Criterion

The BIBO stability of a system is determined by the poles of its transfer function, that is, by the roots of its denominator. The asymptotic stability is determined by the eigenvalues of **A**—that is, by the roots of the characteristic polynomial of **A**. In other words, the BIBO and asymptotic stability are all determined by the roots of polynomials. If the degree of a polynomial is larger than 3, solving for the roots is a very tedious job. Furthermore the knowledge of the exact locations of the roots is not needed in determining the stability. Therefore it is desirable to have some method of determining the stability without solving for the roots. In this section we shall introduce one such method, namely the Routh-Hurwitz criterion.

Definition 4-5

A polynomial with real coefficients is said to be a *Hurwitz polynomial* if all the roots of the polynomial have negative real parts.   ∎

The Routh-Hurwitz criterion is a method of checking whether or not a polynomial is a Hurwitz polynomial. Clearly this criterion can be applied to check the stability of a system. Consider the polynomial

$$D(s) = a_0 s^n + a_1 s^{n-1} + \cdots + a_{n-1} s + a_n \qquad a_0 > 0 \qquad (4\text{-}32)$$

with real coefficients a_i. It is assumed that the leading coefficient a_0 is positive. If a_0 is negative, it can be changed to a positive value by multiplying the whole polynomial by -1, and the roots of the polynomial are not affected. We first give a necessary condition for $D(s)$ to be Hurwitz. If $D(s)$ is Hurwitz—that is, if all the roots of $D(s)$ have negative real parts—then $D(s)$ can be factored as

$$\begin{aligned} D(s) &= a_0 \prod_k (s + \alpha_k) \prod_i (s + \beta_i + j\gamma_i)(s + \beta_i - j\gamma_i) \\ &= a_0 \prod_k (s + \alpha_k) \prod_i (s^2 + 2\beta_i s + \beta_i^2 + \gamma_i^2) \end{aligned} \qquad (4\text{-}33)$$

with $\alpha_k > 0$, $\beta_i > 0$, and $j = \sqrt{-1}$. Since the a_i's are real, complex roots appear in pairs. All the coefficients of the factors in the right-hand side of Equation (4-33) are positive; hence we conclude that if $D(s)$ is Hurwitz, then its coefficients a_i, $i = 1, 2, \ldots, n$, must be all positive. In other words, *given a polynomial with a positive leading coefficient, if some of its coefficients are negative or zero, then the polynomial is not a Hurwitz polynomial.* Conversely, a polynomial with positive coefficients is not necessarily a Hurwitz polynomial. For example, the polynomial with positive coefficients

$$s^3 + s^2 + 11s + 51 = (s + 3)(s - 1 + 4j)(s - 1 - 4j)$$

has roots with positive real parts.

If a polynomial with a positive leading coefficient has a negative or zero coefficient, then we may conclude immediately that it is not Hurwitz. Given a polynomial with all positive coefficients, in order to assert that it is Hurwitz, we have to carry out the following test. For ease of presentation we consider the polynomial of degree 6:

$$a_0 s^6 + a_1 s^5 + a_2 s^4 + a_3 s^3 + a_4 s^2 + a_5 s + a_6 \qquad a_0 > 0 \qquad (4\text{-}34)$$

We form Table 4-1. The first two rows are just the coefficients of the given polynomial. The third row is computed from the first two rows; the fourth row is computed from the second and third rows. In general a row is computed from its immediately preceding two rows. There is a simple pattern in computing each entry. Using this pattern the table can be easily completed. Note that for a polynomial of degree 6, there is a total of seven entries in the first column of the Routh-Hurwitz table.

Theorem 4-8

A polynomial of degree n with a positive leading coefficient is a Hurwitz polynomial if and only if the $n + 1$ entries in the first column of its Routh-Hurwitz table are all positive. ∎

Table 4-1 Routh-Hurwitz Table

s^6	a_0	a_2	a_4	a_6
s^5	a_1	a_3	a_5	0
s^4	$b_1 = \dfrac{a_1 a_2 - a_0 a_3}{a_1}$	$b_2 = \dfrac{a_1 a_4 - a_0 a_5}{a_1}$	$b_3 = \dfrac{a_1 a_6 - a_0 \cdot 0}{a_1}$	0
s^3	$c_1 = \dfrac{b_1 a_3 - a_1 b_2}{b_1}$	$c_2 = \dfrac{b_1 a_5 - a_1 b_3}{b_1}$	$c_3 = \dfrac{b_1 \cdot 0 - a_1 \cdot 0}{b_1}$	0
s^2	$d_1 = \dfrac{c_1 b_2 - b_1 c_2}{c_1}$	$d_2 = \dfrac{c_1 b_3 - b_1 c_3}{c_1}$	$d_3 = 0$	0
s^1	$e_1 = \dfrac{d_1 c_2 - c_1 d_2}{d_1}$	0	0	0
s^0	$f_1 = \dfrac{e_1 d_2 - d_1 \cdot 0}{e_1}$	0	0	0

A proof of this theorem can be found in Reference [1]. If a zero or a negative number appears in the first column before the completion of the table, we may stop there and conclude that the polynomial is not a Hurwitz polynomial.

Example 1

Consider $2s^4 + 2s^3 + 3s + 2$. The coefficient associated with s^2 is zero; hence it is not a Hurwitz polynomial.

Example 2

Consider $2s^4 + 2s^3 + s^2 + 3s + 2$. Form the following table:

s^4	2	1	2
s^3	2	3	
s^2	-2		

Since a negative number appears in the first column, the polynomial is not Hurwitz.

Example 3

Consider $s^3 + s^2 + s + 1$. Form the following table:

s^3	1	1
s^2	1	1
s	0	

Since a zero appears before the completion of the table, the polynomial is not Hurwitz.

We note that in the formation of Table 4-1, the signs of the entries are not affected if we multiply a row by a *positive* number. If we use this fact, the computation of Table 4-1 may often be simplified.

Example 4

Consider $2s^4 + 5s^3 + 5s^2 + 2s + 1$. We form

$$\begin{array}{llll} s^4 & 2 & 5 & 1 \\ s^3 & 5 & 2 & 0 \\ s^2 & 21 & 5 & 0 \quad \text{(after the multiplication by 5)} \\ s^1 & 17 & 0 & 0 \quad \text{(after the multiplication by 21)} \\ s^0 & 5 & 0 & 0 \end{array}$$

Since all the entries in the first column are positive, the polynomial is a Hurwitz polynomial. ∎

A remark is in order concerning the Routh-Hurwitz table. From the signs of the entries in the first column, it is possible, in many cases, to tell the number of poles with positive real parts. Since this information will not be used in the design, we shall not discuss it here.

With the Routh-Hurwitz criterion, the BIBO stability of a transfer function and the asymptotic stability of a dynamical equation can be easily determined.

Example 5

Consider the system shown in Figure 4-12. Is the system BIBO stable? Totally stable?

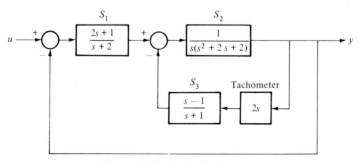

Figure 4-12 A feedback system.

By applying Mason's formula, the overall transfer function can be computed as

$$\hat{g}_f(s) = \frac{\hat{y}(s)}{\hat{u}(s)} = \frac{\dfrac{2s+1}{s+2} \cdot \dfrac{1}{s(s^2+2s+2)}}{1 - \left[\dfrac{-2(s-1)}{(s+1)(s^2+2s+2)} - \dfrac{2s+1}{(s+2)s(s^2+2s+2)}\right]}$$

$$= \frac{(2s+1)(s+1)}{s^5 + 5s^4 + 12s^3 + 14s^2 + 3s + 1}$$

Form the Routh-Hurwitz table for the denominator of $\hat{g}_f(s)$ as follows:

s^5	1	12	3
s^4	5	14	1
s^3	46	14	0 (after the multiplication by 5)
s^2	41 × 14	46	0 (after the multiplication by 46)
s	$\dfrac{(41 \times 196) - (46 \times 46)}{41 \times 14}$	0	0
s^0	46	0	0

Since the entries in the first column are all positive, we conclude that the denominator of $\hat{g}_f(s)$ is a Hurwitz polynomial. Hence the system is BIBO stable.

If system S_i, $i = 1, 2, 3$, is completely characterized by its transfer function, then systems S_1, S_2, and S_3 have, respectively, 1, 3, and 1 state variables. Hence there are totally five state variables in the system. Note that the tachometer is a measuring device; therefore no state variable is introduced in the block denoted as tachometer. The transfer function $\hat{g}_f(s)$ has degree 5; hence the overall system is completely characterized by $\hat{g}_f(s)$. Consequently we conclude that the system is also totally stable and asymptotically stable. ∎

In the design of a control system it is often necessary to find the range of a parameter in which the system is stable. This stability range can be computed by use of the Routh-Hurwitz criterion.

Example 6

Consider the system shown in Figure 4-13. Find the range of α in which the system is BIBO stable.

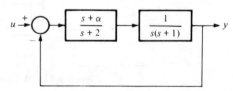

Figure 4-13
A feedback system with a parameter α.

The transfer function of the system is

$$\hat{g}_f(s) = \frac{s + \alpha}{s^3 + 3s^2 + 3s + \alpha}$$

Form the Routh-Hurwitz table for the denominator of $\hat{g}_f(s)$.

$$\begin{array}{c|cc} s^3 & 1 & 3 \\ s^2 & 2 & \alpha \\ s & \dfrac{9-\alpha}{3} & 0 \\ s^0 & \alpha & 0 \end{array}$$

In order for $\hat{g}_f(s)$ to be BIBO stable, it is required that

$$\frac{9-\alpha}{3} > 0$$

and

$$\alpha > 0$$

These two conditions imply $9 > \alpha > 0$. Hence if α is between 0 and 9, the system is BIBO stable. ∎

4-7 Steady-State Responses of Stable Systems

In this section the steady-state responses of a BIBO stable system due to certain inputs will be discussed. By steady-state response we mean the response as the time approaches infinity; that is,

$$y_s(t) \triangleq \lim_{t \to \infty} y(t) \qquad (4\text{-}35)$$

Consider a system with the transfer function

$$\hat{g}(s) = \frac{\beta_0 + \beta_1 s + \cdots + \beta_m s^m}{\alpha_0 + \alpha_1 s + \cdots + \alpha_n s^n} \qquad (4\text{-}36)$$

with $m \leq n$. It is assumed that the system is BIBO stable; that is, all the poles of $\hat{g}(s)$ have negative real parts. If we apply the input

$$u(t) = a \sin \omega_0 t \qquad (4\text{-}37)$$

or

$$\hat{u}(s) = \frac{a\omega_0}{s^2 + \omega_0^2}$$

to the system, then its output is given by

$$\hat{y}(s) = \hat{g}(s)\hat{u}(s) = \hat{g}(s)\left(\frac{a\omega_0}{s^2 + \omega_0^2}\right)$$

which, by partial fraction expansion, can be expanded into

$$\hat{y}(s) = \frac{a\hat{g}(j\omega_0)}{2j(s - j\omega_0)} - \frac{a\hat{g}(-j\omega_0)}{2j(s + j\omega_0)} + \text{terms due to the poles of } \hat{g}(s)$$

Since all the poles of $\hat{g}(s)$ have negative real parts, the time responses resulting from the terms due to the poles of $\hat{g}(s)$ will approach zero as $t \to \infty$. Hence the steady-state response of the system due to the application of the input $u(t) = a \sin \omega_0 t$ is given by

$$y_s(t) = \mathscr{L}^{-1} \left[\frac{a\hat{g}(j\omega_0)}{2j(s - j\omega_0)} - \frac{a\hat{g}(-j\omega_0)}{2j(s + j\omega_0)} \right] \quad (4\text{-}38)$$

For a rational function with real coefficients, it can be shown that the real part of $\hat{g}(j\omega)$, denoted as Re $\hat{g}(j\omega)$, is an even function of ω; and the imaginary part of $\hat{g}(j\omega)$, denoted as Im $\hat{g}(j\omega)$, is an odd function of ω. Hence if we write $\hat{g}(j\omega_0) = |\hat{g}(j\omega_0)|e^{j\theta}$, where the vertical rules denote the magnitude of the quantity within, and $\theta = \tan^{-1} \left[\text{Im } \hat{g}(j\omega_0) / \text{Re } \hat{g}(j\omega_0) \right]$, then $\hat{g}(-j\omega_0) = |\hat{g}(j\omega_0)|e^{-j\theta}$. Consequently Equation (4-38) can be written as

$$y_s(t) = a|\hat{g}(j\omega_0)| \frac{e^{j(\omega_0 t + \theta)} - e^{-j(\omega_0 t + \theta)}}{2j} \quad (4\text{-}39)$$

$$= a|\hat{g}(j\omega_0)| \sin(\omega_0 t + \theta)$$

This is a very important fact. It asserts that if a system is BIBO stable, and if a sinusoidal input is applied to the system, then after the transient dies out the output will approach a sinusoidal function of the same frequency. Its amplitude is equal to $a|\hat{g}(j\omega_0)|$; its phase differs from the phase of the input by $\tan^{-1} \left[\text{Im } \hat{g}(j\omega_0) / \text{Re } \hat{g}(j\omega_0) \right]$. Hence from the amplitude and phase of this steady-state output, the transfer function at $s = j\omega_0$—that is, $\hat{g}(j\omega_0)$—can be measured. By varying the frequency of the input, the transfer function $\hat{g}(j\omega)$ at all frequencies ω can be measured. Once $\hat{g}(j\omega)$ is found, the transfer function $\hat{g}(s)$ can be obtained by substituting $s = j\omega$. This fact is often used in practice to measure the transfer function of a system.

We discuss now the steady-state response of a stable system due to the application of a step function and a ramp function. If we apply the step-function input

$$u(t) = \begin{cases} a & \text{for } t \geq 0 \\ 0 & \text{for } t < 0 \end{cases}$$

or

$$\hat{u}(s) = \frac{a}{s}$$

4-7 STEADY-STATE RESPONSES OF STABLE SYSTEMS

to the system, then the output is given by

$$\hat{y}(s) = \hat{g}(s)\hat{u}(s) = \frac{\beta_0 + \beta_1 s + \cdots + \beta_m s^m}{\alpha_0 + \alpha_1 s + \cdots + \alpha_n s^n} \cdot \frac{a}{s}$$

$$= \hat{g}(0) \frac{a}{s} + \text{terms due to the poles of } \hat{g}(s)$$

$$= \frac{\beta_0}{\alpha_0} \cdot \frac{a}{s} + \text{terms due to the poles of } \hat{g}(s)$$

If the system is BIBO stable, then the responses resulting from the terms due to the poles of $\hat{g}(s)$ will go to zero at $t \to \infty$. Hence the steady-state response of the system due to a step input is

$$y_s(t) = \lim_{t \to \infty} y(t) = \frac{\beta_0}{\alpha_0} \cdot a \qquad (4\text{-}40)$$

Note that this steady-state response depends only on the coefficients associated with s^0 of $\hat{g}(s)$. If the input is a ramp function; that is, if

$$u(t) = at \qquad \text{for } t \geq 0$$

then

$$\hat{u}(s) = \frac{a}{s^2}$$

and

$$\hat{y}(s) = \hat{g}(s) \frac{a}{s^2}$$

$$= \frac{\hat{g}(0)a}{s^2} + \frac{d}{ds}\hat{g}(s)\bigg|_{s=0} \cdot \frac{a}{s} + \text{terms due to poles of } \hat{g}(s) \qquad (4\text{-}41)$$

which is obtained by using Equation (4-15). With $\hat{g}(s)$ given as in (4-36), we have

$$\frac{d}{ds}\hat{g}(s)\bigg|_{s=0} = \frac{(\alpha_0 + \alpha_1 s + \cdots + \alpha_n s^n)(\beta_1 + \cdots + m\beta_m s^{m-1}) - (\beta_0 + \beta_1 s + \cdots + \beta_m s^m)(\alpha_1 + \cdots + \alpha_n s^{n-1})}{(\alpha_0 + \alpha_1 s + \cdots + \alpha_n s^n)^2}\bigg|_{s=0}$$

$$= \frac{\alpha_0 \beta_1 - \beta_0 \alpha_1}{\alpha_0^2}$$

Hence the steady-state response of the system due to the ramp input at is equal to

$$y_s(t) = \frac{\beta_0}{\alpha_0} \cdot at + a \cdot \frac{\alpha_0 \beta_1 - \beta_0 \alpha_1}{\alpha_0^2} \qquad (4\text{-}42)$$

Note that this steady-state response depends only on the coefficients associated with s and s^0 of $\hat{g}(s)$. Equations (4-40) and (4-42) will be used later in the design.

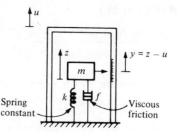

Figure 4-14
A seismometer.

4-8 Simplification

In order to facilitate analyses and designs, simplifications of mathematical descriptions of systems are often made in engineering. For example, the transfer function of an electronic amplifier is $k/(\tau s + 1)$. If the time constant τ is very small, as is often the case in practice, and if the input signals are of low frequency, then the transfer function of the amplifier can be simplified as k. Another example of simplification can be found in the transfer function of a seismometer. Consider the system shown in Figure 4-14. The case is rigidly attached to the ground. Let u and z be, respectively, the displacements of the case and of the mass relative to the inertial space, and let $y = z - u$. Then Newton's law yields

$$m\ddot{z} + f\dot{y} + ky = 0$$

where f is the viscous friction coefficient, and k is the spring constant. By using $z = y + u$, this equation can be written as

$$m\ddot{y} + f\dot{y} + ky = -m\ddot{u}$$

Hence the transfer function from u to y is

$$\hat{g}(s) = \frac{\hat{y}(s)}{\hat{u}(s)} = -\frac{s^2}{s^2 + \dfrac{f}{m}s + \dfrac{k}{m}} \qquad (4\text{-}43)$$

If (4-43) can be approximated by $\hat{g}(s) = -s^2/s^2 = -1$, then the movement of y will record the movement of the ground. Hence the system can be used, under certain approximations, to measure the vibrations of earthquakes. In the following, justifications of these simplifications will be discussed.

If the time constant τ is very small, then for small $s = j\omega$, the amplitude of the transfer function $k/(\tau s + 1)$ is approximately equal to k, and its phase is approximately equal to zero. Hence from (4-39) we may conclude that for input signals with low-frequency spectra, a transfer function $k/(\tau s + 1)$ can be simplified as k as far as the steady-state responses are concerned. In the time domain, the simplification can be justified as follows: For a system with transfer function $k/(\tau s + 1)$ and for the input $\hat{u}(s)$, the output is given by

$$\hat{y}(s) = \frac{k}{\tau s + 1} \hat{u}(s)$$

$$= \frac{k}{\tau s + 1} \hat{u}\left(\frac{-1}{\tau}\right) + \text{terms due to the poles of } \hat{u}(s) \quad (4\text{-}44)$$

If τ is very small and if u has a low frequency spectrum, then $u(-1/\tau)$ is approximately equal to zero (see Figure 3-1). Hence the constant associated with the pole $(\tau s + 1)$ is approximately equal to zero. Furthermore the time response due to the pole $(\tau s + 1)$ dies out rapidly because of the small time constant τ. Hence the first term in the right-hand side of (4-44) can be neglected. On the other hand, the magnitudes of the poles of $\hat{u}(s)$ are small, as is implicitly implied by the low frequency spectrum of $\hat{u}(s)$; hence in computing the constants associated with the poles of $\hat{u}(s)$, the contributions due to $k/(\tau s + 1)$ and due to k are approximately the same. Hence we conclude that for low-frequency signals, Equation (4-44) can be approximated by $\hat{y}(s) = k\hat{u}(s)$, and a transfer function $k/(\tau s + 1)$ with small τ may be simplified to k.

We discuss now the simplification of Equation (4-43). If the coefficients f/m and k/m in (4-43) are very small, then for signals with high frequency spectra, the denominator of the transfer function in (4-43) is approximately equal to s^2; hence the transfer function in (4-43) can be approximated by -1. We use a numerical example to verify this approximation.

Example

Consider the system shown in Figure 4-14. Let $k/m = 10^{-5}$ and $f/m = 10^{-2} + 10^{-3}$. Then the transfer function from u to y is

$$\hat{g}(s) = \frac{-s^2}{s^2 + (10^{-2} + 10^{-3})s + 10^{-5}} = \frac{-s^2}{(s + 10^{-2})(s + 10^{-3})}$$

If the input $u_1(t) = 10^{-2} e^{-0.1t} \sin 10t$ is applied to the system, then the output is given by

$$\hat{y}(s) = \frac{-s^2}{(s + 10^{-2})(s + 10^{-3})} \frac{10^{-2} \times 10}{(s + 0.1)^2 + 100}$$

$$= \frac{10^{-5}}{s + 10^{-2}} - \frac{10^{-7}}{s + 10^{-3}} - \frac{10^{-2} \times 10}{(s + 0.1)^2 + 100}$$

which implies that

$$y(t) = 10^{-5} e^{-10^{-2} t} - 10^{-7} e^{-10^{-3} t} - 10^{-2} e^{-0.1t} \sin 10t$$

If the sensitivity of the indicator is 10^{-4} meter—that is, the pointer can only respond to signals larger than 10^{-4} meter (0.1 mm)—then the picked-up signal

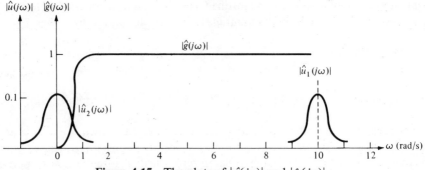

Figure 4-15 The plots of $|\hat{g}(j\omega)|$ and $|\hat{u}_i(j\omega)|$.

will be $-10^{-2}e^{-0.1t}\sin 10t$, which is identical to the input signal except for the sign. This verifies that the transfer function in (4-43) can indeed be simplified to -1. The frequency spectrum of u_1 and the plot of $|\hat{g}(j\omega)|$ are shown in Figure 4-15. We note that in the frequency range of u_1, the plot of $|\hat{g}(j\omega)|$ is essentially equal to 1.

If the input $u_2(t) = 10^{-2}e^{-0.1t}$ is applied to the system, the output is given by

$$\hat{y}(s) = \frac{-s^2}{(s+10^{-2})(s+10^{-3})}\frac{10^{-2}}{s+0.1}$$

$$= \frac{10^{-3}}{s+10^{-2}} - \frac{10^{-5}}{s+10^{-3}} - \frac{10^{-2}}{s+0.1}$$

which implies that

$$y(t) = 10^{-3}e^{-10^{-2}t} - 10^{-5}e^{-10^{-3}t} - 10^{-2}e^{-0.1t}$$

If the sensitivity of the indicator is still 10^{-4} meter, then the detected signal is $10^{-3}e^{-10^{-2}t} - 10^{-2}e^{-0.1t}$, in which the first term becomes dominant after $t \geq 26$ seconds. Hence for the input signal u_2, the transfer function cannot be simplified to -1. This fact can also be concluded from the frequency spectrum of u_2 shown in Figure 4-15. ∎

From the foregoing discussion, we see that the transfer function of a system can often be simplified for a certain class of input signals. The signals in most

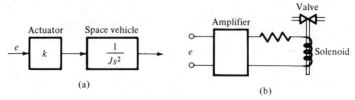

Figure 4-16 (a) Model of a gas-jet-controlled space vehicle. (b) An actuator whose transfer function can be approximated as k.

control systems have low frequency spectra; hence from the discussion of (4-44), we know that a pole with a very small time constant can often be neglected. If so, the design of a control system can be simplified. As an example, in the attitude control of a space vehicle using a gas jet, the system can be modeled as shown in Figure 4-16(a), where the actuator is a mechanism for controlling the gas-jet valve. The actuator may be a solenoid, as shown in Figure 4-16(b), or another more-complicated device. However if it is designed so that its time constants are much smaller than the other time constants of the system, then its transfer function can be approximated simply by a gain k. Another example of an actuator is shown in Figure 4-17. Its transfer function can be shown to be roughly equal to $\hat{g}(s) = 1$ if the input signal u has a magnitude smaller than 1 and a frequency spectrum smaller than 10 rad/s. See Reference [3].

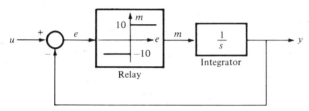

Figure 4-17 An actuator.

4-9 Remarks and Review Questions

The analysis techniques introduced in this chapter are applicable to any linear time-invariant systems if their overall mathematical descriptions are first computed. For a single-variable composite system, the overall transfer functions can be easily obtained by applying Mason's formula. There is however no similar formula for the multivariable systems and for the state-variable description. In these cases the overall mathematical descriptions can be computed by eliminating the internal variables, as in Section 4-2.

What is the general procedure for the computation of the output of a system by using the Laplace transform?

If the Laplace transform of an input is not a rational function, is it easy to use the Laplace transform in computing the response?

Partial fraction expansion can be obtained systematically by using Equations (4-15) and (4-16). For some special cases we may avoid the use of the formula and employ the following type of manipulation:

$$\frac{s^2 + 3s + 6}{s^3} = \frac{s^2}{s^3} + \frac{3s}{s^3} + \frac{6}{s^3} = \frac{1}{s} + \frac{3}{s^2} + \frac{6}{s^3}$$

$$\frac{2s + 5}{(s + 1)^2} = \frac{2(s + 1) + 3}{(s + 1)^2} = \frac{2(s + 1)}{(s + 1)^2} + \frac{3}{(s + 1)^2} = \frac{2}{(s + 1)} + \frac{3}{(s + 1)^2}$$

How many methods have we introduced in computing e^{At}? List each of them. Why do we have to compute e^{At}?

What is a time constant? In practice it is often said that a response dies out in 4 time constants. Verify this by showing that the magnitude of a response reduces to 2 percent of its original magnitude in 4 time constants.

What is the open right-half s-plane? The closed right-half s-plane? Why do we need to make such a fine distinction?

How many stability definitions have we introduced? What are their conditions?

There are many stability definitions in the literature. Therefore when we talk about stability, we must be specific; otherwise confusion may arise. For example, there is a stability definition that permits simple poles on the $j\omega$-axis of the s-plane.

Does BIBO stability imply asymptotic stability? How about the converse?

What is total stability?

Under what condition will BIBO stability, total stability, and asymptotic stability imply each other?

What condition on a system do we need in the discussion of the steady-state response?

Can you use the final-value theorem to establish Equation (4-39)? Why?

Approximation is very useful in engineering. By proper approximation analysis and synthesis can often be simplified. What is a good approximation is often difficult to define precisely. Good reasoning and engineering judgment should be employed in carrying out approximations.

All the results except Mason's formula can be extended directly to the multivariable case. For example, if all the poles of every entry of the transfer-function matrix of a multivariable system have negative real parts, then the multivariable system is BIBO stable. There is no modification required in checking the asymptotic stability of a multivariable system.

References

[1] Chen, C. T., *Introduction to Linear System Theory*. New York: Holt, Rinehart and Winston, 1970.
[2] Cooper, G. R., and C. D. McGillem, *Methods of Signal and System Analysis*. New York: Holt, Rinehart and Winston, 1967.
[3] Hsu, J. C., and A. U. Mayer, *Modern Control Principles and Applications*. New York: McGraw-Hill, 1968.
[4] Mason, S. J., and H. J. Zimmerman, *Electronic Circuits, Signals, and Systems*. New York: Wiley, 1960.

[5] Ogata, K., *Modern Control Engineering*. Englewood Cliffs, N.J.: Prentice-Hall, 1970.
[6] Truxal, J. G., *Control System Synthesis*. New York: McGraw-Hill, 1955.

Problems

4-1 Given two single-variable systems S_i, for $i = 1$ and 2, that are described as follows: For S_1,

$$\begin{bmatrix} \dot{x}_1 \\ \dot{x}_2 \end{bmatrix} = \begin{bmatrix} 0 & 1 \\ 0 & -1 \end{bmatrix} \begin{bmatrix} x_1 \\ x_2 \end{bmatrix} + \begin{bmatrix} 0 \\ 1 \end{bmatrix} u_1$$

$$y_1 = \begin{bmatrix} 2 & 1 \end{bmatrix} \begin{bmatrix} x_1 \\ x_2 \end{bmatrix}$$

and for S_2,

$$\dot{x} = -x + 2u_2$$

$$y_2 = -x - u_2$$

Find the state-variable and the transfer-function descriptions of the tandem connection of S_1 followed by S_2.

4-2 Find the state-variable and the transfer-function descriptions of the feedback system shown in Figure 4-1 with S_1 and S_2 given as in Problem 4-1.

4-3 Find the transfer functions from u to y of the systems shown in Figure P4-1.

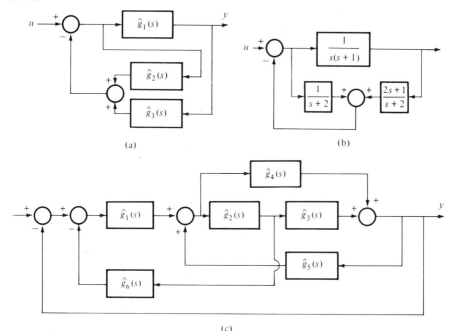

Figure P4-1

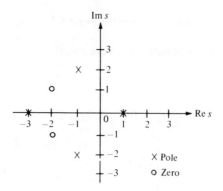

Figure P4-2

4-4 Given a system with poles and zeros as shown in Figure P4-2. It is also known that $\hat{g}(2) = 2.1$. What is the transfer function of the system?

4-5 Find the zero-state response of the system with transfer function $2(s - 1)/s(s + 1)$ due to the application of the input $u(t) = e^{-t}$.

4-6 Given $\hat{g}(s) = \omega_n^2/(s^2 + 2\delta\omega_n s + \omega_n^2)$ with $0 < \delta < 1$. Show that the output due to a unit step input is given by

$$y(t) = 1 + \frac{e^{-\delta\omega_n t}}{\sqrt{1 - \delta^2}} \sin\left(\omega_n\sqrt{1 - \delta^2}\, t + \tan^{-1}\frac{\sqrt{1 - \delta^2}}{\delta}\right) \qquad t \geq 0$$

4-7 Find the output of

$$\dot{\mathbf{x}} = \begin{bmatrix} -1 & 1 & 0 \\ 0 & -1 & 0 \\ 0 & 0 & 0 \end{bmatrix} \mathbf{x} + \begin{bmatrix} 0 \\ 1 \\ 2 \end{bmatrix} u$$

$$y = \begin{bmatrix} 1 & -1 & 0 \end{bmatrix} \mathbf{x}$$

due to the initial condition $[9\ 1\ 0]'$ and a unit-step-function input. Solve it by using the Laplace transform *and* directly in the time domain.

4-8 Find the output of the system shown in Figure P4-3 due to the initial state $\mathbf{x}(0) = [1\ 0]'$, where S is described by

$$\dot{\mathbf{x}} = \begin{bmatrix} 0 & 1 \\ 1 & -1 \end{bmatrix} \mathbf{x} + \begin{bmatrix} 0 \\ 1 \end{bmatrix} u$$

$$y = \begin{bmatrix} 1 & 0 \end{bmatrix} \mathbf{x}$$

4-9 Find the output of the system shown in Figure P4-4 due to a unit-step-function input. Can you conclude the BIBO stability of the system from this input-output pair?

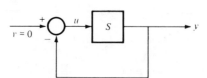

Figure P4-3

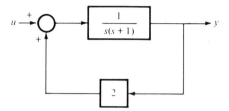

Figure P4-4

4-10 Consider an ac motor. It is assumed that its transfer function from the electrical voltage input to the angular position of the motor shaft is $2/s(s + 1)$. Is the motor BIBO stable? If the angular velocity, rather than the displacement, of the motor shaft is considered as the output, what is its transfer function? With respect to this input and output, is the system BIBO stable?

4-11 Check the BIBO stability of the following systems:

a. $\hat{g}(s) = \dfrac{s^3 - 1}{s^5 + s^4 + 2s^3 + 2s^2 + 5s + 5}$

b. $\hat{g}(s) = \dfrac{s^3 - 1}{s^4 + 14s^3 + 71s^2 + 154s + 120}$

c. The system shown in Figure P4-5.

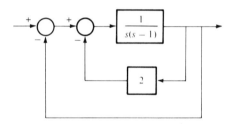

Figure P4-5

4-12 Study the asymptotic stability of the zero-input response of the system in Problem 4-7. Study also the BIBO stability of its zero-state response.

4-13 Determine whether each of the following statements is true or false.
a. If an overall system is BIBO stable, then all of its subsystems must be BIBO stable.
b. If all subsystems are BIBO stable, then the overall system must be BIBO stable.
c. If all subsystems are not BIBO stable, then the overall system is not BIBO stable.
d. If all the poles of a transfer function lie in the closed left-half plane, then its impulse response is bounded.
e. If the zero-input response of a system is asymptotically stable, then the system is totally stable.

4-14 Consider the systems shown in Figure P4-6. Find the ranges of k in which the systems are BIBO stable?

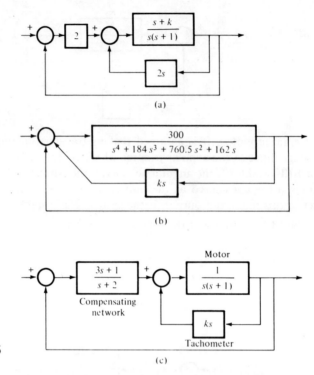

Figure P4-6

4-15 What are the steady-state responses of the system with transfer function $1/(s^2 + 2s + 1)$ due to the application of the following inputs:
a. $u_1(t) = 2 \sin 2\pi t$, for $t \geq 0$
b. $u_2(t) = $ a unit step function
c. $u_3(t) = $ a ramp function
d. $u_4 = u_1 + u_2$

4-16 Consider a system with transfer function $\hat{g}(s)$. It is assumed that $\hat{g}(s)$ has no poles in the closed right-half s-plane except a simple pole at the origin. Show that if the input $u(t) = a \sin \omega_0 t$ is applied, the steady-state response excluding the dc part is given by Equation (4-39).

4-17 Derive Equation (4-40) by using the final-value theorem (see Appendix A). Can you use the theorem in deriving Equation (4-42)?

4-18 Study the BIBO and total stability of the system shown in Figure P4-7. It is assumed that every subsystem is completely characterized by its transfer function.

4-19 An armature-controlled dc motor is used to drive a load. Let the transfer function from the applied input voltage to the angular displacement of the motor shaft be of the form $k_m/s(\tau_m s + 1)$. If an input of 100 volts is applied, it is measured that the motor shaft reaches a speed (*not* displacement) of 2 rad/s at 1.5 second. The steady-state motor shaft speed is measured as 3 rad/s. What is the transfer function of the system?

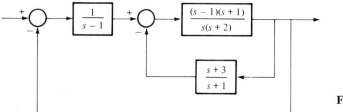

Figure P4-7

4-20 In a modern rapid transit system, a train can be controlled manually or automatically. The block diagram shown in Figure P4-8 is a possible way to

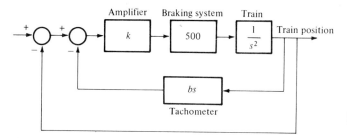

Figure P4-8

control the train automatically. If the tachometer is not used in the feedback—that is, if $b = 0$—is it possible for the system to be totally stable for some k? If $b = 0.2$, find the range of k so that the system is totally stable.

4-21 Maintaining the liquid level at a fixed height is an important task in many process control systems. Such a system and its block diagram are shown in Figure P4-9, with $\tau = 10$, $k_1 = 10$, $k_2 = 0.95$, and $k_3 = 2$. The variables h

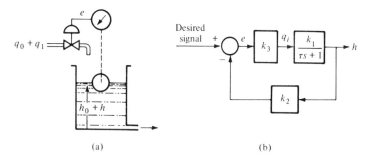

Figure P4-9

and q_i in the block diagram are the deviations from the nominal values h_0 and q_0; hence the desired or command signal for this system is zero. This kind of system is called a *regulating system*. If h at $t = 0$ is -1, what is the behavior of $h(t)$ for $t > 0$?

CHAPTER

5

Computer Simulations and Realizations

In recent years computers have become indispensable in the design and control of sophisticated control systems. They are used in space vehicles, industry process controls, auto pilots in airplanes, numerical controls in machine tools, and other applications. Computers can be employed to carry out complicated computations, collect data, and then generate control signals, serve as control components, or simulate mathematical equations. In this chapter computer simulations of mathematical equations will be discussed.

Computers can be divided into analog and digital computers. An interconnection of a digital and an analog computer is called a hybrid computer. The signals in an analog computer are defined over the entire time interval of interest, whereas the signals in a digital computer are defined only at discrete instants of time. In other words, a digital computer accepts only sequences of numbers, and its outputs again consist of only sequences of numbers. In this chapter both analog and digital computer simulations will be discussed. We discuss only the fundamental idea in the simulation. Technical details will not be discussed.

5-1 Analog Computer Simulations of Dynamical Equations

In order to unify the presentation of analog and digital computer simulations of linear dynamical equations, we shall first develop a block diagram for a

5-1 ANALOG COMPUTER SIMULATIONS OF DYNAMICAL EQUATIONS

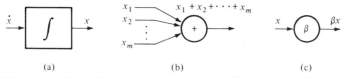

Figure 5-1 Basic blocks. (a) Integrator. (b) Summer. (c) Amplifier with gain $-\infty < \beta < \infty$.

dynamical equation. The basic elements used in the block diagram are the integrator, summer, and amplifier (or attenuator), as shown in Figure 5-1. By interconnection of these three basic elements, a block diagram can be built for any linear dynamical equation. We shall call this kind of diagram *a basic block diagram*. We give in Figure 5-2 the basic block diagram of the following dynamical equation:

$$\begin{bmatrix} \dot{x}_1 \\ \dot{x}_2 \end{bmatrix} = \begin{bmatrix} a_{11} & a_{12} \\ a_{21} & a_{22} \end{bmatrix} \begin{bmatrix} x_1 \\ x_2 \end{bmatrix} + \begin{bmatrix} b_{11} & b_{12} \\ b_{21} & b_{22} \end{bmatrix} \begin{bmatrix} u_1 \\ u_2 \end{bmatrix} \quad \textbf{(5-1a)}$$

$$\begin{bmatrix} y_1 \\ y_2 \end{bmatrix} = \begin{bmatrix} c_{11} & c_{12} \\ c_{21} & c_{22} \end{bmatrix} \begin{bmatrix} x_1 \\ x_2 \end{bmatrix} + \begin{bmatrix} d_{11} & d_{12} \\ d_{21} & d_{22} \end{bmatrix} \begin{bmatrix} u_1 \\ u_2 \end{bmatrix} \quad \textbf{(5-1b)}$$

Note that for a two-dimensional dynamical equation we need two integrators. By assigning the output of each integrator as a state variable, a basic block diagram can be readily obtained from (5-1). We also note that every branch is unidirectional, and the signal transmits only in one direction.

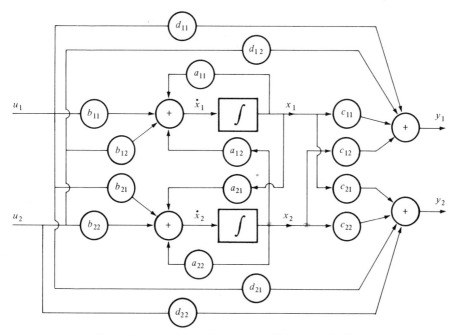

Figure 5-2 Basic block diagram of Equation (5-1).

With this background, we are ready to discuss analog computer simulations of dynamical equations. The actual analog computer elements are listed in Figure 5-3. They are built by using operational amplifiers, resistors, and capacitors (see Section 3-6 and Problem 3-3). Most analog computers use standardized resistors and capacitors, for example, 10^6, 0.25×10^6, and 0.1×10^6 ohms for resistors and 10^{-6} farad for capacitors. Hence we have no freedom in choosing continuous ranges of gains for a_i and b_i in Figure 5-3. The often available a_i and b_i are 1 and 10.

The basic elements introduced in Figure 5-1 are just mathematical entities. They can be constructed however by using actual analog computer elements, as shown in Figure 5-4. We note that the β in Figure 5-4(c) can be any positive or negative value by connecting amplifiers and a potentiometer. Now since

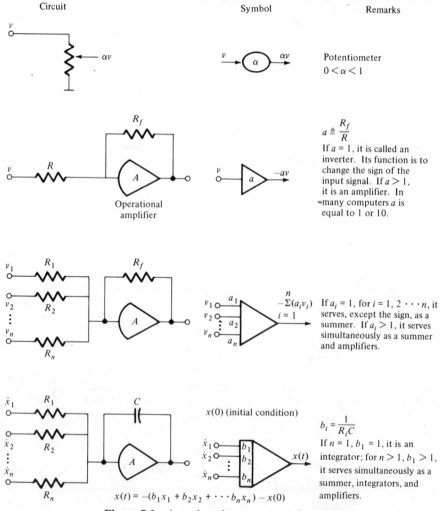

Figure 5-3 Actual analog computer elements.

5-1 ANALOG COMPUTER SIMULATIONS OF DYNAMICAL EQUATIONS

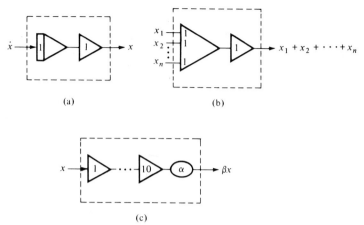

Figure 5-4 The construction of basic blocks by using actual analog computer elements.

every dynamical equation can be set up by an interconnection of the basic elements shown in Figure 5-1, and since every basic element can be constructed by using actual analog computer elements, we conclude that every linear time-invariant dynamical equation can be simulated in an analog computer. A block diagram built by using actual analog computer elements is called an *analog simulation block diagram*.

Example 1

Simulate the following dynamical equation in an analog computer:

$$\begin{bmatrix} \dot{x}_1 \\ \dot{x}_2 \end{bmatrix} = \begin{bmatrix} -0.9 & 81 \\ -1 & 2 \end{bmatrix} \begin{bmatrix} x_1 \\ x_2 \end{bmatrix} + \begin{bmatrix} 1.2 \\ 1 \end{bmatrix} u \qquad \text{(5-2a)}$$

$$y = \begin{bmatrix} 2.5 & 0.7 \end{bmatrix} \begin{bmatrix} x_1 \\ x_2 \end{bmatrix} \qquad \text{(5-2b)}$$

A basic block diagram of (5-2) is drawn in Figure 5-5(a). Each basic block in Figure 5-5(a) is replaced by actual analog computer elements in Figure 5-5(b). The resulting diagram is an analog simulation block diagram for (5-2). ∎

The purpose of this example is to demonstrate that every linear time-invariant dynamical equation can indeed be simulated on an analog computer. The block diagram in this example is unsatisfactory however because it uses excessive numbers of analog computer elements. Although the block diagram can be simplified, it is easier to set up a new analog simulation block diagram by a different procedure. The procedure is very simple: For an n-dimensional dynamical equation, we need n integrators. We then assign the output of each integrator as either x_i or $-x_i$. If we *assign* it as $-x_i$, then the input of the integrator is equal to $\dot{x}_i$. If we assign it as x_i, then the input of the integrator is

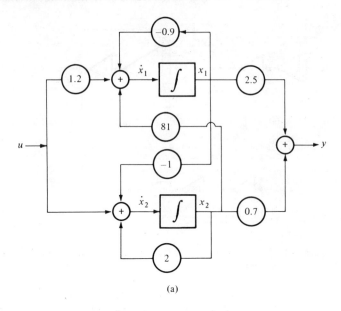

(a)

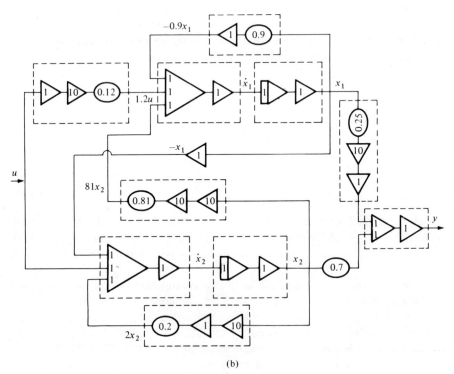

(b)

Figure 5-5 A simulation of Equation (5-2).

5-1 ANALOG COMPUTER SIMULATIONS OF DYNAMICAL EQUATIONS

$-\dot{x}_i$. By generating $\dot{x}_i$ or $-\dot{x}_i$, using amplifiers, potentiometers, and the given dynamical equation, at the input of each integrator, the analog simulation block diagram is completed.

Example 2

Find an analog simulation block diagram for Equation (5-2). The equation has dimension 2; hence we need two integrators. Their outputs are assigned as $-x_1$ and $-x_2$, as shown in Figure 5-6(a). The input of integrator 1 is equal to $-0.9x_1 + 81x_2 + 1.2u$. The input of integrator 2 is equal to $-x_1 + 2x_2 + u$. They are generated as shown in the figure.

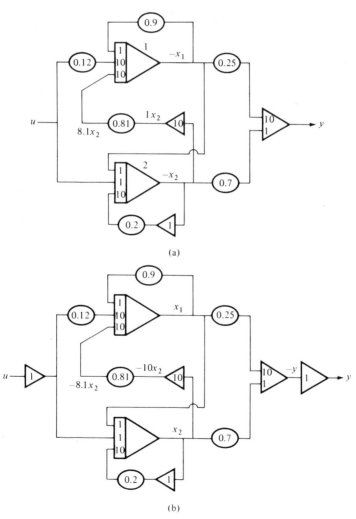

Figure 5-6 An analog simulation block diagram of Equation (5-2).

If the outputs of the integrators are assigned as x_1 and x_2, then the inputs of the integrators are $-\dot{x}_1 = 0.9x_1 - 81x_2 - 1.2u$ and $-\dot{x}_2 = x_1 - 2x_2 - u$. The block diagram with this choice of state variables is shown in Figure 5-6(b). Note that if the outputs of the integrators are assigned as x_1 and $-x_2$, we will obtain a different block diagram. ∎

Magnitude scaling. The signals in analog computers are measured in terms of voltage. The variables of the dynamical equations to be simulated are however not necessarily voltage; they can be current, temperature, or other variables. Therefore, strictly speaking, we have to carry out unit changes before any simulation. This step can be avoided if we use just the numerical values of an analog computer and then add the appropriate unit to the variables. For example, if x_1 in (5-2) stands for displacement and is expressed in meters, and if x_1 reads 0.2 volt on an analog computer, then it should mean 0.2 meter in the dynamical equation. Therefore in the following, we discuss only computer variables.

The magnitude scaling[1] is required in analog computer simulations for the following two reasons: First, the value a variable can assume in an analog computer is limited to a range, typically ± 100 volts, ± 50 volts, or ± 10 volts. Hence in order to avoid saturation of the operational amplifier, the magnitude of every variable must be limited. Second, if the value of a variable is very small, then the accuracy of its readout, or plot, will be poor. Hence in order to increase the accuracy, the magnitude of every variable should be increased to make use of the full scale. Clearly it is not possible to achieve these two objectives simultaneously by changing the magnitude of the input alone. To achieve these objectives, we have to perform the following magnitude scaling: Consider the linear time-invariant dynamical equation

$$\dot{\mathbf{x}} = \mathbf{A}\mathbf{x} + \mathbf{B}\mathbf{u} \qquad (5\text{-}3a)$$

$$\mathbf{y} = \mathbf{C}\mathbf{x} + \mathbf{D}\mathbf{u} \qquad (5\text{-}3b)$$

Let

$$\bar{x}_1 = p_1 x_1$$
$$\bar{x}_2 = p_2 x_2$$
$$\vdots$$
$$\bar{x}_n = p_n x_n$$

and let

$$\bar{y}_1 = q_1 y_1$$
$$\bar{y}_2 = q_2 y_2$$
$$\vdots$$
$$\bar{y}_m = q_m y_m$$

---

[1] In addition to the magnitude scaling to be discussed, there is a scaling called *unit*, or *normalized*, scaling. The concept and the scaling process are the same as in magnitude scaling. See References [4] and [5].

or, in matrix form,

$$\bar{\mathbf{x}} = \mathbf{P}\mathbf{x} \qquad (5\text{-}4)$$

$$\bar{\mathbf{y}} = \mathbf{Q}\mathbf{y} \qquad (5\text{-}5)$$

where

$$\mathbf{P} = \begin{bmatrix} p_1 & 0 & \cdots & 0 \\ 0 & p_2 & \cdots & 0 \\ \vdots & \vdots & & \vdots \\ 0 & 0 & \cdots & p_n \end{bmatrix} \qquad \mathbf{P}^{-1} = \begin{bmatrix} 1/p_1 & 0 & \cdots & 0 \\ 0 & 1/p_2 & \cdots & 0 \\ \vdots & \vdots & & \vdots \\ 0 & 0 & \cdots & 1/p_n \end{bmatrix}$$

$$\mathbf{Q} = \begin{bmatrix} q_1 & 0 & \cdots & 0 \\ 0 & q_2 & \cdots & 0 \\ \vdots & \vdots & & \vdots \\ 0 & 0 & \cdots & q_m \end{bmatrix} \qquad \mathbf{Q}^{-1} = \begin{bmatrix} 1/q_1 & 0 & \cdots & 0 \\ 0 & 1/q_2 & \cdots & 0 \\ \vdots & \vdots & & \vdots \\ 0 & 0 & \cdots & 1/q_m \end{bmatrix}$$

The constant p_i and q_i are *nonzero* real constants.[2] They are chosen so that the new variable $\bar{x}_i$ and $\bar{y}_i$ will not saturate and will take the full scale. For example, if the maximum magnitude of x_1, x_2, and y_1 are known to be 25 volts, 0.1 volt, and 40 volts, and if the allowable range of the computer is from -10 volts to $+10$ volts, then p_1, p_2, and q_1 can be chosen as $p_1 = 10/25 = 0.4$, $p_2 = 10/0.1 = 100$, and $q_1 = 10/40 = 0.25$. Then the new variables $\bar{x}_1$, $\bar{x}_2$, and $\bar{y}_1$ can vary over the entire allowable range. By the substitution of $\mathbf{x} = \mathbf{P}^{-1}\bar{\mathbf{x}}$, $\mathbf{y} = \mathbf{Q}^{-1}\bar{\mathbf{y}}$, and $\dot{\mathbf{x}} = \mathbf{P}^{-1}\dot{\bar{\mathbf{x}}}$, (5-3) becomes

$$\dot{\bar{\mathbf{x}}} = \mathbf{P}\mathbf{A}\mathbf{P}^{-1}\bar{\mathbf{x}} + \mathbf{P}\mathbf{B}\mathbf{u} \qquad (5\text{-}6\text{a})$$

$$\bar{\mathbf{y}} = \mathbf{Q}\mathbf{C}\mathbf{P}^{-1}\bar{\mathbf{x}} + \mathbf{Q}\mathbf{D}\mathbf{u} \qquad (5\text{-}6\text{b})$$

If this equation is used in the simulation, the variables will not saturate and will range over the full scale. By using $\mathbf{x} = \mathbf{P}^{-1}\bar{\mathbf{x}}$ and $\mathbf{y} = \mathbf{Q}^{-1}\bar{\mathbf{y}}$, the responses of the original equation (5-3) can then be obtained.

Example 3

Consider the dynamical equation

$$\begin{bmatrix} \dot{x}_1 \\ \dot{x}_2 \end{bmatrix} = \begin{bmatrix} 0 & 1 \\ -2 & -2 \end{bmatrix} \begin{bmatrix} x_1 \\ x_2 \end{bmatrix} + \begin{bmatrix} 1 \\ 4 \end{bmatrix} u$$

$$y = \begin{bmatrix} 2 & -1 \end{bmatrix} \mathbf{x}$$

Suppose it is known that the maximum magnitudes of x_1, x_2, and y are, respectively, 25 volts, -10 volts, and 40 volts for the class of inputs of interests and suppose that the analog computer used has a range from -10 volts to 10 volts. Then the matrices $\mathbf{P}$ and $\mathbf{Q}$ can be chosen as

$$\mathbf{P} = \begin{bmatrix} 10/25 & 0 \\ 0 & 10/10 \end{bmatrix} \qquad \mathbf{Q} = 0.25$$

---

[2] Note that p_i and q_i can be either positive or negative. If positive, the polarity of the variable will not change. If negative, then the polarity will change.

For these **P** and **Q**, Equation (5-6) becomes

$$\dot{\mathbf{x}} = \begin{bmatrix} 0 & 0.4 \\ -5 & -2 \end{bmatrix} \mathbf{x} + \begin{bmatrix} 0.4 \\ 400 \end{bmatrix} u$$

$$\bar{y} = \begin{bmatrix} 1.25 & -0.25 \end{bmatrix} \mathbf{x}$$

This is the equation after magnitude scaling. ∎

In magnitude scaling, in order to choose **P** and **Q** properly, the maximum magnitudes of x_i and y_i must be known. Unfortunately this is generally not the case. Hence the choices of **P** and **Q** must be proceeded by trial and error. This trial-and-error process can be eliminated if the equation is first solved or simulated in a digital computer (see Section 5-2). Otherwise the magnitude scaling could constitute a problem in an analog computer simulation.

Time scaling. In an analog computer simulation, it is sometimes desirable to increase or decrease the speed in which the computer solves the equation. For example, it takes three days for a spaceship to travel from the earth to the moon; however its entire trajectory can be simulated in a computer in a matter of hours. This is achieved by the change of time scale. Consider the linear time-invariant dynamical equation

$$\frac{d}{dt} \mathbf{x}(t) = \mathbf{A}\mathbf{x}(t) + \mathbf{B}\mathbf{u}(t) \tag{5-7a}$$

$$\mathbf{y}(t) = \mathbf{C}\mathbf{x}(t) + \mathbf{D}\mathbf{u}(t) \tag{5-7b}$$

with $\mathbf{x}(0) = \mathbf{x}_0$. Now let $\tau = \alpha t$, and define $\bar{\mathbf{u}}(\tau) \triangleq \mathbf{u}(\tau/\alpha)$, $\bar{\mathbf{x}}(\tau) \triangleq \mathbf{x}(\tau/\alpha)$, $\bar{\mathbf{y}}(\tau) \triangleq \mathbf{y}(\tau/\alpha)$. Then by using $dt = (1/\alpha) d\tau$, Equation (5-7) becomes

$$\alpha \frac{d}{d\tau} \bar{\mathbf{x}}(\tau) = \mathbf{A}\bar{\mathbf{x}}(\tau) + \mathbf{B}\bar{\mathbf{u}}(\tau) \tag{5-8a}$$

$$\bar{\mathbf{y}}(\tau) = \mathbf{C}\bar{\mathbf{x}}(\tau) + \mathbf{D}\bar{\mathbf{u}}(\tau) \tag{5-8b}$$

or

$$\frac{d}{d\tau} \bar{\mathbf{x}}(\tau) = \frac{1}{\alpha} \mathbf{A}\bar{\mathbf{x}}(\tau) + \frac{1}{\alpha} \mathbf{B}\bar{\mathbf{u}}(\tau) \tag{5-9a}$$

$$\bar{\mathbf{y}}(\tau) = \mathbf{C}\bar{\mathbf{x}}(\tau) + \mathbf{D}\bar{\mathbf{u}}(\tau) \tag{5-9b}$$

If α in $\tau = \alpha t$ is greater than 1, then the solutions of (5-8) and (5-9) are, compared with that of (5-7), slowed down. On the other hand, if α is less than 1, then the solutions are speeded up. Since $\bar{x}(\tau) = x(\tau/\alpha)$, $\bar{y}(\tau) = y(\tau/\alpha)$, the solutions of (5-7), (5-8), and (5-9) are the same, except for time scaling.

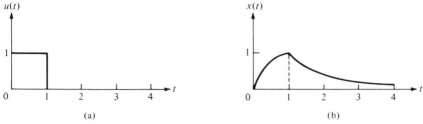

Figure 5-7 A time response.

Example 4

Consider the one-dimensional dynamical equation

$$\frac{dx(t)}{dt} = -x(t) + u(t) \tag{5-10}$$

The solution of (5-10) due to the initial state $x(0) = 0$ and the input $u(t)$ shown in Figure 5-7(a) is computed in Figure 5-7(b). If we introduce $\tau = \alpha t$, then Equation (5-10) becomes

$$\frac{d\bar{x}(\tau)}{d\tau} = -\frac{1}{\alpha}\bar{x}(\tau) + \frac{1}{\alpha}\bar{u}(\tau) \tag{5-11}$$

The solutions of Equation (5-11) for $\alpha = 3$ and 0.2 are shown in Figure 5-8. ∎

We see from (5-8) and (5-9) that there are two possible ways to achieve the change of time scale. After Equation (5-7) is simulated, if we reduce the gains of all integrators by a factor of α (this corresponds to the increase of the capacitance of the capacitor in Figure 5-3 by a factor of α), then the change of time scale is

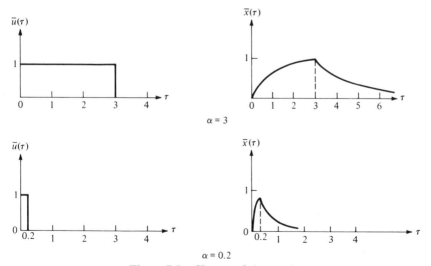

Figure 5-8 Change of time scale.

achieved. If the gains of integrators cannot be changed arbitrarily, as is the case in most analog computers, then in order to change the time scale, we have to simulate the new equation (5-9).

Every present-day analog computer has two or more different built-in speed operations. This is achieved by use of two or more different capacitors in all integrators. As we see from Figure 5-3 that, if the capacitance changes, say from 10^{-6} farad to 10^{-8} farad, then the gains of all integrators will increase by 100 times. Consequently the solution can be speeded up by 100 times. These different speeds of operation can expand the applications of analog computers. See Reference [6].

5-2 Digital Computer Simulations of Dynamical Equations

A digital computer operates only on sequences of numbers. Hence when we use a digital computer to solve a differential equation, the equation must first be discretized. Consider the dynamical equation

$$\dot{\mathbf{x}} = \mathbf{A}\mathbf{x} + \mathbf{B}\mathbf{u} \tag{5-12}$$

or, in integral form,

$$\mathbf{x}(t + \Delta t) = \mathbf{x}(t) + \int_{t}^{t+\Delta t} [\mathbf{A}\mathbf{x}(\tau) + \mathbf{B}\mathbf{u}(\tau)] \, d\tau \tag{5-13}$$

If Δt is very small and if x and u are roughly constant in the interval $[t, t + \Delta t]$, then Equation (5-13) can be approximated by

$$\mathbf{x}(t + \Delta t) = \mathbf{x}(t) + [\mathbf{A}\mathbf{x}(t) + \mathbf{B}\mathbf{u}(t)] \, \Delta t \tag{5-14}$$

This equation can be used to solve for $\{\mathbf{x}(0), \mathbf{x}(\Delta t), \mathbf{x}(2\,\Delta t), \ldots\}$ in a digital computer. It is clear that the smaller the Δt, the more accurate the solution. Δt is called the *integration step size*. For a given step size, Equation (5-14) is the simplest and crudest method for carrying out the integration in (5-13). There are many other more complicated and accurate methods for carrying out the integration, including the trapezoidal, the Runge-Kutta method, and the predictor-corrector method, among others. These methods are readily available as subroutines in most computing centers. These subroutines are well tested and can be easily employed to solve dynamical equations.

In addition to the aforementioned subroutines, there are many specialized digital computer simulation programs for solving dynamical equations, such as MIDAS (*M*odified *I*ntegration *D*igital *A*nalog *S*imulator), MIMIC (an improved version of MIDAS), IBM 1130 CSMP, System/360 CSMP (*C*ontinuous *S*ystem *M*odeling *P*rogram), and TELSIM (*Tele*type *Sim*ulator). These programs can be easily prepared from the basic block diagram of a dynamical equation or directly from a dynamical equation.

5-2 DIGITAL COMPUTER SIMULATIONS OF DYNAMICAL EQUATIONS

We give a simple example of the application of System/360 CSMP. Consider the dynamical equation

$$\begin{bmatrix} \dot{x}_1 \\ \dot{x}_2 \end{bmatrix} = \begin{bmatrix} 2 & -1 \\ 1 & 5 \end{bmatrix} \begin{bmatrix} x_1 \\ x_2 \end{bmatrix} + \begin{bmatrix} 0 \\ 1.5 \end{bmatrix} u$$

$$y = \begin{bmatrix} 1 & 0.5 \end{bmatrix} \begin{bmatrix} x_1 \\ x_2 \end{bmatrix}$$

Find the output y and state variable x_1 from 0 to 20 seconds due to the initial condition $x_1(0) = 1$, $x_2(0) = 0$, and a unit-step-function input.

The CSMP input statements for this problem are listed in the following. It is written in Fortran IV language.

```
*
DYNAMIC
   PARAMETER U=1.0
      X1DT=2*X1-X2
      X2DT=X1+5*X2+1.5*U
      X1=INTGR(1.0,X1DT)
      X2=INTGR(0.0,X2DT)
      Y=X1+0.6*X2
   TIMER DELT=0.001,FINTIM=20.0,OUTDEL=0.10
   PRTPLT X1,Y
END
STOP
```

The first part of the program is self-explanatory. "DELT" is the integration step size. "FINTIM" is the final time of computation. "OUTDEL" is the interval in which the responses will be printed. In order to have an accurate result, DELT is usually chosen to be very small. It is however unnecessary to print out every computed result; therefore the printout interval is chosen much larger than DELT. In employing CSMP the user has to decide the sizes of DELT and OUTDEL. For this program we have asked the computer to *print* as well as *plot* the output y and the state variable x_1.

Comparisons of analog computer simulations and digital computer simulations are in order. In analog computer simulation, magnitude scaling might constitute a difficult problem. This difficulty however will not arise on a digital computer because the range of numbers it can handle is very large—for example, from -10^{38} to 10^{38} in the IBM 7090. The accuracy of an analog computer is often limited to 0.1 percent of its full scale; a digital computer may have a precision of eight or more decimal digits. Therefore the result from a digital computer simulation is much more accurate than that from an analog computer. The generation of nonlinear functions is easier on a digital computer than on an analog computer. However the interaction between an analog computer and the user is very good. By this we mean that the parameters of a simulation can be easily adjusted, and the consequence of the adjustments can be immediately observed. The interaction between a general-purpose digital computer and the user is generally not very satisfactory. But this interaction has been improved in recent years because of the introduction of time sharing and of remote terminals. In any case, the use of an analog or a digital computer will make the design much easier.

5-3 Realization Problem

Network synthesis is one of the important disciplines in electrical engineering. It is concerned mainly with the determination of a network that has a prescribed impedance or transfer function. The realization problem we shall introduce in the remainder of this chapter is along the same line, that is, to determine a linear time-invariant dynamical equation that has a prescribed rational transfer-function matrix. Hence this problem could be viewed as a branch of network synthesis.

If we are given a transfer-function matrix $\hat{G}(s)$ and if it is possible to find a linear time-invariant dynamical equation that has $\hat{G}(s)$ as its transfer function, then $\hat{G}(s)$ is said to be *realizable*, and the dynamical equation is said to be a *realization* of $\hat{G}(s)$. We study the realization problem for the following reasons: First, a dynamical equation can be readily simulated on an analog or a digital computer; hence if a realization is found, the transfer function can then be simulated by using the realization. Second, the realization provides a method of synthesizing a transfer function by use of operational-amplifier circuits. Finally, there are many design techniques and computational algorithms developed exclusively for dynamical equations. In order to apply these techniques and algorithms, transfer-function matrices must be realized.

If a transfer-function matrix is realizable at all, then it is possible to have infinitely many different dynamical equation realizations. These dynamical equations may have the same dimensions or different dimensions. However their dimensions can never be smaller than the degree of the transfer-function matrix (see Definition 2-7). The dynamical equation realization whose dimension is equal to the degree of the transfer-function matrix is called a *minimal-dimensional* (or *irreducible*) *realization*. It can be shown that a minimal-dimensional realization is always controllable and observable. In this chapter we discuss only minimal-dimensional realizations. The significance of a minimal-dimensional realization is that, if the transfer function is simulated on an analog computer by using this realization, then the number of integrators used will be the smallest possible.

Not every transfer function is realizable by a linear time-invariant dynamical equation. The condition for a transfer-function matrix to be realizable by a finite-dimensional dynamical equation of form

$$\dot{x} = Ax + Bu$$

$$y = Cx + Du$$

is that the transfer-function matrix must be a proper rational matrix. If a transfer function is *not* rational (the system is not a lumped system but rather a distributed system), then it cannot be realized by a finite-dimensional dynamical equation. If a transfer function is rational but not proper, then the realization will be of the form

$$\dot{\mathbf{x}} = \mathbf{Ax} + \mathbf{Bu}$$

$$\mathbf{y} = \mathbf{Cx} + \mathbf{Du} + \mathbf{D}_1\dot{\mathbf{u}} + \mathbf{D}_2\ddot{\mathbf{u}} + \cdots$$

In this case, differentiators are needed to generate $\dot{\mathbf{u}}, \ddot{\mathbf{u}}, \ldots$. Hence if differentiators are not permitted, as is often the case in practice, then a rational function must be proper.

In the subsequent two sections minimal-dimensional realizations of scalar proper rational functions and vector proper rational transfer functions will be discussed. The minimal-dimensional realizations of general multivariable proper rational transfer matrices are too complicated to be discussed here; the interested reader is referred to Reference [1].

5-4 Realizations of Proper Rational Functions

In this section various minimal-dimensional realizations of scalar proper rational transfer functions will be introduced. We first introduce a realization of $\beta/D(s)$, then three different realizations of the same $N(s)/D(s)$.

Realizations of $\beta/D(s)$. Consider the transfer function

$$\hat{g}(s) = \frac{\beta'}{\alpha_0' s^n + \alpha_1' s^{n-1} + \cdots + \alpha_{n-1}' s + \alpha_n'} \qquad \text{with } \alpha_0' \neq 0$$

which can be written as

$$\hat{g}(s) = \frac{\beta}{s^n + \alpha_1 s^{n-1} + \cdots + \alpha_{n-1} s + \alpha_n} \triangleq \frac{\beta}{D(s)} \qquad (5\text{-}15)$$

with $\beta = \beta'/\alpha_0'$, $\alpha_i = \alpha_i'/\alpha_0'$. It is assumed that β and α_i are real constants. In the time domain, Equation (5-15) can be written as

$$(p^n + \alpha_1 p^{n-1} + \cdots + \alpha_n) y(t) = \beta u(t) \qquad (5\text{-}16)$$

where p^i stands for d^i/dt^i. By taking the Laplace transform of (5-16) and assuming zero initial conditions, it is easy to verify that the transfer function from u to y in (5-16) is indeed the one given in (5-15). It is well known that in order for an nth-order differential equation to have a unique solution for any u, we need n number of initial conditions. Hence the state vector will consist of n

components. For Equation (5-16), it turns out that the output y and its derivatives up to the $(n-1)$th order qualify as state variables. Let

$$\begin{aligned} x_1(t) &\triangleq y(t) \\ x_2(t) &\triangleq y'(t) = py(t) = \dot{x}_1(t) \\ x_3(t) &\triangleq y''(t) = p^2 y(t) = \dot{x}_2(t) \\ &\vdots \\ x_n(t) &\triangleq y^{(n-1)}(t) = p^{n-1} y(t) = \dot{x}_{n-1}(t) \end{aligned} \qquad (5\text{-}17)$$

Differentiating $x_n(t)$ once and using (5-16), we obtain

$$\dot{x}_n(t) = p^n y(t) = -\alpha_n x_1 - \alpha_{n-1} x_2 - \cdots - \alpha_1 x_n + \beta u \qquad (5\text{-}18)$$

If we define

$$\mathbf{x}(t) \triangleq \begin{bmatrix} x_1(t) \\ x_2(t) \\ \vdots \\ x_n(t) \end{bmatrix}$$

then Equations (5-17) and (5-18) can be arranged in matrix form as

$$\dot{\mathbf{x}} = \begin{bmatrix} 0 & 1 & 0 & \cdots & 0 \\ 0 & 0 & 1 & \cdots & 0 \\ 0 & 0 & 0 & \cdots & 0 \\ \vdots & \vdots & \vdots & & \vdots \\ 0 & 0 & 0 & & 1 \\ -\alpha_n & -\alpha_{n-1} & -\alpha_{n-2} & \cdots & -\alpha_1 \end{bmatrix} \mathbf{x} + \begin{bmatrix} 0 \\ 0 \\ 0 \\ \vdots \\ 0 \\ \beta \end{bmatrix} u \qquad (5\text{-}19\text{a})$$

$$y = \begin{bmatrix} 1 & 0 & 0 & \cdots & 0 \end{bmatrix} \mathbf{x} \qquad (5\text{-}19\text{b})$$

This is a minimal-dimensional realization of the transfer function in Equation (5-15), for its dimension is equal to the degree of the transfer function. We note that the realization can be read out directly from the coefficients of $\hat{g}(s)$ in (5-15). Since (5-19) is derived from $\hat{g}(s)$ in (5-15), its transfer function $\mathbf{c}(s\mathbf{I} - \mathbf{A})^{-1}\mathbf{b}$ should be equal to $\hat{g}(s)$. This can be verified algebraically or, more easily, by using Mason's formula. The basic block diagram of (5-19) is shown in Figure 5-9. There is one forward path, p_1, from u to y, with gain β/s^n. The Δ in Mason's formula is

$$\Delta = 1 + \frac{\alpha_1}{s} + \frac{\alpha_2}{s^2} + \cdots + \frac{\alpha_{n-1}}{s^{n-1}} + \frac{\alpha_n}{s^n}$$

It is clear that $\Delta_1 = 1$. Hence the transfer function of (5-19) is

$$\hat{g}(s) = \frac{\beta/s^n}{1 + \frac{\alpha_1}{s} + \frac{\alpha_2}{s^2} + \cdots + \frac{\alpha_n}{s^n}} = \frac{\beta}{s^n + \alpha_1 s^{n-1} + \alpha_2 s^{n-2} + \cdots + \alpha_n}$$

This verifies that (5-19) is indeed a realization of $\hat{g}(s)$ in (5-15).

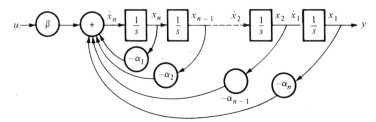

Figure 5-9 A basic block diagram of Equation (5-19).

We study now the realizations of general proper rational transfer functions. It is clear that every proper rational transfer function can be written as

$$\hat{g}_1(s) = \frac{\beta_0' s^n + \beta_1' s^{n-1} + \cdots + \beta_n'}{\alpha_0' s^n + \alpha_1' s^{n-1} + \cdots + \alpha_n'} = d' + \frac{\beta_1 s^{n-1} + \beta_2 s^{n-2} + \cdots + \beta_n}{s^n + \alpha_1 s^{n-1} + \cdots + \alpha_n}$$

$$\triangleq d' + \frac{N(s)}{D(s)} \qquad (5\text{-}20)$$

If the dynamical equation

$$\dot{\mathbf{x}} = \mathbf{A}\mathbf{x} + \mathbf{b}u$$

$$y = \mathbf{c}\mathbf{x} + du$$

is a realization of $\hat{g}_1(s)$ in Equation (5-20), then we have

$$\mathbf{c}(s\mathbf{I} - \mathbf{A})^{-1}\mathbf{b} + d = d' + \frac{N(s)}{D(s)} \qquad (5\text{-}21)$$

This equality holds for any s; in particular, $s = \infty$. Since $N(s)/D(s) = 0$ and $\mathbf{c}(s\mathbf{I} - \mathbf{A})^{-1}\mathbf{b} = 0$ at $s = \infty$, Equation (5-21) implies that $d = d'$. Hence if a proper rational function is expanded into the form of (5-20), the constant term gives immediately the d in a realization. Hence in the following, we need to consider only the realization of the strictly proper rational function

$$\hat{g}(s) = \frac{\hat{y}(s)}{\hat{u}(s)} = \frac{N(s)}{D(s)} \triangleq \frac{\beta_1 s^{n-1} + \beta_2 s^{n-2} + \cdots + \beta_n}{s^n + \alpha_1 s^{n-1} + \cdots + \alpha_n} \qquad (5\text{-}22)$$

or, equivalently, the differential equation

$$D(p)y(t) = N(p)u(t) \qquad (5\text{-}23)$$

where $D(p) = p^n + \alpha_1 p^{n-1} + \cdots + \alpha_n$, $N(p) = \beta_1 p^{n-1} + \cdots + \beta_n$, and $p^i \triangleq d^i/dt^i$. We now introduce three different realizations of $\hat{g}(s)$.

Observable-form realization of $N(s)/D(s)$. Consider Equation (5-23), an nth-order differential equation. It is well known that its solution is uniquely determinable if n initial conditions are known. However if we choose $y(t)$, $y'(t), \ldots, y^{(n-1)}(t)$ as the state variables, as we did in (5-17), then the resulting equation will be of the form $\dot{\mathbf{x}} = \mathbf{Ax} + \mathbf{b}u$, $y = \mathbf{cx} + d_0 u + d_1 \dot{u} + d_2 \ddot{u} + \cdots$. This realization requires differentiation; hence it is not acceptable in the presence of noise. In order to realize (5-22) or (5-23) in the form $\dot{\mathbf{x}} = \mathbf{Ax} + \mathbf{b}u$, $y = \mathbf{cx}$, a different set of state variables has to be chosen.

Taking the Laplace transform of (5-23) and grouping the terms associated with the same power of s, we obtain

$$\hat{g}(s) = \frac{N(s)}{D(s)} \hat{u}(s) + \frac{1}{D(s)} \{y(0) s^{n-1}$$
$$+ [y'(0) + \alpha_1 y(0) - \beta_1 u(0)] s^{n-2} + \cdots \qquad (5\text{-}24)$$
$$+ [y^{(n-1)}(0) + \alpha_1 y^{(n-2)}(0) - \beta_1 u^{(n-2)}(0) + \alpha_2 y^{(n-3)}(0)$$
$$- \beta_2 u^{(n-3)}(0) + \cdots + \alpha_{n-1} y(0) - \beta_{n-1} u(0)] s^0 \}$$

The first term on the right-hand side of (5-24) gives the response due to the input $\hat{u}(s)$; the remainder gives the response due to the initial conditions. Therefore if all the coefficients associated with $s^{n-1}, s^{n-2}, \ldots, s^0$ in (5-24) are known, then for any u, a unique output y can be determined. Consequently if we choose the state-variables as

$$x_n(t) = y(t)$$
$$x_{n-1}(t) = y'(t) + \alpha_1 y(t) - \beta_1 u(t)$$
$$x_{n-2}(t) = y''(t) + \alpha_1 y'(t) - \beta_1 u'(t) + \alpha_2 y(t) - \beta_2 u(t) \qquad (5\text{-}25)$$
$$\vdots$$
$$x_1(t) = y^{(n-1)}(t) + \alpha_1 y^{(n-2)}(t) - \beta_1 u^{(n-2)}(t) + \cdots + \alpha_{n-1} y(t) - \beta_{n-1} u(t)$$

then $\mathbf{x} = [x_1 \; x_2 \; \cdots \; x_n]'$ qualifies as a state vector, where the "prime" stands for the transpose. The set of equations in (5-25) yields

$$y = x_n$$
$$x_{n-1} = \dot{x}_n + \alpha_1 x_n - \beta_1 u$$
$$x_{n-2} = \dot{x}_{n-1} + \alpha_2 x_n - \beta_2 u$$
$$\vdots$$
$$x_1 = \dot{x}_2 + \alpha_{n-1} x_n - \beta_{n-1} u$$

Differentiating x_1 in (5-25) once and using (5-23), we obtain

$$\dot{x}_1 = -\alpha_n x_n + \beta_n u$$

The foregoing equations can be arranged in matrix form as

$$\begin{bmatrix} \dot{x}_1 \\ \dot{x}_2 \\ \dot{x}_3 \\ \vdots \\ \dot{x}_{n-1} \\ \dot{x}_n \end{bmatrix} = \begin{bmatrix} 0 & 0 & 0 & \cdots & 0 & -\alpha_n \\ 1 & 0 & 0 & \cdots & 0 & -\alpha_{n-1} \\ 0 & 1 & 0 & \cdots & 0 & -\alpha_{n-2} \\ \vdots & \vdots & \vdots & & \vdots & \vdots \\ 0 & 0 & 0 & \cdots & 0 & -\alpha_2 \\ 0 & 0 & 0 & \cdots & 1 & -\alpha_1 \end{bmatrix} \begin{bmatrix} x_1 \\ x_2 \\ x_3 \\ \vdots \\ x_{n-1} \\ x_n \end{bmatrix} + \begin{bmatrix} \beta_n \\ \beta_{n-1} \\ \beta_{n-2} \\ \vdots \\ \beta_2 \\ \beta_1 \end{bmatrix} u \qquad (5\text{-}26)$$

$$y = [0 \; 0 \; 0 \; \cdots \; 0 \; 1] \, \mathbf{x}$$

5-4 REALIZATIONS OF PROPER RATIONAL FUNCTIONS 139

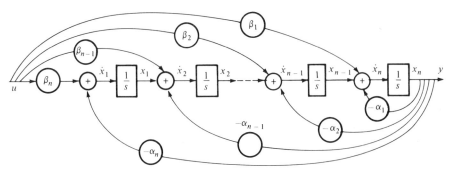

Figure 5-10 A basic block diagram of Equation (5-26).

This is a realization of the transfer function in (5-22), as can be verified by using Mason's formula. The basic block diagram of (5-26) is shown in Figure 5-10. There are n loops with loop gains $-\alpha_1/s$, $-\alpha_2/s^2$, ..., $-\alpha_{n-1}/s^{n-1}$ and $-\alpha_n/s^n$. All the loops touch each other; hence the Δ in Mason's formula is given by

$$\Delta = 1 + \frac{\alpha_1}{s} + \frac{\alpha_2}{s^2} + \cdots + \frac{\alpha_n}{s^n} = \frac{1}{s^n}(s^n + \alpha_1 s^{n-1} + \alpha_2 s^{n-2} + \cdots + \alpha_n)$$

There are n forward paths, with gains β_1/s, β_2/s^2, ..., β_n/s^n. Every forward path touches all the loops; hence $\Delta_i = 1$, for $i = 1, 2, \ldots, n$. Consequently the transfer function from u to y of Figure 5-10 or, equivalently, of (5-26), is

$$\hat{g}(s) = \frac{\frac{\beta_1}{s} + \frac{\beta_2}{s^2} + \cdots + \frac{\beta_n}{s^n}}{\frac{1}{s^n}(s^n + \alpha_1 s^{n-1} + \cdots + \alpha_n)} = \frac{\beta_1 s^{n-1} + \beta_2 s^{n-2} + \cdots + \beta_n}{s^n + \alpha_1 s^{n-1} + \cdots + \alpha_n}$$

This verifies that Equation (5-26) is indeed a realization of $\hat{g}(s)$ in (5-22). Note that this realization can be obtained immediately from the coefficients of $\hat{g}(s)$.

Recall that a realization of a transfer function is called a *minimal-dimensional* realization if its dimension is equal to the degree of the transfer function. The dimension of (5-26) is n; the degree of $\hat{g}(s)$ in (5-22) is also n if $N(s)$ and $D(s)$ have no common factor. Hence we conclude that if $N(s)$ and $D(s)$ have no common factor, then (5-26) is a minimal-dimensional realization of (5-22). It can also be shown that (5-26) is both controllable and observable. However if $N(s)$ and $D(s)$ have a common factor, then although (5-26) is still observable, it is not controllable. This is the reason we call (5-26) the observable-form realization of (5-22).

Controllable-form realization of $N(s)/D(s)$. We introduce now a different realization called the *controllable-form* dynamical equation realization. Consider $\hat{g}(s) = N(s)/D(s)$ or, equivalently, the differential equation

$$D(p)y(t) = N(p)u(t) \qquad (6\text{-}27)$$

Let us introduce a new variable $v(t)$ satisfying

$$D(p)v(t) = u(t) \tag{5-28}$$

By substituting (5-28) into (5-27) and using the fact that $N(p)D(p) = D(p)N(p)$, we obtain

$$y(t) = N(p)v(t) \tag{5-29}$$

Observe that Equation (5-29) is in the form of (5-16) with $\beta = 1$. Hence by defining, as in (5-17), $x_1 = v$, $x_2 = v'$, ..., $x_n = v^{(n-1)}$, we immediately obtain the equation, as in (5-19a),

$$\dot{\mathbf{x}} = \begin{bmatrix} 0 & 1 & 0 & \cdots & 0 \\ 0 & 0 & 1 & \cdots & 0 \\ 0 & 0 & 0 & \cdots & 0 \\ \vdots & \vdots & \vdots & & \vdots \\ 0 & 0 & 0 & & 1 \\ -\alpha_n & -\alpha_{n-1} & -\alpha_{n-2} & \cdots & -\alpha_1 \end{bmatrix} \mathbf{x} + \begin{bmatrix} 0 \\ 0 \\ 0 \\ \vdots \\ 0 \\ 1 \end{bmatrix} u \tag{5-30a}$$

Substituting $x_i = v^{(i-1)}$, for $i = 1, 2, \ldots, n$, into (5-29) yields

$$\begin{aligned} y &= \beta_1 v^{(n-1)} + \beta_2 v^{(n-2)} + \cdots + \beta_{(n-1)} v' + \beta_n v \\ &= \beta_n x_1 + \beta_{n-1} x_2 + \cdots + \beta_1 x_n \\ &= [\beta_n \ \beta_{n-1} \ \cdots \ \beta_2 \ \beta_1] \mathbf{x} \end{aligned} \tag{5-30b}$$

This is a novel way to realize a transfer function. That the transfer function of (5-30) is equal to $N(s)/D(s)$ can again be verified by using Mason's formula. As in the preceding subsection, if $N(s)$ and $D(s)$ have no common factor, then (5-30) is a different minimal-dimensional realization of $\hat{g}(s)$ in (5-22) and is both controllable and observable. If $N(s)$ and $D(s)$ have a common factor, then (5-30) is not observable but is still controllable. This is the reason we call it the *controllable-form realization*.

Example

Find the controllable-form and observable-form dynamical equation realizations of

$$\hat{g}(s) = \frac{3s^5 + 6s^4 - 10s^3 - 6s + 2}{2s^5 + 4s^4 - 8s^3 + 2s + 2} = \frac{s^3 + 1.5s - 0.5}{s^5 + 2s^4 - 4s^3 + s + 1} + 1.5$$

From the coefficients of $\hat{g}(s)$, the observable-form realization is obtained as

$$\begin{bmatrix} \dot{x}_1 \\ \dot{x}_2 \\ \dot{x}_3 \\ \dot{x}_4 \\ \dot{x}_5 \end{bmatrix} = \begin{bmatrix} 0 & 0 & 0 & 0 & -1 \\ 1 & 0 & 0 & 0 & -1 \\ 0 & 1 & 0 & 0 & 0 \\ 0 & 0 & 1 & 0 & 4 \\ 0 & 0 & 0 & 1 & -2 \end{bmatrix} \mathbf{x} + \begin{bmatrix} -0.5 \\ 1.5 \\ 0 \\ 1 \\ 0 \end{bmatrix} u \tag{5-31}$$

$$y = [0 \ \ 0 \ \ 0 \ \ 0 \ \ 1] \mathbf{x} + 1.5u$$

The controllable-form realization is

$$\begin{bmatrix} \dot{x}_1 \\ \dot{x}_2 \\ \dot{x}_3 \\ \dot{x}_4 \\ \dot{x}_5 \end{bmatrix} = \begin{bmatrix} 0 & 1 & 0 & 0 & 0 \\ 0 & 0 & 1 & 0 & 0 \\ 0 & 0 & 0 & 1 & 0 \\ 0 & 0 & 0 & 0 & 1 \\ -1 & -1 & 0 & 4 & -2 \end{bmatrix} \mathbf{x} + \begin{bmatrix} 0 \\ 0 \\ 0 \\ 0 \\ 1 \end{bmatrix} u \qquad (5\text{-}32)$$

$$y = \begin{bmatrix} -0.5 & 1.5 & 0 & 1 & 0 \end{bmatrix} \mathbf{x} + 1.5u$$

We note that although we use the same **x** in (5-31) and (5-32), they represent entirely different quantities. ∎

Jordan-form realization of $N(s)/D(s)$. We introduce in this subsection another realization called the *Jordan-form* dynamical equation realization. We use an example to illustrate the realization procedure. The idea can be easily extended to the general case. Consider a strictly proper rational function $\hat{g}(s)$, with three distinct poles λ_1, λ_2, and λ_3, with multiplicities 3, 1, and 1. It is assumed that $\hat{g}(s)$ can be expanded, by partial fraction expansion, into

$$\hat{g}(s) = \frac{e_{11}}{(s-\lambda_1)^3} + \frac{e_{12}}{(s-\lambda_1)^2} + \frac{e_{13}}{(s-\lambda_1)} + \frac{e_2}{(s-\lambda_2)} + \frac{e_3}{(s-\lambda_3)}$$

$$(5\text{-}33)$$

with $e_{11} \neq 0$, $e_2 \neq 0$, and $e_3 \neq 0$. Block diagrams are drawn in Figure 5-11 for (5-33) in terms of each pole. We note that each block in Figure 5-11 can be viewed as consisting of an integrator, as shown in Figure 5-12. Hence the output of each block is qualified as a state variable. By assigning the output of each block as a state variable, and referring to Figure 5-12, we can easily obtain the mathematical description of each block in Figure 5-11 as shown. By grouping these equations, we can obtain the dynamical equation description of the block diagram in Figure 5-11(a) as

$$\begin{bmatrix} \dot{x}_{11} \\ \dot{x}_{12} \\ \dot{x}_{13} \\ \dot{x}_2 \\ \dot{x}_3 \end{bmatrix} = \begin{bmatrix} \lambda_1 & 1 & 0 & 0 & 0 \\ 0 & \lambda_1 & 1 & 0 & 0 \\ 0 & 0 & \lambda_1 & 0 & 0 \\ 0 & 0 & 0 & \lambda_2 & 0 \\ 0 & 0 & 0 & 0 & \lambda_3 \end{bmatrix} \mathbf{x} + \begin{bmatrix} 0 \\ 0 \\ 1 \\ 1 \\ 1 \end{bmatrix} u \qquad (5\text{-}34)$$

$$y = \begin{bmatrix} e_{11} & e_{12} & e_{13} & e_2 & e_3 \end{bmatrix} \mathbf{x}$$

The dynamical equation description of Figure 5-11(b) is

$$\begin{bmatrix} \dot{x}_{11} \\ \dot{x}_{12} \\ \dot{x}_{13} \\ \dot{x}_2 \\ \dot{x}_3 \end{bmatrix} = \begin{bmatrix} \lambda_1 & 0 & 0 & 0 & 0 \\ 1 & \lambda_1 & 0 & 0 & 0 \\ 0 & 1 & \lambda_1 & 0 & 0 \\ 0 & 0 & 0 & \lambda_2 & 0 \\ 0 & 0 & 0 & 0 & \lambda_3 \end{bmatrix} \mathbf{x} + \begin{bmatrix} e_{11} \\ e_{12} \\ e_{13} \\ e_2 \\ e_3 \end{bmatrix} u \qquad (5\text{-}35)$$

$$y = \begin{bmatrix} 0 & 0 & 1 & 1 & 1 \end{bmatrix} \mathbf{x}$$

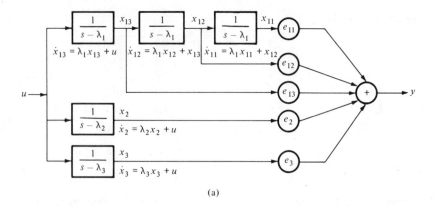

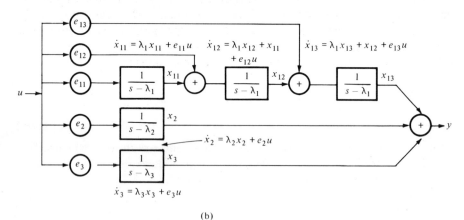

Figure 5-11 Block diagram of Equation (5-33).

These equations are two different realizations of $\hat{g}(s)$ in (5-33); both are said to be in Jordan form.[3] These dynamical equations are minimal-dimensional realizations of $\hat{g}(s)$, because if the denominator and numerator of $\hat{g}(s)$ have common factors, the common factors will be automatically eliminated in the process of partial fraction expansion.

Comparing with the controllable-form and observable-form realizations, there are two disadvantages in using Jordan-form realizations. First, the denominator of a transfer function must be first factored. This is a very tedious job for a transfer function with degree larger than 3. Second, if the transfer function has complex poles, as often might be the case, then the Jordan-form realization will consist of complex numbers. The complex numbers can be changed to real numbers by the so-called equivalence transformation; the

---

[3] The name Jordan is borrowed from linear algebra. See References [1] and [9].

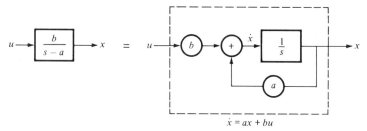

Figure 5-12 Internal structure of block $b/(s - a)$.

interested reader is referred to Reference [1], pages 231 and 257. The Jordan-form realization, on the other hand, has one advantage over the controllable-form or observable-form realizations. Its eigenvalues are less sensitive to parameter variations. See References [2] and [9].

5-5 Realizations of Vector Proper Rational Transfer Functions

In this section realizations of vector proper rational transfer functions will be studied. By a vector rational function we mean either a $1 \times p$ rational-function matrix or a $q \times 1$ rational-function matrix. Consider the $q \times 1$ proper rational-function matrix

$$\hat{\mathbf{G}}(s) = \begin{bmatrix} \hat{g}_1'(s) \\ \hat{g}_2'(s) \\ \vdots \\ \hat{g}_q'(s) \end{bmatrix} \qquad (5\text{-}36)$$

It is assumed that each $\hat{g}_i'(s)$ is an irreducible proper rational function. We first expand $\hat{\mathbf{G}}$ into

$$\hat{\mathbf{G}}(s) = \begin{bmatrix} d_1 \\ d_2 \\ \vdots \\ d_q \end{bmatrix} + \begin{bmatrix} \hat{g}_1(s) \\ \hat{g}_2(s) \\ \vdots \\ \hat{g}_q(s) \end{bmatrix}$$

where $d_i = \hat{g}_i'(\infty)$, and $\hat{g}_i(s) \triangleq g_i'(s) - d_i$ is a strictly proper rational function. Next find the least common denominator of $\hat{g}_i$, for $i = 1, 2, \ldots, p$, say $s^n + \alpha_1 s^{n-1} + \cdots + \alpha_n$; then express $\hat{\mathbf{G}}(s)$ as

$$\hat{\mathbf{G}}(s) = \begin{bmatrix} d_1 \\ d_2 \\ \vdots \\ d_q \end{bmatrix} + \frac{1}{s^n + \alpha_1 s^{n-1} + \cdots + \alpha_n} \begin{bmatrix} \beta_{11} s^{n-1} + \cdots + \beta_{1n} \\ \beta_{21} s^{n-1} + \cdots + \beta_{2n} \\ \vdots \\ \beta_{q1} s^{n-1} + \cdots + \beta_{qn} \end{bmatrix}$$

$$(5\text{-}37)$$

It is claimed that the dynamical equation

$$\dot{\mathbf{x}} = \begin{bmatrix} 0 & 1 & 0 & \cdots & 0 \\ 0 & 0 & 1 & \cdots & 0 \\ \vdots & \vdots & \vdots & & \vdots \\ 0 & 0 & 0 & \cdots & 1 \\ -\alpha_n & -\alpha_{n-1} & -\alpha_{n-2} & \cdots & -\alpha_1 \end{bmatrix} \mathbf{x} + \begin{bmatrix} 0 \\ 0 \\ \vdots \\ 0 \\ 1 \end{bmatrix} u \qquad (5\text{-}38a)$$

$$\begin{bmatrix} y_1 \\ y_2 \\ \vdots \\ y_q \end{bmatrix} = \begin{bmatrix} \beta_{1n} & \beta_{1(n-1)} & \cdots & \beta_{11} \\ \beta_{2n} & \beta_{2(n-1)} & \cdots & \beta_{21} \\ \vdots & \vdots & & \vdots \\ \beta_{qn} & \beta_{q(n-1)} & \cdots & \beta_{qn} \end{bmatrix} \mathbf{x} + \begin{bmatrix} d_1 \\ d_2 \\ \vdots \\ d_q \end{bmatrix} u \qquad (5\text{-}38b)$$

is a realization of (5-37). This can be proved by using the controllable-form realization of $\hat{g}(s)$ in (5-22). By comparing (5-38) with (5-30), we see that the transfer function from u to y_i is equal to

$$d_i + \frac{\beta_{i1}s^{n-1} + \cdots + \beta_{in}}{s^n + \alpha_1 s^{n-1} + \cdots + \alpha_n}$$

which is the ith component of $\hat{\mathbf{G}}(s)$. This proves the assertion. Since $\hat{g}_i'(s)$ for $i = 1, 2, \ldots, q$ are assumed to be irreducible, the degree of $\hat{\mathbf{G}}(s)$ is equal to n. The dynamical equation (5-38) also has dimension n; hence it is a minimal-dimensional realization of $\hat{\mathbf{G}}(s)$ in (5-37).

For single-input single-output transfer functions, we have both the controllable-form and the observable-form realizations. But for column rational functions it is *not* possible to have the observable-form realization.

Example 1

Consider

$$\hat{\mathbf{G}}(s) = \begin{bmatrix} \dfrac{s+3}{(s+1)(s+2)} \\ \\ \dfrac{s+4}{s+3} \end{bmatrix} = \begin{bmatrix} 0 \\ 1 \end{bmatrix} + \begin{bmatrix} \dfrac{s+3}{(s+1)(s+2)} \\ \\ \dfrac{1}{s+3} \end{bmatrix}$$

$$= \begin{bmatrix} 0 \\ 1 \end{bmatrix} + \frac{1}{(s+1)(s+2)(s+3)} \begin{bmatrix} (s+3)^2 \\ (s+1)(s+2) \end{bmatrix}$$

$$= \begin{bmatrix} 0 \\ 1 \end{bmatrix} + \frac{1}{s^3 + 6s^2 + 11s + 6} \begin{bmatrix} s^2 + 6s + 9 \\ s^2 + 3s + 2 \end{bmatrix}$$

Hence a minimal-dimensional realization of $\hat{\mathbf{G}}(s)$ is given by

$$\begin{bmatrix} \dot{x}_1 \\ \dot{x}_2 \\ \dot{x}_3 \end{bmatrix} = \begin{bmatrix} 0 & 1 & 0 \\ 0 & 0 & 1 \\ -6 & -11 & -6 \end{bmatrix} \mathbf{x} + \begin{bmatrix} 0 \\ 0 \\ 1 \end{bmatrix} u$$

$$\mathbf{y} = \begin{bmatrix} 9 & 6 & 1 \\ 2 & 3 & 1 \end{bmatrix} \mathbf{x} + \begin{bmatrix} 0 \\ 1 \end{bmatrix} u$$

■

We study now the realizations of $1 \times p$ proper rational-function matrices. Since its development is similar to the one for the $q \times 1$ case, we present only the result. Consider the $1 \times p$ proper rational matrix

$$\hat{\mathbf{G}}(s) = [\hat{g}_1'(s) \vdots \hat{g}_2'(s) \vdots \cdots \vdots \hat{g}_p'(s)]$$

$$= [d_1 \vdots d_2 \vdots \cdots \vdots d_p] + [\hat{g}_1(s) \vdots \hat{g}_2(s) \vdots \cdots \vdots \hat{g}_p(s)]$$

$$= [d_1 \vdots d_2 \vdots \cdots \vdots d_p] + \frac{1}{s^n + \alpha_1 s^{n-1} + \cdots + \alpha_n}$$

$$\times [\beta_{11} s^{n-1} + \beta_{12} s^{n-2} + \cdots + \beta_{1n} \vdots \beta_{21} s^{n-1} + \beta_{22} s^{n-2} + \cdots$$

$$+ \beta_{2n} \vdots \cdots \vdots \beta_{p1} s^{n-1} + \beta_{p2} s^{n-2} + \cdots + \beta_{pn}]$$

(5-39)

Then the dynamical equation

$$\begin{bmatrix} \dot{x}_1 \\ \dot{x}_2 \\ \dot{x}_3 \\ \vdots \\ \dot{x}_n \end{bmatrix} = \begin{bmatrix} 0 & 0 & \cdots & 0 & -\alpha_n \\ 1 & 0 & \cdots & 0 & -\alpha_{n-1} \\ 0 & 1 & \cdots & 0 & -\alpha_{n-2} \\ \vdots & \vdots & & \vdots & \vdots \\ 0 & 0 & \cdots & 1 & -\alpha_1 \end{bmatrix} \mathbf{x} + \begin{bmatrix} \beta_{1n} & \beta_{2n} & \cdots & \beta_{pn} \\ \beta_{1(n-1)} & \beta_{2(n-1)} & \cdots & \beta_{p(n-1)} \\ \beta_{1(n-2)} & \beta_{2(n-2)} & \cdots & \beta_{p(n-2)} \\ \vdots & \vdots & & \vdots \\ \beta_{11} & \beta_{21} & \cdots & \beta_{p1} \end{bmatrix} \mathbf{u}$$

(5-40)

$$y = [0 \ 0 \ \cdots \ 0 \ 0 \ 1] \mathbf{x} + [d_1 \ d_2 \ \cdots \ d_p] \mathbf{u}$$

is a realization of (5-39). ∎

It is also possible to find Jordan-form realizations for vector proper rational functions. The procedure is similar to the one for the scalar case. We use an example to illustrate the procedure.

Example 2

Find a Jordan-form realization of the 2×1 rational function

$$\hat{\mathbf{G}}(s) = \begin{bmatrix} \dfrac{s+3}{(s+1)(s+2)} \\ \dfrac{s+4}{s+1} \end{bmatrix} = \begin{bmatrix} \dfrac{2}{s+1} - \dfrac{1}{s+2} \\ 1 + \dfrac{3}{s+1} \end{bmatrix} \quad (5\text{-}41)$$

$$= \begin{bmatrix} 0 \\ 1 \end{bmatrix} + \frac{1}{s+1}\begin{bmatrix} 2 \\ 3 \end{bmatrix} + \frac{1}{s+2}\begin{bmatrix} -1 \\ 0 \end{bmatrix}$$

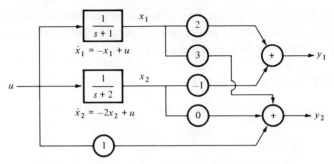

Figure 5-13 Block diagram of Equation (5-41).

A block diagram of Equation (5-41) is given in Figure 5-13. With the state variables chosen as shown, the dynamical equation

$$\begin{bmatrix} \dot{x}_1 \\ \dot{x}_2 \end{bmatrix} = \begin{bmatrix} -1 & 0 \\ 0 & -2 \end{bmatrix} \begin{bmatrix} x_1 \\ x_2 \end{bmatrix} + \begin{bmatrix} 1 \\ 1 \end{bmatrix} u$$

$$\begin{bmatrix} y_1 \\ y_2 \end{bmatrix} = \begin{bmatrix} 2 & -1 \\ 3 & 0 \end{bmatrix} \begin{bmatrix} x_1 \\ x_2 \end{bmatrix} + \begin{bmatrix} 0 \\ 1 \end{bmatrix} u$$

can be found to describe the block diagram. Hence it is a realization of (5-41).

Example 3

Find an analog simulation block diagram for the control system shown in Figure 5-14.

There are two ways to simulate the system shown in Figure 5-14. One is to find the overall transfer function and then simulate it. The other is to simulate each block and then connect them together. The second method is recommended

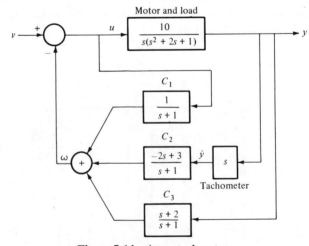

Figure 5-14 A control system.

because, by simulating each block separately, not only are the state variables identifiable with each block, but also it is easier to adjust the parameters of each block.

If the transfer function $10/(s^3 + 2s^2 + 2s)$ is realized as

$$\begin{bmatrix} \dot{x}_1 \\ \dot{x}_2 \\ \dot{x}_3 \end{bmatrix} = \begin{bmatrix} 0 & 1 & 0 \\ 0 & 0 & 1 \\ 0 & -2 & -2 \end{bmatrix} \mathbf{x} + \begin{bmatrix} 0 \\ 0 \\ 10 \end{bmatrix} u$$

$$y = \begin{bmatrix} 1 & 0 & 0 \end{bmatrix} \mathbf{x}$$

then the block can be simulated as shown in Figure 5-15. We note that the outputs of two integrators are chosen as $-x_1$ and $-x_3$. The output of the third integrator is chosen as $+x_2$. These choices are purely for convenience. Since $x_2 = \dot{x}_1$ and $y = x_1$, the tachometer need not be simulated; the signal $\dot{y}$ can be obtained directly from x_2. The compensators, C_1, C_2, and C_3, all have the same pole. By simulating them as a vector proper rational function, the number of integrators used can be reduced from 3 to 1. Consider

$$\hat{\omega}(s) = \frac{1}{s+1}\hat{u}(s) + \frac{-2s+3}{s+1}\hat{y}(s) + \frac{s+2}{s+1}\hat{\dot{y}}(s)$$

$$= \begin{bmatrix} \dfrac{1}{s+1} & \dfrac{-2s+3}{s+1} & \dfrac{s+2}{s+1} \end{bmatrix} \begin{bmatrix} \hat{u}(s) \\ \hat{y}(s) \\ \hat{\dot{y}}(s) \end{bmatrix}$$

The 1×3 proper rational function can be written as

$$\begin{bmatrix} \dfrac{1}{s+1} & \vdots & \dfrac{-2s+3}{s+1} & \vdots & \dfrac{s+2}{s+1} \end{bmatrix} = \begin{bmatrix} 0 & \vdots & -2 & \vdots & 1 \end{bmatrix} + \frac{1}{s+1} \begin{bmatrix} 1 & \vdots & 5 & \vdots & 1 \end{bmatrix}$$

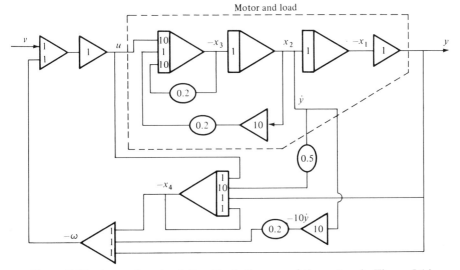

Figure 5-15 An analog simulation block diagram of the system in Figure 5-14.

Hence, from (5-40), it has the following one-dimensional realization:

$$\dot{x}_4 = -x_4 + \begin{bmatrix} 1 & 5 & 1 \end{bmatrix} \begin{bmatrix} u(t) \\ \dot{y}(t) \\ y(t) \end{bmatrix}$$

$$\omega = -x_4 + \begin{bmatrix} 0 & -2 & 1 \end{bmatrix} \begin{bmatrix} u(t) \\ \dot{y}(t) \\ y(t) \end{bmatrix}$$

If we use this dynamical equation, the complete analog simulation block diagram of the system in Figure 5-14 can be obtained as shown in Figure 5-15. We use four integrators in the simulation. The system cannot be simulated by using a smaller number of integrators. ∎

To conclude this chapter, let us remark on the realizations of general proper rational matrices. Consider the 2 × 2 rational matrix

$$\hat{\mathbf{G}}(s) = \begin{bmatrix} \hat{g}_{11}(s) & \hat{g}_{12}(s) \\ \hat{g}_{21}(s) & \hat{g}_{22}(s) \end{bmatrix} \quad\quad\quad (5\text{-}42)$$

A block diagram of Equation (5-42) is shown in Figure 5-16. After obtaining the realizations of each $\hat{g}_{ij}(s)$ and then computing the overall dynamical-equation description of the composite system in Figure 5-16, we can obtain a realization of (5-42). However this realization is generally not a minimal-dimensional realization. In other words, its dimension is generally larger than the degree of $\hat{\mathbf{G}}(s)$. Finding a minimal-dimensional realization of $\hat{\mathbf{G}}(s)$ is rather complicated; the interested reader is referred to Reference [1], Chapter 6.

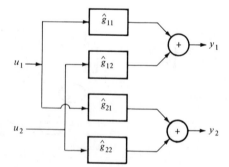

Figure 5-16
A realization of a 2 × 2 rational matrix.

5-6 Remarks and Review Questions

In most texts analog computer simulations of transfer functions are carried out directly, rather than through dynamical equation realizations as is suggested in this text. There are however many reasons for the latter approach. First, the procedure is more systematic. Second, the magnitude and time scalings can be easily carried out through dynamical equations. Finally, the procedure can be extended to the multivariable case.

Do you have freedom in assigning variables to either the input or the output of an integrator? To both the input and output?

What is the main difficulty in using analog computer simulation? Do you have the same difficulty in digital computer simulation?

How do you carry out magnitude scaling?

Why must the constants p_i and q_i in Equations (5-4) and (5-5) be different from zero?

An analog computer usually has a normal mode and a first mode of operation. How can this be designed?

What are the reasons for the study of the realization problem?

What is a minimal-dimensional realization? Is it unique?

Can you always assign the output $y(t)$ and its derivatives $\dot{y}(t)$, $\ddot{y}(t)$, ... as state variables? For what type of transfer functions are you permitted to do so?

How do you compare, in terms of ease of realizing, the controllable-form realization, the observable-form realization, and the Jordan-form realization?

Is it possible to obtain both the controllable-form realization and the observable-form realization for a vector rational transfer matrix? Why?

References

[1] Chen, C. T., *Introduction to Linear System Theory*. New York: Holt, Rinehart and Winston, 1970.
[2] Chi, H. H., and C. T. Chen, "A sensitivity study of analog computer simulation," *Proc. Allerton Conf.*, pp. 845–854, 1969.
[3] Chu, Y., *Digital Simulation of Continuous Systems*. New York: McGraw-Hill, 1969.
[4] Hausner, A., *Analog and Analog-Hybrid Computer Programming*. Englewood Cliffs, N.J.: Prentice-Hall, 1971.
[5] Jackson, A. S., *Analog Computation*. New York: McGraw-Hill, 1960.
[6] Korn, G. A., and T. M. Korn, *Electronic Analog and Hybrid Computers*. New York: McGraw-Hill, 1964.
[7] Levine, L., *Methods for Solving Engineering Problems using Analog Computers*. New York: McGraw-Hill, 1964.
[8] Mantey, P. E., "Eigenvalue sensitivity and state-variable selection," *IEEE Trans. Automatic Control*, vol. AC-13, pp. 263–269, 1968.
[9] Nering, E. D., *Linear Algebra and Matrix Theory*. New York: Wiley, 1963.

Problems

5-1 Draw analog computer simulation block diagrams for the dynamical equations

a. $\dot{\mathbf{x}} = \begin{bmatrix} 9 & -0.1 & 1 \\ 1.9 & 0 & 4.5 \\ 1 & 2 & 5 \end{bmatrix} \mathbf{x} + \begin{bmatrix} 1.1 \\ 0 \\ 2 \end{bmatrix} u$

$y = \begin{bmatrix} 2.5 & 1 & 1.2 \end{bmatrix} \mathbf{x}$

b. $\dot{\mathbf{x}} = \begin{bmatrix} -1 & -2 \\ 81 & -0.9 \end{bmatrix} \mathbf{x} + \begin{bmatrix} 1.5 \\ 1.1 \end{bmatrix} u$

$y = \begin{bmatrix} 0.7 & 2.1 \end{bmatrix} \mathbf{x}$

5-2 If it is known that the maximum magnitudes of x_1, x_2, and x_3, and y in Problem 5-1a are, respectively, 15 volts, 0.1 volt, -8 volts, and 40 volts for a certain class of inputs, how do you simulate the problem in an analog computer whose variables are limited to ± 10 volts?

5-3 How do you simulate the dynamical equation in Problem 5-1b if it is desired to increase the speed of response by a factor of 5?

5-4 Find two different realizations for the transfer function

$$\hat{g}(s) = \frac{s^2 + 2}{s^3}$$

one in Jordan form, the other in controllable or observable canonical form.

5-5 Find realizations for the transfer functions

a. $\hat{g}_1(s) = \dfrac{3s^4 + 1}{2s^4 + 3s^3 + 4s^2 + s + 5}$

b. $\hat{g}_2(s) = \dfrac{(s+3)^2}{(s+1)^2(s+2)}$

c. $\hat{\mathbf{G}}_1(s) = \begin{bmatrix} \dfrac{2s+1}{s^2+2s+1} & \dfrac{s+1}{3s+1} \end{bmatrix}$

d. $\hat{\mathbf{G}}_2(s) = \begin{bmatrix} \dfrac{s+3}{(s+1)(s+2)} \\ 1 \\ \dfrac{2}{s+1} \end{bmatrix}$

5-6 Write dynamical equation descriptions for the block diagrams shown in Figure P5-1 with the chosen state variables.

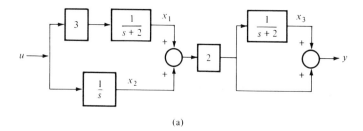

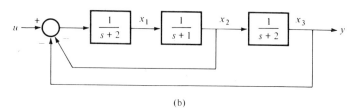

Figure P5-1

5-7 Given the network shown in Figure P5-2. Draw an analog computer simulation block diagram for it.

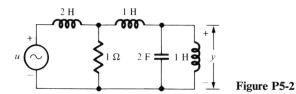

Figure P5-2

5-8 Write analog computer simulation block diagrams for the systems shown in Figure P5-3.

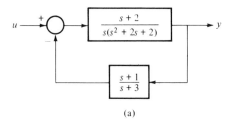

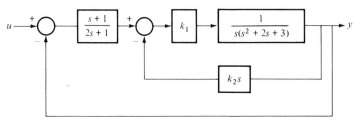

Figure P5-3

CHAPTER

6

Introduction to Design

With the background introduced in the previous chapters, we are ready to study the design of control systems. Before plunging into specific design techniques, it is important to obtain a total picture of the design problem. In this chapter we shall discuss the choice of a plant, the performance criterion, and the possible difficulties in the design. Two basic approaches in the analytical design will also be discussed.

6-1 The Choice of a Plant

At a large airport, in order to direct incoming airplanes entering the traffic area, it is necessary to point a radar antenna toward the airplanes and follow their movements. In order to meet these requirements, control systems must be designed. The reasons for using control systems are as follows: First, the antenna may be too heavy to be moved by an operator. By using a control system, the power required to move the antenna is reduced to that of turning a knob. Second, the antenna may be remotely located, and hence a remote control is needed. The first step in the design is the selection of an actuator to move the antenna. Depending on the available power supply, or on space or economical limitations, it could be a dc motor, an ac motor, or a hydraulic servomotor.

6-1 THE CHOICE OF A PLANT

The size of the motor is determined by the inertia, velocity, and acceleration ranges of the antenna. In order to drive the antenna at its maximum angular velocity $(\dot{\theta}_L)_{max}$ and its maximum acceleration $(\ddot{\theta}_L)_{max}$, the motor should deliver at least the following power to the shaft of the antenna:

$$P_L = J_L(\ddot{\theta}_L)_{max}(\dot{\theta}_L)_{max} + f_L(\dot{\theta}_L)^2_{max} \tag{6-1}$$

where J_L is the moment of inertia of the antenna and f_L is the viscous friction coefficient. Since the maximum angular velocity $(\dot{\theta}_L)_{max}$ and the maximum angular acceleration $(\ddot{\theta}_L)_{max}$ may not occur at the same instant, the power computed in Equation (6-1) may be too large. On the other hand, the inertias of the gears and the motor that are yet to be chosen are not included in (6-1). Furthermore the power to overcome the Coulomb and static frictions is not considered. Hence as a first try, we may select a motor with a rated power that is twice P_L in (6-1). Since the motor's rated speed is usually larger than the required load speed, a gear train with gear ratio, say N_m/N_L = (average load speed)/(rated motor speed), must be used. With the chosen motor and gear train, the power computed in (6-1) must be modified to

$$P_L' = \left[J_L' + \left(\frac{N_L}{N_m} \right)^2 J_m' \right] (\ddot{\theta}_L)_{max}(\dot{\theta}_L)_{max} + \left[f_L' + \left(\frac{N_L}{N_m} \right)^2 f_m' \right] (\dot{\theta}_L)^2_{max} \tag{6-2}$$

where J_L' and J_m' are, respectively, the total moments of inertia on the load and motor shafts, and f_L' and f_m' are, respectively, the total viscous friction coefficients on the load and motor shafts. Note that J_m' and f_m' are transferred to the load shaft by multiplying $(N_L/N_m)^2$. In Equation (6-2), the power to overcome the Coulomb and static friction are not included. Hence the rated power of the selected motor should be slightly larger but not greatly larger than the value of P_L' in (6-2). Otherwise the same process has to be repeated until a suitable motor is chosen. For convenience of discussion, an armature-controlled dc motor is chosen, rather arbitrarily, to drive the antenna. It is also decided to use a dc generator as a power amplifier, as shown in Figure 6-1. This collection of devices, including the load, is called the *plant* of the control system. We see from the foregoing discussion that the choice of a plant is not unique. Different designers often choose different plants; even the same designer may choose different plants at different times. However the various design techniques we shall introduce in the remainder of this book are equally applicable.

Once a plant is chosen, the remainder of the design is concerned with trying to make the best use of the plant. If the design is to be carried out analytically, either the transfer-function description or the state-variable description, depending on the design techniques used, of the plant must be found. If after exhausting all design methods, we are still not able to design a satisfactory system from the plant, then we have to select a new plant and repeat the design process.

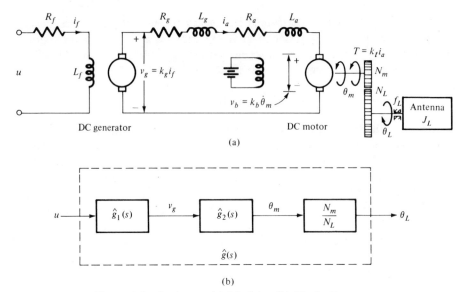

Figure 6-1 A plant. (a) Model. (b) Block diagram.

6-2 Performance Criterion

Every control system is designed for a specific application and therefore should satisfy a certain set of specifications in order to be acceptable for practical use. This set of specifications is called the *performance criterion*. Before introducing the criterion, we need to examine the concept of *test functions*.

A control system is often designed for a certain class of command signals. The command, or desired, signals of some control systems, such as home heating systems or the system in Example 2 of Section 3-9, are known to the designer. This may not be the case for other systems however. For example, the desired antenna direction in an airport is dictated by the incoming airplanes; hence the command signals of the control system may not always be the same and are not exactly known to the designer. In this case standard test signals are used in the design. The use of standard test signals is justifiable for the following two reasons: First, there is a strong correlation between the response of a system to a test signal and the responses of the system to the actual command signals. Second, the command signals of many control systems are often very close to test signals.

The standard test signals used in practice are the step function, the ramp function, and the accelerating (or parabolic) function, as shown in Figure 6-2. The command signal of a home heating system is a step function whose magnitude is the desired temperature. If an antenna is used to track a communication satellite, then the command signal is very close to a ramp function. An accelerating function is not used as often as a step or ramp function.

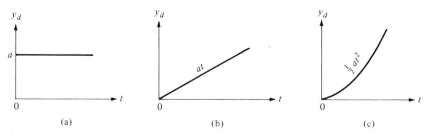

Figure 6-2 Test functions. (a) Step function. (b) Ramp function. (c) Accelerating function.

The specifications of a control system are usually stated in terms of the test functions. They can be divided into two parts: the steady-state performance and the transient performance.

Steady-state performance. The steady-state response of a system due to the application of an input is, by definition, the response as time approaches infinity. Consider the system shown in Figure 6-3. Let $y_d(t)$ be the command or desired signal, and let $y(t)$ be the corresponding output. If y_d is a step function with magnitude a [that is, $y_d(t) = a$, for $t \geq 0$] or a ramp function at [that is, $y_d(t) = at$, for $t \geq 0$], then the percentage steady-state error is defined as

$$e(t) = \lim_{t \to \infty} \left| \frac{y_d(t) - y(t)}{a} \right| \times 100 \text{ percent} \qquad (6\text{-}3)$$

The physical meaning of this steady-state error can be easily seen from Figure 6-4. If the steady-state error is zero, then the actual output will be, after the transient dies out, identical to the desired, or command, signal. For different

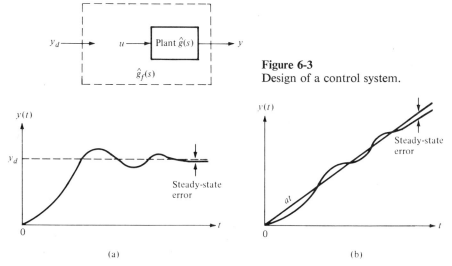

Figure 6-3 Design of a control system.

Figure 6-4 Steady-state performance.

control systems, the specifications on the steady-state error may be different. For example, the steady-state error of a home heating system is not as critical as the one of pointing a radiotelescope to a remote star.

The response of the system shown in Figure 6-3 is determined by the transfer function from y_d to y. Let

$$\hat{g}_f(s) = \frac{y(s)}{y_d(s)} = \frac{\beta_0 + \beta_1 s + \cdots + \beta_m s^m}{\alpha_0 + \alpha_1 s + \cdots + \alpha_n s^n} \qquad m \leq n \qquad (6\text{-}4)$$

be its transfer function, which is yet to be designed. In order for the overall system to be operative, it is assumed that the system is totally stable. This implies that all the poles of $\hat{g}_f(s)$ have negative real parts. Now if the step function $y_d(t) = a$, for $t \geq 0$, is applied to the system, the steady-state response of the system, as derived in (4-40), is given by

$$y_s(t) = \frac{\beta_0}{\alpha_0} a$$

Hence the steady-state error, e_1, due to a step function is equal to

$$e_1 = \left| \frac{a - \frac{\beta_0}{\alpha_0} a}{a} \right| = \left| 1 - \frac{\beta_0}{\alpha_0} \right| = \left| \frac{\alpha_0 - \beta_0}{\alpha_0} \right| \qquad (6\text{-}5)$$

Hence if the specification on the steady-state error is given, the constraint on the overall system can be immediately obtained. For example, if the steady-state error due to a step function is required to be less than γ (or 100γ percent), then it is required that

$$\left| 1 - \frac{\beta_0}{\alpha_0} \right| < \gamma$$

or

$$1 - \gamma < \frac{\beta_0}{\alpha_0} < 1 + \gamma$$

or

$$(1 - \gamma)\alpha_0 < \beta_0 < (1 + \gamma)\alpha_0 \qquad (6\text{-}6)$$

Note that α_0 is a positive number because of the stability assumption on $\hat{g}_f(s)$. Hence if β_0, the constant term in the numerator of $\hat{g}_f(s)$, is in the range specified in (6-6), then the steady-state error due to a step function will be less than γ. We also see that the requirement on the steady-state error due to a step function imposes constraints only on α_0 and β_0; no constraints are imposed on other coefficients.

We discuss now the constraints imposed on $\hat{g}_f(s)$ due to the specification on the steady-state error with respect to a ramp function. As derived in Equation

(4-42), the steady-state response due to the application of $y_d = at$, for $t \geq 0$, is given by

$$y_s(t) = \frac{\beta_0}{\alpha_0} at + \frac{\alpha_0 \beta_1 - \beta_0 \alpha_1}{\alpha_0^2} a$$

Hence the steady-state error, e_2, due to a ramp function is

$$e_2(t) = \left| \left(1 - \frac{\beta_0}{\alpha_0}\right) t - \frac{\alpha_0 \beta_1 - \beta_0 \alpha_1}{\alpha_0^2} \right| \tag{6-7}$$

We see that if $\beta_0 \neq \alpha_0$, then the steady-state error e_2 will increase with time. Hence a necessary condition for having a finite e_2 is that $\alpha_0 = \beta_0$. If $\alpha_0 = \beta_0$, then $e_2(t)$ reduces to

$$e_2(t) = \left| \frac{\beta_1 - \alpha_1}{\alpha_0} \right| \quad \text{if } \alpha_0 = \beta_0 \tag{6-8}$$

Hence if the steady-state error due to a ramp function is required to be less than γ (or 100γ percent), then it is required that

$$\alpha_0 = \beta_0$$

and

$$\left| \frac{\beta_1 - \alpha_1}{\alpha_0} \right| < \gamma$$

This inequality can be simplified to

$$\alpha_1 - \gamma \alpha_0 < \beta_1 < \alpha_1 + \gamma \alpha_0 \tag{6-9}$$

Hence if $\beta_0 = \alpha_0$ and if β_1 is in the range expressed in (6-9), then the steady-state error due to a ramp input will be less than γ. It is clear that if the steady-state error due to a ramp function is finite, then the steady-state error due to a step function is zero. We see that the steady-state error due to a ramp function imposes constraints on only the coefficients of $\hat{g}_f(s)$ associated with s^0 and s^1; no constraints are imposed on other coefficients.

The foregoing discussions can be easily extended to the steady-state error due to an acceleration function. This extension is left as an exercise.

In addition to the steady-state performance defined for the step, ramp, and acceleration test functions, there is another type of steady-state performance defined for sinusoidal functions. This performance specification is used in the frequency-domain design and will be discussed in Chapter 10.

Transient performance. The steady-state performance is defined for the response as $t \to \infty$ and is defined for the step or ramp test function, whereas the transient performance is defined for the response right after the application of an input and is defined with respect to a step test function. Consider the responses due to the application of step functions, as shown in Figure 6-5. The speed of the response is measured by t_r, called the *rise time*. There are many

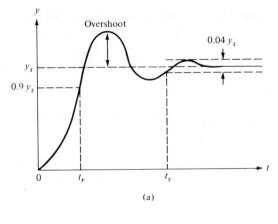

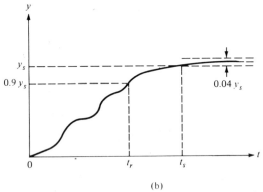

Figure 6-5 Transient performance.

ways to define a rise time. We shall define it as the time required for the response to rise from 0 to 90 percent of its steady-state value, as shown in Figure 6-5. It is clear that the smaller the rise time, the faster the response. The time denoted by t_s in Figure 6-5 is called the *settling time*. It is the time required for the responses to reach and stay within a certain percentage, say 2 percent, of the steady-state value. Let y_{max} be the maximum value of $|y(t)|$; that is,

$$y_{max} \triangleq \max_{0 < t \leq \infty} |y(t)|$$

If y_{max} is smaller than the steady-state value y_s, then the response is said to have no overshoot. If $y_{max} \geq y_s$, the percentage *overshoot* is defined as

$$\text{overshoot} = \frac{y_{max} - y_s}{y_s} \times 100 \text{ percent}$$

The transient performance of a control system is generally specified in terms of the rise time, the settling time, and the overshoot.

In the steady-state performance, the specification can be easily translated into the overall transfer function, as shown in Equations (6-5) and (6-7). This however cannot be done in the transient performance. Except for a very special case, there seems no simple relationship between these three specifications and the overall transfer function. This problem will be discussed in Chapter 9 when we use the specifications in actual designs.

A control system is inherently a time-domain system, and hence the introduced specifications are natural and have simple physical interpretations. For example, in pointing a telescope to a star, the steady-state performance (accuracy) is the main concern; the specifications on the rise time, overshoot, and settling time are not critical. However in aiming a gun at an aircraft, both the accuracy and the speed of response are important. In the design of an aircraft, the specification is often given as shown in Figure 6-6. It is required that the step response of the system be confined to the region shown. This region is obtained by a compromise between the comfort or physical limitation of the pilot and the swiftness of the manipulation of the aircraft. In the design of an elevator, any appreciable overshoot is undesirable. From these examples we see that the specifications on steady-state and transient performance are indeed meaningful.

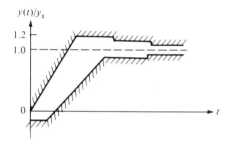

Figure 6-6
Allowable step response.

Although these specifications are reasonable, they cannot always be directly used in the design. Depending on the design technique employed, these specifications have to be properly translated. For example, if the design is carried out in the frequency domain, then these specifications in the time domain must be translated into the frequency domain (see Chapter 10). If optimization techniques are used, then these specifications must again be modified. This will be discussed in Chapter 8.

6-3 Two Basic Approaches in the Design

As discussed earlier, the first step in the design is the choice of a plant. The choice of a plant is dictated not only by the performance specifications but also by size, weight, available power supply, and cost. For example, size and weight are very important factors in choosing actuators for control systems in space vehicles. Therefore the plant chosen generally cannot meet the specifications discussed in the previous section. Although there is no reason that the designer cannot choose a new plant, it is however much easier and much cheaper to introduce some compensation to force the chosen plant to meet the specifications. Therefore once a plant is chosen, the design problem is always posed as follows: Given a plant and a set of specifications, design compensators so that the resulting overall system will meet the specifications. There are basically two approaches to this design problem:

1. From the specifications and the plant, find an overall system that meets the specifications and then compute the necessary compensators.

2. Introduce compensators to the plant, and then choose the parameters of the compensators so that the resulting system will meet the specifications.

These two approaches are quite different philosophically. In the first approach, the designer first finds an overall system and then forces the plant to meet it. In the second approach the designer works inside the plant and slices things up so that the overall system will hopefully do the job required. In the next two chapters we shall be concerned mainly with the problems associated with the first approach. The second approach will be studied in the remainder of the book. Comparisons of these two approaches will also be discussed.

6-4 Difficulties in the Design

Before introducing specific design techniques, we shall discuss some difficulties that may be encountered in the design of most control systems.

Saturation. Most physical systems are, strictly speaking, nonlinear. Their linear mathematical descriptions are all obtained under linearization assumptions. Nonlinearities may take the forms of saturation, backlash, Coulomb friction, dead zone, or others. Among them, saturation is most often encountered in control systems. It occurs in electronic amplifiers, generators, dc and ac motors, and hydraulic motors. We shall first use an example to illustrate the effect of saturation.

Example 1

Consider the system shown in Figure 6-7(a). The element N is an amplifier with gain 2. The overall transfer function is $(2s + 4)/(s^2 + s + 4)$. The system is clearly BIBO stable. Since the constants in the numerator and denominator are the same, the system has zero steady-state error due to a step function. The responses of the system due to two step functions with different magnitudes are shown in Figure 6-8.

In reality, the amplifier may have the characteristic shown in Figure 6-7(b). For ease of simulation, the saturation curve is approximated by the dashed lines shown. The responses of the system with the approximated saturation

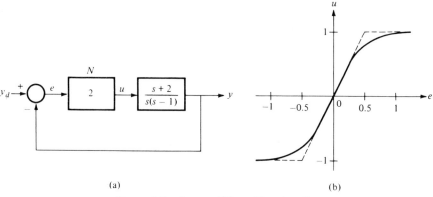

Figure 6-7 An amplifier with saturation.

curve due to the application of the two step functions are shown in Figure 6-8. This is obtained by using IBM System/360 CSMP. We see that if the element N is not saturated, the linear analysis gives a very accurate result. On the other hand, if the element N is saturated, then the response of the linear model and that of the actual physical system are quite different. In fact, the system with saturation will become unstable if $y_d(t) = a$, with $a \geq 1.15$, although the linear model is always BIBO stable. ∎

We see from this example that in the design of a control system by using linear models, it is important to keep the components from saturating; otherwise the system will not function as expected. To keep the component from saturating is however a very difficult problem because, in the process of design, the knowledge of the exact response of the resulting system is not available.

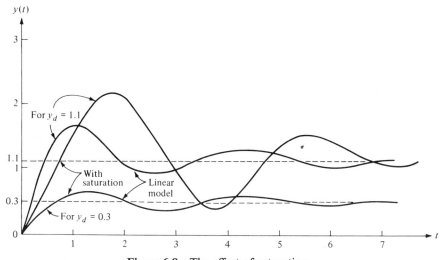

Figure 6-8 The effect of saturation.

Hence the saturation problem can be checked only after the design is completed. If saturation indeed occurs in the resulting system, the system has to be redesigned to improve its performance.

Noise. Noise is present almost everywhere in most control systems. For example, in the antenna tracking problem discussed in Section 6-1, because of steady winds or gusts, an external noise force is applied to the antenna. If a potentiometer is used as a transducer, because of brush jumps, wire irregularity, or variations of contact resistance, there are always unwanted spurious voltages. Shot noises are also unavoidable in electronic amplifiers and motors. Therefore the problem of noise cannot be neglected in control system design. The formal way to attack this problem requires statistical descriptions of noises. This is beyond the scope of this text, and the reader is referred to References [3] and [5]. In the following we discuss the constraints imposed by noises on compensating networks.

The noises in control systems usually have high frequency spectra. These signals will be greatly amplified after passing through systems—in particular, differentiators—with improper transfer functions.[1] For example, the signal $0.1 \sin 100t$ becomes $10 \cos 100t$ after passing through a differentiator. On the other hand, signals with high frequency spectra will be attenuated after passing through systems whose transfer functions are strictly proper and stable. [See Equation (4-39) and note that $|\hat{g}(j\omega_0)|$ is generally very small at high frequencies if $\hat{g}(s)$ is strictly proper.] Hence systems with strictly proper transfer functions are often said to have *low-pass characteristics*.

The transfer functions of most control systems are strictly proper. Hence if the noises have high frequency spectra and if they are not further amplified by systems with improper transfer functions, then, because of the low-pass characteristics of most control systems, the noises will be automatically suppressed. This is demonstrated in the following example.

Example 2

Consider the control system shown in Figure 6-9. It is assumed that the noise generated by the amplifier can be represented by $n(t)$ as shown. The compensator $C(s)$ is given as $(s + 1)/(s + 2)$. The transfer functions from θ_d to θ_o and from n to θ_o can be computed as

$$\hat{g}(s) = \frac{\hat{\theta}_o(s)}{\hat{\theta}_d(s)} = \frac{2}{s^2 + 2s + 2}$$

and

$$\hat{g}_n(s) = \frac{\hat{\theta}_o(s)}{\hat{n}(s)} = \frac{1}{s^2 + 2s + 2}$$

---

[1] A transfer function is called *improper* if the degree of the numerator is larger than that of the denominator.

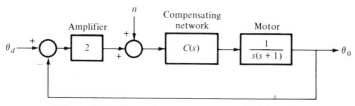

Figure 6-9 Control system corrupted with noise.

For convenience of discussion, we assume that $\theta_d(t) = \sin t$ and $n(t) = 0.1 \sin 10t$. Then the steady-state outputs due to $\theta_d(t)$ and $n(t)$ are, respectively,

$$\theta_o \text{ (due to } \theta_d) = |\hat{g}(j)| \sin (t - \underline{/\hat{g}(j)}) = 0.87 \sin (t - \underline{/\hat{g}(j)})$$
$$\theta_o \text{ (due to } n) = 0.1|\hat{g}_n(10j)| \sin (10t - \underline{/\hat{g}_n(10j)})$$
$$\doteq 10^{-3} \sin (t - \underline{/\hat{g}_n(10j)})$$

We see that the original signal-to-noise amplitude ratio is $1/0.1 = 10$; however the signal-to-noise amplitude ratio at the output becomes $0.87/10^{-3} = 870$. Hence the noise is indeed automatically suppressed in the control system. ∎

The analysis in this example is, strictly speaking, incorrect, because the noise is a random variable and cannot be represented by $0.1 \sin 10t$. However the analysis does imply that if the noise has a high frequency spectrum and if the system has low-pass characteristics, then the noise will be automatically suppressed in the system.

If the compensating network in Figure 6-9 is chosen as $1 + s$, then it can be computed that, for the same θ_d and n, the signal-to-noise amplitude ratio at the output is approximately equal to 46. This is much smaller than 870 for $C(s) = (s + 1)/(s + 2)$. Furthermore the differentiator may cause the motor to become saturated. For example, if the noise contains a very-high-frequency component, say $0.1 \sin 10^3 t$, then after differentiation the signal becomes $10^2 \cos 10^3 t$, which may drive the motor into the saturating region. Hence compensators with improper transfer functions are generally avoided in control systems.

Because of the presence of noise in control systems, *the compensating networks used in this text will be required to have proper transfer functions.* Otherwise the noises will be accentuated. This not only will obscure information-bearing signals but also will saturate the systems.

The loading problem. As discussed in Section 3-9, the transfer function of a system often changes after a second system is connected to it. This change of transfer function due to loading is an annoying problem in the design of control systems. For example, consider the plant with transfer function $\hat{g}_p(s)$ shown in Figure 6-1. Suppose after a lengthy computation based on $\hat{g}_p(s)$ it is decided to use potentiometer, tachometer, and network feedbacks, as shown in Figure 6-10, to improve the performance of the system. After the introduction of these feedbacks, the transfer function of the plant has changed to $\hat{g}_p'(s)$, because the

164 INTRODUCTION TO DESIGN

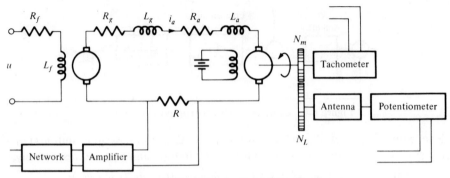

Figure 6-10 Introduction of feedback.

motor must now also drive the potentiometer and tachometer, and an additional resistor is added to the armature circuit. Consequently the feedbacks are actually introduced for $\hat{g}_p'(s)$, although they were intended for $\hat{g}_p(s)$. Hence we may not obtain the expected performance from the resulting system. Therefore the loading problem has to be taken into consideration in the design of control systems.

Nonuniqueness. As discussed in Section 6-1, even for the same control problem it is unlikely for two different designers to choose exactly the same plant. A designer generally prefers to use the devices that he is most familiar with or has access to. Even starting from the same plant, the resulting final systems may again be different. One person may design a system without tachometer feedback; another may design a system without a network feedback. Depending on the transducers used and the complexities of compensating networks, it is possible to design many different working systems starting from the same plant.

We recapitulate what we have discussed in this section. Saturation and noise are two inherent problems in control systems. Presence of noise prohibits us from using compensating networks whose transfer functions are not proper. The loading problem, though annoying, can always be resolved with proper care. Nonuniqueness is an advantage rather than a drawback in the design; it permits the designer ample room for manipulation. As far as the saturation problem is concerned, there is presently no complete solution for it. It remains a difficult problem in the design of control systems.

In conclusion, we mention that a good supplement to the material in this chapter can be found in Chapter 8 of Reference [4].

6-5 Review Questions

What information do you need in choosing the size of a motor?

Is the choice of a plant unique?

If a plant is found not to satisfy a set of specifications, why not just change the plant rather than design compensators to improve it?

Why do we have to introduce test functions? Is it justifiable to use test functions in a design?

What performance specifications have we introduced? Are they in the time domain or frequency domain?

How do you translate the steady-state performance into an overall transfer function?

What are the two basic approaches in a design?

What are the difficulties in a design?

How do we take care of the loading problem in a design?

Why, in the design of control systems, are compensators required to have proper transfer functions?

References

[1] Baeck, H. S., *Practical Servomechanical Design*. New York: McGraw-Hill, 1963.
[2] Chestnut, H., and R. W. Mayer, *Servomechanisms and Regulating System Design*, vols. 1 and 2. New York: Wiley, 1951.
[3] Meditch, J. S., *Stochastic Optimal Linear Estimation and Control*. New York: McGraw-Hill, 1969.
[4] Thaler, G. J., and R. G. Brown, *Analysis and Design of Feedback Control Systems*, 2d ed. New York: McGraw-Hill, 1960.
[5] Truxal, J. G., *Control Systems Synthesis*. New York: McGraw-Hill, 1955.

Problems

6-1 Find the ranges of k so that the systems shown in Figure P6-1 will be BIBO stable and so that the steady-state error due to a step input will be less than 5 percent.

6-2 Find the ranges of k so that the systems in Problem 6-1 have a steady-state error due to a ramp function of less than 10 percent.

6-3 It is suggested that the testing signals be extended to include signals of the type $y_d(t) = a_0 + a_1 t + a_2 t^2$. Is there any difference between using $y_d(t) = t^2$ and $y_d = a_0 + a_1 t + a_2 t^2$ as far as the steady-state performance is concerned?

6-4 For the system shown in Figure P6-2, show that if $\hat{g}(s)$ has one pole at the origin, then the steady-state error due to a step function is zero for any stable k. Show also that if $\hat{g}(s)$ has two poles at the origin, then the steady-state error due to a ramp function is zero for any stable k. Do the above two statements hold for any configuration of feedback system? (*Answer:* no.)

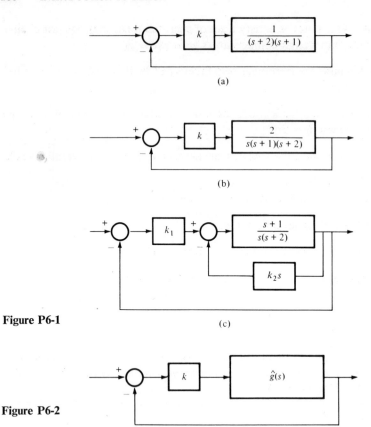

Figure P6-1

Figure P6-2

6-5 Derive Equation (6-7) by using the final-value theorem? Check carefully the applicability condition of the theorem.

6-6 Find the conditions on $\hat{g}_f(s)$ if the steady-state error due to an acceleration input is required to be less than γ (or 100γ percent).

6-7 Consider the control system studied in Figure 3-27. Its block diagram is reproduced in Figure P6-3, with various constants replaced by numerical values. It is assumed that the noise generated in the electronic amplifier can be represented by n, as shown. If $\theta_d = \sin t$ and $n = 0.1 \sin 10t$, find the steady-state outputs due to θ_d and n, respectively. What is the signal-to-noise amplitude ratio at the output?

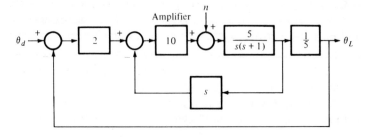

Figure P6-3

6-8 The movement of the pen of a recorder can be controlled by the system shown in Figure P6-4(a). Its block diagram is shown in Figure P6-4(b). Find the range of k so that the steady-state error due to a ramp input is smaller than 1 percent.

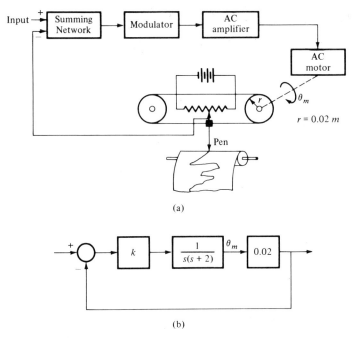

Figure P6-4

CHAPTER 7
Synthesizable Transfer Functions and Feedback

The problem of the design of a control system can be stated as follows: Given a plant with transfer function $\hat{g}(s)$ and a set of specifications, design compensators so that the resulting overall system will meet the specifications. This problem can be depicted as shown in Figure 7-1. As discussed in the previous chapter, one approach in the design is first to find an overall system with transfer function $\hat{g}_f(s)$ that meets the specifications and then to compute the required compensators. Because of the presence of noise in control systems, the compensators used are required to have proper transfer rational functions. This restriction on compensators conceivably imposes certain constraints on $\hat{g}_f(s)$. In this chapter these constraints will be studied. We shall also discuss the reasons for introducing feedback in control systems.

7-1 Pole-Zero Excesses of an Overall Transfer Function

Let $\hat{g}(s)$ be the transfer function of the plant shown in Figure 7-1 and let $\hat{g}_f(s)$ be the transfer function of an overall system yet to be designed. What we are going to establish is that if

$$\hat{g}(s) = \frac{N(s)}{D(s)} \qquad (7\text{-}1)$$

7-1 POLE-ZERO EXCESSES OF AN OVERALL TRANSFER FUNCTION

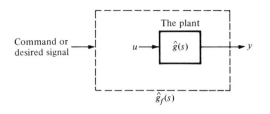

Figure 7-1
A design problem.

and

$$\hat{g}_f(s) = \frac{N_f(s)}{D_f(s)} \qquad (7\text{-}2)$$

and if the compensators used have proper transfer functions, then in most practical situations we have

$$\delta D_f(s) - \delta N_f(s) \geq \delta D(s) - \delta N(s)$$

where $\delta(\ )$ denotes the degree of a polynomial. In words, the pole-zero excess of an overall system must be equal to or larger than the pole-zero excess of the plant, if only compensators with proper rational functions are used. To be more specific, we state it formally as a theorem.

Theorem 7-1

Consider the control problem shown in Figure 7-1. No matter how many compensators are used or how they are connected, if the following conditions are met:

1. all the forward paths from y_d to y pass through the plant,
2. all compensators have proper rational transfer functions, and
3. there is no loop with loop gain 1 or loops the sum or the product of whose loop gains are ± 1,

then the overall system must satisfy the following pole-zero excess inequality:

$$\delta D_f(s) - \delta N_f(s) \geq \delta D(s) - \delta N(s) \qquad (7\text{-}3)$$

∎

Before proving the theorem, we shall make several comments concerning the three conditions. Since the plant consists of an actuator and the object to be controlled, it is clear that condition 1 should be met in every control system. Condition 2 is required because of the presence of noise in control systems. To explain condition 3, consider the single-loop system shown in Figure 7-2. If the loop gain is equal to 1, then the overall transfer function is

$$\frac{C_1(s)\hat{g}(s)}{1 - [-C_1(s)C_2(s)\hat{g}(s)]} = \frac{C_1(s)\hat{g}(s)}{1 - 1}$$

Figure 7-2
A feedback control system.

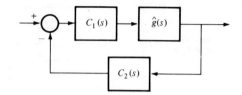

which is not well defined because its denominator is equal to zero. Hence condition 3 should be met for single-loop systems in order for the system to be well defined. For more complicated configurations of compensators, the physical meaning of condition 3 is not clear. But it is believed that condition 3 is met in most practical systems.

Proof

We shall prove Theorem 7-1 by using Mason's formula. It is clear that under condition 1, every overall transfer function $\hat{g}_f(s)$ is of the form

$$\frac{\hat{g}(s)(\sum_i \bar{P}_i \Delta_i)}{\Delta}$$

where $\bar{P}_i$ is the gain excluding $\hat{g}(s)$ of the ith forward path, and Δ and Δ_i are defined as in Section 4-1. Now condition 3 implies that the constant 1 in Δ cannot be canceled completely; hence the denominator and numerator of Δ are of the same degree. Consequently we have

$$\delta D_f(s) - \delta N_f(s) = [\delta D(s) - \delta N(s)] + [\delta D_c(s) - \delta N_c(s)] \qquad (7\text{-}4)$$

where D_c and N_c denote the denominator and numerator of $(\sum_i \bar{P}_i \Delta_i)$. Condition 3 implies that Δ_i is proper for each i. $\bar{P}_i$ is also proper because of condition 2; hence we have $\delta D_c(s) - \delta N_c(s) \geq 0$. This proves the theorem. ∎

Theorem 7-1 is very general in the sense that no specific configuration is mentioned. An important implication of this theorem is that if conditions 1 and 3 are met and if the pole-zero excess inequality is not satisfied, then the overall control system may not be constructed by using exclusively compensators with proper rational transfer functions. Therefore the inequality $\delta D_f(s) - \delta N_f(s) \geq \delta D(s) - \delta N(s)$ is a basic constraint, imposed by the permissible compensators in the design of every control system.

In the design of control systems it is sometimes desirable to introduce state feedback (see Chapter 8). By *state feedback* we mean that compensators are connected to the state variables of the plant. We shall show that inequality (7-3) still holds for state feedback. Consider the plant shown in Figure 7-3(a), where

$\hat{x}_i(s)$ is the ith state variable of the plant and $\phi_i(s)$ is the transfer function from u to x_i. The transfer function $\phi_i(s)$ can be computed from

$$\hat{\mathbf{x}}(s) = (s\mathbf{I} - \mathbf{A})^{-1}\mathbf{b}\hat{u}(s) = \frac{1}{\det(s\mathbf{I} - \mathbf{A})}[\text{Adj}(s\mathbf{I} - \mathbf{A})]\mathbf{b}\hat{u}(s) \triangleq \begin{bmatrix} \phi_1(s) \\ \phi_2(s) \\ \vdots \\ \phi_n(s) \end{bmatrix} \hat{u}(s)$$

(7-5)

which is obtained from Equation (2-27) by assuming $\mathbf{x}(0) = \mathbf{0}$. Since every entry of Adj $(s\mathbf{I} - \mathbf{A})$ has degree $(n - 1)$ or less, whereas $\det(s\mathbf{I} - \mathbf{A})$ has degree n, hence $\phi_i(s)$ is strictly proper. Therefore if compensators with proper rational functions are connected to the state variables, condition 2 of Theorem 7-1 is not violated. Consequently we conclude that the inequality $\delta D_f(s) - \delta N_f(s) \geq \delta D(s) - \delta N(s)$ still holds for state feedback.

The transfer functions encountered in control systems are almost exclusively proper or strictly proper rational functions. The only exception occurs in some transducers, such as tachometers and rate gyros. The transfer function of a tachometer is ks, which is not proper. However if it is connected to a plant with a strictly proper rational transfer function, as is the case in practice, then from Figure 7-3(b) we may conclude that inequality (7-3) still holds.

Inequality (7-3) inevitably imposes a limit on what we can achieve in the design of a control system. Suppose that the overall transfer function $\hat{g}_f(s)$ in Figure 7-1 is 1; then the output y will follow the desired signal y_d faithfully. The overall system will also satisfy any specifications discussed in Section 6-2.

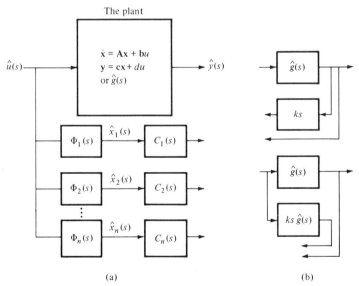

Figure 7-3 (a) State feedback. (b) Tachometer feedback.

Hence the overall transfer function $\hat{g}_f(s) = 1$ is the best we can achieve in the design. However if the transfer function of a plant is strictly proper, as is always the case in practice, then the inequality $\delta D_f(s) - \delta N(s) = 0 \geq \delta D(s) - \delta N(s) = 1$ cannot hold. Hence it is impossible, due to the class of compensators used, to build the best possible system $\hat{g}_f(s) = 1$.

7-2 Zeros and Poles of an Overall Transfer Function

Consider again the control problem in Figure 7-1. In Section 7-1 we have shown that the pole-zero excess of the plant imposes a constraint on the overall transfer function. We shall study another constraint in this section.

Let $\hat{g}(s) = N(s)/D(s)$ be the transfer function of the plant. The roots of $N(s)$ are called the zeros of $\hat{g}(s)$.

Theorem 7-2

Consider the control system shown in Figure 7-1. If all the forward paths from y_d to y pass through the plant, then the zeros of the plant will not be affected, no matter how the compensators are introduced. Hence the zeros of the plant will appear as zeros of the overall transfer function except those canceled by either the poles of the compensators or the missing poles of the overall transfer function. ■

This theorem can again be proved by using Mason's formula. If all the forward paths from y_d to y pass through the plant, then the overall transfer function is of the form $\hat{g}(s)(\sum_i \bar{P}_i \Delta_i)/\Delta$, where $\bar{P}_i$, Δ_i, and Δ are defined as in the proof of Theorem 7-1. From this formula, the assertion in this theorem can be deduced.

Theorem 7-2 has a very important implication. It implies that if the overall transfer function $\hat{g}_f(s)$ does not contain all the zeros of $\hat{g}(s)$, the missing zero must be canceled by a pole. If the missing zero has a negative real part, the pole that is introduced to cancel it is a stable pole.[1] If the missing zero has a zero or positive real part, the pole that cancels it must be an unstable pole. To introduce a stable pole to cancel out an open left-half s-plane zero may not affect seriously the response. It is however unacceptable in practice to introduce an unstable pole to cancel a right-half s-plane zero. As discussed earlier, the pole-zero cancellation does not really get rid of the undesired pole or zero. The undesired pole or zero merely does not appear in the overall transfer function; its effect still remains inside the system. Even without this theoretical difficulty, it is very expensive in practice to build an *exact* pole to cancel a zero. Hence *in designing a control system, if the plant has right-half s-plane zeros,*

---

[1] A pole in the open left-half s-plane is called a *stable* pole; otherwise it is called an *unstable* pole.

then these zeros should appear as the zeros of the overall transfer function $\hat{g}_f(s)$. Otherwise it is not practically advisable to synthesize $\hat{g}_f(s)$.

The zeros of $\hat{g}(s)$ impose, as discussed above, some constraints on the zeros of $\hat{g}_f(s)$. One may wonder whether or not the poles of $\hat{g}(s)$ impose any constraint on the poles of $\hat{g}_f(s)$. Surprisingly there is no constraint. In fact, if compensators are properly introduced, no matter what poles $\hat{g}(s)$ has, all the poles of the overall transfer function $\hat{g}_f(s)$ can be *arbitrarily* assigned. We state this fact as a theorem.

Theorem 7-3

Consider a plant with a transfer function of degree n. If compensators of degree $(n-1)$ are properly introduced, then the $(2n-1)$ poles of the overall transfer function can be arbitrarily assigned. ∎

The proof of this theorem can be found in References [1] and [2]. This theorem will be used in the next chapter to design compensators. The purpose of introducing this theorem here is to point out that, although a plant imposes some constraint on the zeros of an overall system, it imposes no constraint at all on the poles of an overall system.

We summarize what we have discussed so far in this chapter. In the design of a control system, the plant imposes two constraints on an overall transfer function. First, the pole-zero excess of an overall transfer function cannot be less than that of the plant transfer function. Second, the closed right-half s-plane zeros of the plant transfer function must appear as the zeros of an overall transfer function. Any overall transfer function satisfying these two constraints will be said to be *synthesizable*. A synthesizable transfer function can be built by using compensators with proper rational transfer functions. Furthermore the resulting system will not have hidden unstable poles.

Example

If the plant transfer function is $1/s(s+4)$, then the transfer functions $2/(s^2+2s+2)$, $(s+2)/(s+3)(s^2+2s+2)$, and $1/(s+2)(s^2+2s+2)$ are all synthesizable, but $1/(s+1)$ is not. If the plant transfer function is $(s-1)/s(s+1)$, then the transfer functions $(s-1)/(s^2+2s+2)$ and $(s^2-1)/(s+3)(s^2+2s+2)$ are synthesizable, but $1/(s+1)$ is not. If the plant transfer function is $(s+2)/s(s+1)$, then $1/(s+1)$ is synthesizable. ∎

7-3 Why Feedback?

If we are given a plant with transfer function $\hat{g}(s)$ and a synthesizable $\hat{g}_f(s)$, there are generally many ways to synthesize $\hat{g}_f(s)$ by using different configurations of compensators. Therefore it is natural to study the relative merits of various configurations. A general study of this problem is out of the question because of the almost infinitely many possible configurations of compensators. Therefore we shall study in this section only two of the most typical configurations.

Consider the configurations of compensators shown in Figure 7-4. The one in Figure 7-4(a) is called an *open-loop* system, and the one in Figure 7-4(b) is called a *closed-loop*, or *feedback*, system. The signal u is called the *actuating* signal. The major difference between these two systems is that one of the actuating signals depends on the actual output y, whereas the other does not. In the open-loop system, the actuating signal is predetermined; it will not change no matter what happens to the actual output. In the closed-loop, or feedback, system, the actuating signal is a function of the actual output. If $C_3(s) = 1$ and if the actual output does not differ very much from the desired signal, then the actuating signal will be small; otherwise a large actuating signal will be generated to bring the actual output close to the desired one as fast as possible. In other words, a properly designed closed-loop system has a self-correcting property. Hence a closed-loop system is generally better than an open-loop system.

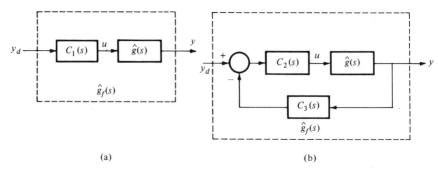

Figure 7-4 (a) Open-loop system. (b) Closed-loop, or feedback, system.

The above comparison between the two configurations in Figure 7-4 is based on physical arguments. We shall establish it more quantitatively in the following: Given a plant with transfer function $\hat{g}(s)$ and given a synthesizable overall transfer function $\hat{g}_f(s)$, then

1. It is not always possible in practice to synthesize $\hat{g}_f(s)$ by using the open-loop configuration.

2. If all the poles of $\hat{g}_f(s)$ are different from the poles of $\hat{g}(s)$, then the degree of the compensator used in the open-loop configuration will be larger than or equal to that used in the closed-loop configuration.

3. A closed-loop system, if properly designed, is less sensitive to the parameter variations of the plant than an open-loop system.

4. A properly designed closed-loop system is less sensitive to disturbances than an open-loop system.

The first two statements can be demonstrated by examples.

Example 1

Given a plant with transfer function $2/(s - 1)(s + 3)$. Let the overall transfer function be $\hat{g}_f(s) = 2/(s^2 + 2s + 2)$. If the open-loop configuration in Figure 7-4(a) is chosen, the required compensator is $C_1(s) = (s - 1)(s + 3)/(s^2 + 2s + 2)$. This open-loop system has an unstable pole cancellation, and therefore it is not acceptable in practice. The degree of the compensator is 2.

If the closed-loop configuration in Figure 7-4(b) is chosen, it can be verified that the design is completed by choosing $C_2(s) = 1$, $C_3(s) = 2.5$. There is no pole-zero cancellation, and the degree of the compensators is zero.

Example 2

Given a plant with transfer function $1/s(s + 1)$. Let the overall transfer function be $\hat{g}_f(s) = 2/(s^2 + 2s + 2)$. If the open-loop configuration is chosen, the compensator required is equal to $2s(s + 1)/(s^2 + 2s + 2)$. This open-loop system has an unstable pole-zero cancellation (the pole at the origin); hence this is not desirable in practice. The degree of the compensator is 2.

If $C_3(s)$ is chosen as 1, then $C_2(s)$ in Figure 7-4(b) for this example can be computed as $C_2(s) = 2(s + 1)/(s + 2)$. Although there is still a pole-zero cancellation in this configuration, the missing pole $(s + 1)$ is a stable pole. Furthermore the degree of the compensator is 1. ∎

We discuss now the assertion that a closed-loop system, if properly designed, is less sensitive to the parameter variations of the plant than an open-loop system. Consider the systems shown in Figure 7-5. The systems shown on the left-hand side are designed for the nominal transfer function $\hat{g}(s)$ and for

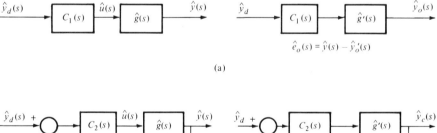

Figure 7-5 (a) Open-loop system. (b) Closed-loop system.

the same overall transfer function; hence their output should be equal; that is

$$\hat{y}(s) = \hat{g}(s)C_1(s)\hat{y}_d(s) = \frac{C_2(s)\hat{g}(s)}{1 + C_2(s)C_3(s)\hat{g}(s)}\hat{y}_d(s) \quad (7\text{-}6)$$

Because of inaccuracy in modeling or because of load variations, the actual plant transfer function often differs from the one on which the design is carried out. Let $\hat{g}'(s)$ be the actual plant transfer function. The difference $\hat{g}(s) - \hat{g}'(s)$ will be called the *plant variation*. Because of the plant variation, the actual outputs in both systems may no longer be the same. Let $\hat{y}_o(s)$ and $\hat{y}_c(s)$ be, respectively, the actual outputs of the open-loop and closed-loop systems. If the errors due to the plant variation are defined as

$$\hat{e}_o(s) = \hat{y}(s) - \hat{y}_o(s) \quad (7\text{-}7)$$

$$\hat{e}_c(s) = \hat{y}(s) - \hat{y}_c(s) \quad (7\text{-}8)$$

then they can be computed as

$$\hat{e}_o(s) = C_1(s)[\hat{g}(s) - \hat{g}'(s)]\hat{y}_d(s) \quad (7\text{-}9)$$

$$\hat{e}_c(s) = \left[\frac{C_2(s)\hat{g}(s)}{1 + C_2(s)C_3(s)\hat{g}(s)} - \frac{C_2(s)\hat{g}'(s)}{1 + C_2(s)C_3(s)\hat{g}'(s)}\right]\hat{y}_d(s)$$

$$= \frac{C_2(s)[\hat{g}(s) - \hat{g}'(s)]}{[1 + C_2(s)C_3(s)\hat{g}(s)][1 + C_2(s)C_3(s)\hat{g}'(s)]}\hat{y}_d(s) \quad (7\text{-}10)$$

Now by using (7-6) and (7-9), Equation (7-10) can be written as

$$\hat{e}_c(s) = \frac{1}{1 + C_2(s)C_3(s)\hat{g}'(s)}\hat{e}_o(s) \doteq \frac{1}{1 + C_2(s)C_3(s)\hat{g}(s)}\hat{e}_o(s) \quad (7\text{-}11)$$

where we have implicitly assumed that the plant variation is small, or equivalently, $\hat{g}'(s)$ is very close to $\hat{g}(s)$. We shall call

$$S(s) \triangleq \frac{1}{1 + C_2(s)C_3(s)\hat{g}(s)} \quad (7\text{-}12)$$

the *comparison sensitivity*. We see that if $|S(j\omega)| \ll 1$ in the frequency spectra of control signals, then the error $\hat{e}_c(j\omega)$ will be much smaller than $\hat{e}_o(j\omega)$. This justifies the assertion that a properly designed closed-loop system is less sensitive to plant variations.

Example 3

Consider the design of transistor amplifiers. It is assumed that the single-stage amplifier, an open-loop system, shown in Figure 7-6(a) has a voltage gain of -10. To design a feedback amplifier with the same voltage gain, three stages of amplifiers, each of voltage gain $-A$, are used as shown in Figure

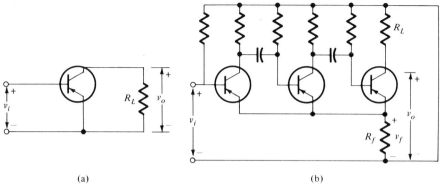

Figure 7-6 (a) Single-stage amplifier. (b) Three-stage amplifier with feedback.

7-6(b). The voltage gain, $-A_f$, of the feedback amplifier can be computed, under certain approximations, as

$$A_f = \frac{A^3}{1 + kA^3}$$

where $k = v_f/v_o = R_f/R_L$ (see Reference [3]). If $A = 10$, in order to have $A_f = 10$, k is chosen as 0.099.

We compare now the sensitivities of the gains of the open-loop and feedback amplifiers with respect to gain variations. If the voltage gain of each stage of the transistor amplifier decreases by 10 percent each year, then the voltage gain of the open-loop amplifier will become -9 at the end of one year. The voltage gain of the feedback amplifier at the end of one year is

$$A_f = \frac{(9)^3}{1 + 0.099 \times (9)^3} = \frac{729}{73.5} = 9.9$$

This shows that the feedback amplifier is indeed less sensitive to parameter variations than the open-loop amplifier.

Although the feedback amplifier needs three times more hardware, it actually is more economical than the open-loop amplifier. To show this, we compute the time span needed for the feedback amplifier to have gain of -9. Let A_x be the gain of each stage of transistor amplifier at which $A_f = 9$. Then we have

$$9 = \frac{A_x^3}{1 + 0.099 \times A_x^3}$$

or

$$A_x = 4.35$$

In other words, the voltage gain of the feedback amplifier reduces to -9 only when all three stages of the transistor amplifiers have gains reduced to -4.35. If the voltage gain reduces by 10 percent each year, it will take y years, where

$10 \cdot (0.9)^y = 4.35$, or $y = 7.9$. It means that the feedback amplifier can last 7.9 times longer than the open-loop amplifier, even though it needs three times more hardware. Hence the feedback amplifier is more economical. ∎

The signals in control systems generally have low frequency spectra. Hence if a plant has one or more poles at the origin as is often the case in practice, then we have $|S(s)| \ll 1$ at low frequencies. Hence even without any effort, a closed-loop system is often less sensitive to plant variations than an open-loop system.

The comparison discussed above is not completely fair because, in designing closed-loop systems, we have to use transducers in the feedback connection. These transducers, such as potentiometers or tachometers, will certainly introduce additional noise into the system, as shown in Figure 7-7. Hence the effect of this noise $\hat{n}(s)$ should be studied. The undesirable output $\hat{y}(s)$ generated by the transducer noise can be computed as

$$\hat{y}(s) = \frac{-C_2(s)C_3(s)\hat{g}(s)}{1 + C_2(s)C_3(s)\hat{g}(s)} \hat{n}(s)$$

If the noise has a high frequency spectrum, and if $\hat{g}(s)$ is strictly proper, then at high frequencies we have $|C_2(s)C_3(s)\hat{g}(s)| \ll 1$, and $\hat{y}(j\omega) \doteq -C_2(j\omega)C_3(j\omega) \cdot \hat{g}(j\omega)\hat{n}(j\omega)$ is very small. Hence noise with a high frequency spectrum, as is often the case, is again suppressed by feedback. See Section 6-4.

We discuss now the assertion that a properly designed closed-loop system is less sensitive to disturbances than an open-loop system. In the systems shown in Figure 7-8, it is assumed that an unwanted disturbance z entered the plant, as indicated. Clearly the output excited by the disturbance should be kept as small as possible. Let y_o and y_c be, respectively, the output excited by the disturbance in the open- and closed-loop configurations shown in Figure 7-8. Then we have

$$\hat{y}_o(s) = \hat{g}_2(s)\hat{z}(s)$$

and

$$\hat{y}_c(s) = \frac{\hat{g}_2(s)}{1 + C_2(s)C_3(s)\hat{g}_1(s)\hat{g}_2(s)} \hat{z}(s)$$

We see that $\hat{y}_o(s)$ and $\hat{y}_c(s)$ are related by

$$\hat{y}_c(s) = \frac{1}{1 + C_2(s)C_3(s)\hat{g}_1(s)\hat{g}_2(s)} \hat{y}_o(s) = S(s)\hat{y}_o(s) \qquad (7\text{-}13)$$

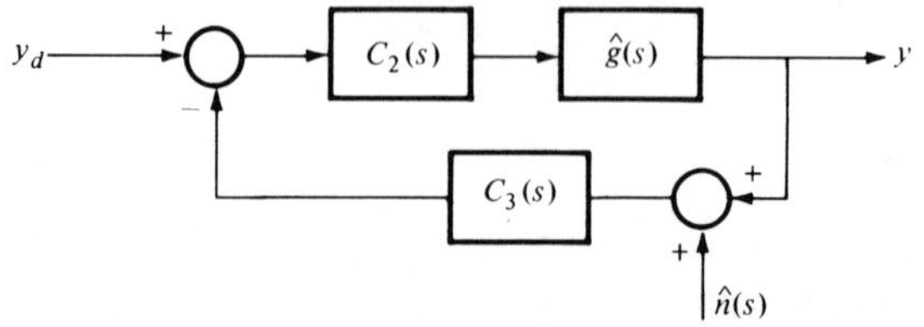

Figure 7-7 A system with noise from the transducer.

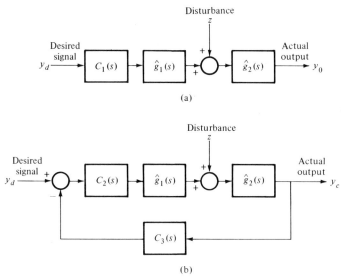

Figure 7-8 Design of a control system with a load disturbance. (a) Open-loop configuration. (b) Closed-loop configuration.

which is similar to Equation (7-11). Hence if the magnitude of $S(s)$ in (7-13) is very small in the frequency spectrum of the disturbance, then $\hat{y}_c(s)$ is much smaller than $\hat{y}_o(s)$. This establishes the assertion that a properly designed closed-loop system is less sensitive to disturbances than an open-loop system.

Example 4

Consider the speed control of rollers in an aluminum factory. Heated aluminum ingots are rolled through the rollers repeatedly to be pressed into sheets. The rollers are assumed to be driven by armature-controlled dc motors. Because the ingots cannot all have the same thickness, the system is modeled as shown in Figure 7-9. The transfer function of the motor and roller is computed for the average size of ingots; the discrepancy in size of an ingot is then represented by the disturbance shown. The disturbance for this system can be approximated by step functions.

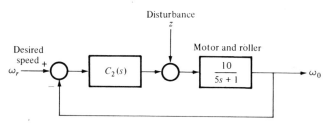

Figure 7-9 Control system with an unwanted disturbance.

It is assumed that we want to design an overall system with $2/(s+2)$ as the transfer function from ω_d to ω_0. If the open-loop configuration is used, the required compensator $C_1(s)$ can be easily computed as $C_1(s) = (5s+1)/5(s+2)$. For this configuration, the transfer function from z to ω_0 remains as $10/(5s+1)$, no matter what C_1 is introduced. Hence if there is a disturbance of type a/s, then the steady-state speed deviation from the desired one is

$$\lim_{s \to 0} s \frac{a}{s} \frac{10}{5s+1} = 10a$$

If the configuration in Figure 7-9 is used, the compensator can be computed as

$$C_2(s) = \frac{5s+1}{5s}$$

The transfer function from z to ω_0 in Figure 7-9 is

$$\hat{g}(s) = \frac{\dfrac{10}{5s+1}}{1 + \dfrac{10}{5s+1} \cdot \dfrac{5s+1}{5s}} = \frac{10s}{(5s+1)(s+2)}$$

Now for any disturbance of type $\hat{z}(s) = a/s$, the steady-state output due to this disturbance is

$$\lim_{s \to 0} s\hat{g}(s) \frac{a}{s} = \lim_{s \to 0} a\hat{g}(s) = 0$$

This means that, for the feedback configuration shown in Figure 7-9, if there is any disturbance, the actual speed ω_0 will return to the desired one after transient. Hence a closed-loop system is indeed less sensitive to disturbances than an open-loop system. ∎

For this example, the disturbance is automatically suppressed in the feedback loop. This may not always be the case for every system. In general, the compensators $C_2(s)$ and $C_3(s)$ in Figure 7-8(b) must be chosen so that the overall system will meet the specifications from y_d to y and from z to y. This is discussed in Reference [5]. In any case, a properly designed closed-loop system is less sensitive to disturbances than an open-loop system.

The closed-loop configuration introduced in Figure 7-6(b) is just one of many possible closed-loop configurations. A discussion of all possible closed-loop configurations is very difficult and hence will not be attempted. However it is generally true that most closed-loop configurations are better than open-loop systems. Hence in the remainder of this book, only closed-loop configurations of compensators will be used in the design of control systems.

7-4 Remarks and Review Questions

What is the pole-zero excess inequality? Under what conditions is it derived?

What is the transfer function of the best system one would like to design if no physical constraints were imposed?

Do the unstable poles of a plant impose any constraint in a design?

Do the right-half s-plane zeros of a plant impose any constraint in a design?

What is a synthesizable transfer function?

Can you use pole-zero cancellations in a design? If yes, what kind of caution is needed?

What advantages do we have in choosing a closed-loop configuration in a design?

In addition to aging and inaccuracy in modeling, plant variations occur in many control systems. For example, consider the control system in a steel mill studied in Example 2 of Section 3-9. Because of the differences in thickness and the inhomogeneity in density of steel sheets, the moment of inertia, J_m, in Equation (3-50) cannot be a constant. Hence a variation of transfer function $\hat{g}_2(s)$ is not avoidable in practice. Another example is the control of an antenna, as discussed in Example 5 of Section 3-9. In deriving the block diagram shown in Figure 3-30(b), wind gust is not considered. Depending on the wind speed, the torque generated by the wind on the antenna will be different. Consequently we have a different plant transfer function. Hence the sensitivity problem due to plant variations is an important one in practice.

The comparison of compensators between the open-loop and the closed-loop configurations by their degrees alone is open to argument. Consider the systems shown in Figure 7-10. The plant has transfer function $0.5/(s + 1)$; the overall systems all have transfer function $2/(s + 2)$. The degrees of compensators of the closed-loop configurators are indeed equal to or smaller than that of the open-loop configuration. We compare now the dc gains of the compensators. The dc gain of a compensator is defined as the gain at $s = 0$. The dc gain of $C_1(s)$ in Figure 7-10(a) is 2; the dc gains of $C_2(s)$ in Figure 7-10(b) through (e) are, respectively, 2, 4, 20, and ∞. Although the dc gain of $C_2(s)$ in Figure 7-10(a) and (b) is the same, its comparison sensitivity, as defined in Equation (7-12), is $(s + 1)/(0.5s + 1)$, the magnitude of which is larger than 1 for all ω. Hence the closed-loop system in Figure 7-10(b) is *not* less sensitive to plant variation than the open-loop configuration in Figure 7-10(a). The comparison sensitivity of the system in Figure 7-10(c) is $(s + 1)/(s + 2)$, which is smaller than 1, but the dc gain of $C_2(s)$ is twice as large as in Figure 7-10(a). The comparison sensitivity in Figure 7-10(d) is smaller than the one in (c), but the dc gain of $C_2(s)$ is larger. The comparison sensitivity of Figure

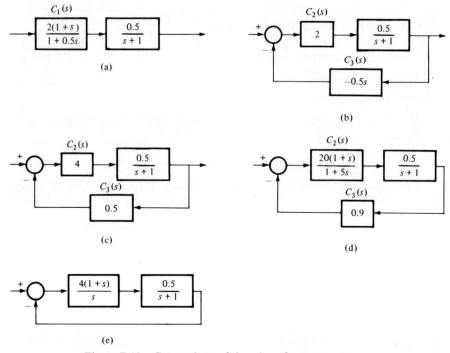

Figure 7-10 Comparison of dc gains of compensators.

7-10(e) at $\omega = 0$ is 0, but its dc gain is infinity. Hence we conclude from these systems that in order to reduce sensitivity comparisons, the dc gains of compensators have to increase. This is the price we often have to pay in the introduction of feedback.

References

[1] Chen, C. T., "Design of feedback control systems," *Proc. Nat. Electron. Conf.*, vol. 57, pp. 46–51, 1969.
[2] Chen, C. T., *Introduction to Linear System Theory*. New York: Holt, Rinehart and Winston, 1970.
[3] Fitchen, F. C., *Transistor Circuit Analysis and Design*. New York.: Van Nostrand, 1960.
[4] Perkins, W. R., and J. B. Cruz, Jr., *Engineering of Dynamic Systems*. New York: Wiley, 1969.
[5] Truxal, J. G., *Control System Synthesis*. New York: McGraw-Hill, 1955.

Problems

7-1 Given a plant with transfer function

$$\hat{g}(s) = \frac{(s-1)(s+2)}{s^2(s+1)}$$

determine which of the following overall transfer functions are synthesizable

$$\hat{g}_{f1}(s) = \frac{s+1}{s^2+2s+3}$$

$$\hat{g}_{f2}(s) = \frac{-s+1}{s^2+2s+1}$$

$$\hat{g}_{f3}(s) = \frac{(s+3)(1-s)}{s^3+2s^2+3s+3}$$

$$\hat{g}_{f4}(s) = \frac{-s+1}{s^3+2s^2+3s+1}$$

7-2 Consider the problem shown in Figure 7-1. Show that if $\hat{g}_f(s)$ is synthesizable and BIBO stable and if $\hat{g}(s)$ has at least one pole at the origin, then the actuating signal $u(t)$ approaches zero as $t \to \infty$, if y_d is a step function.

7-3 Given a plant with transfer function $1/s(s^2 + 2s + 3)$ and an overall transfer function $3/(s^3 + 3.5s^2 + 5s + 3)$. If the configurations of compensators in Figure 7-4 are chosen, what are the required compensators? How do you compare these two configurations?

7-4 Consider the systems shown in Figure P7-1.

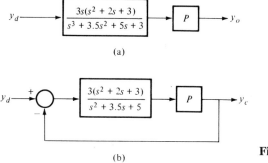

Figure P7-1

a. If the plant, denoted by P, has the following nominal transfer function

$$\frac{1}{s(s^2+2s+3)}$$

show that both systems have the same steady-state output due to $y_d = \sin 0.1t$.

b. If, because of aging or some other reason, the transfer function of the plant changes to

$$\frac{1}{s(s^2+2.1s+3.06)}$$

what are the steady-state outputs of both systems due to $y_d = \sin 0.1t$?

c. Which system has a steady-state output closer to the one computed in part a?

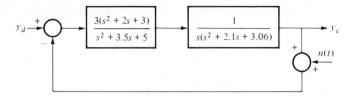

Figure P7-2

7-5 Consider the system shown in Figure P7-2, in which $n(t)$ denotes the noise due to a transducer.

a. Compute the steady-state y_c due to $n(t) = 0.1 \sin 10t$.
b. What is the steady-state y_c due to $y_d = \sin 0.1t$ and $n(t) = 0.1 \sin 10t$?
c. Compare the steady-state error resulting in Figure P7-1(a) with the one resulting in Figure P7-2. Is the reduction in the steady-state error due to feedback large enough to offset the increase of the steady-state error due to noise?

7-6 Compare the sensitivities with respect to plant variations of the configurations shown in Figure P7-3. Is it possible to assert which configuration is better?

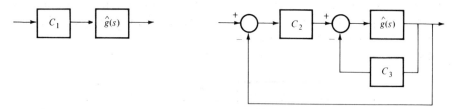

Figure P7-3

7-7 Consider the speed control of rollers discussed in Example 4 of Section 7-3. Instead of modeling the system as shown in Figure 7-9, we shall model it in this problem as having a transfer function $10/(\tau s + 1)$, with $\tau = 5$ as the nominal value and τ ranging between 3 and 6. Find compensators so that the overall open- and closed-loop systems shown in Figure 7-4 have a transfer function $2/(s + 2)$. Now if τ changes to 3 because of load variation, what are the responses of these two systems due to a unit step input? Plot the responses and see which configuration is less sensitive to the plant variation.

CHAPTER

8
Optimal Control Systems

There are, as discussed in Chapter 6, basically two approaches to the design of control systems. In one approach we first find an overall system that satisfies the specifications and then compute the required compensators. In the other approach we search for compensators so that the resulting overall system will satisfy the specifications. In this chapter the design techniques taking the first approach will be introduced. This approach consists of two separate problems: searching for an overall transfer function and solving for compensators. The next four sections will be concerned with the first problem; the remainder, with the second problem.

In this chapter we shall study plants having a single input and a single output. The design of multivariable systems is out of the scope of this text.

8-1 Quadratic Performance Criterion

The performance of a control system, as discussed in Section 6-2, is specified in terms of rise time, settling time, overshoot, and steady-state error. Given these specifications, it is possible, by using computer simulation, to find a synthesizable overall transfer function $\hat{g}_f(s)$ to satisfy these specifications. This simulation approach is however a trial-and-error method. Furthermore if the chosen $\hat{g}_f(s)$ does not meet the specifications, the approach does not provide a direction for improvement. Hence a more systematic approach to choosing $\hat{g}_f(s)$ is desirable.

As engineers, we are often interested in designing not only a working system but the best possible system. In order for us to talk about the best system, a measure of comparisons must be introduced. The measure or criterion could be chosen as

$$J_1 \triangleq k_1 \text{ (rise time)} + k_2 \text{ (settling time)} + k_3 \text{ (overshoot)} \\ + k_4 \text{ (steady-state errors)} \quad (8\text{-}1)$$

where the k_i's are weighting factors. We note that J_1 is a real number; otherwise a comparison is not possible (see Problem 8-1). The system that has the smallest J_1 is called the *optimal* system with respect to criterion J_1. Although the criterion seems reasonable, it is not trackable analytically. Hence a different criterion has to be used.

In fact, the *only* criterion that can be solved analytically and will yield a linear time-invariant overall system is of the form[1]

$$J_2 \triangleq \int_0^\infty [y_d(t) - y(t)]^2 \, dt \quad (8\text{-}2)$$

where $y_d(t)$ is the command, or desired, signal and $y(t)$ is the actual output. Hence $(y_d - y)$ is an error of the system. Although Equation (8-2) is chosen for mathematical trackability, it turns out to be acceptable in practice. The criterion penalizes positive and negative errors equally. It penalizes heavily on large error; hence a small J_2 usually results in a system with a small overshoot. Since the integration in J_2 is carried out over $[0,\infty)$, if $y(t)$ does not approach $y_d(t)$ as $t \to \infty$, then J_2 will be infinity. This implies that if J_2 is finite, the steady-state error is zero. Therefore we conclude that the criterion in (8-2) is a reasonable one.

Unfortunately no matter what the plant transfer function $\hat{g}(s)$ is, the overall system that minimizes the criterion in (8-2) always has a transfer function of 1. This is, as discussed in Section 7-1, not necessarily synthesizable. Hence the performance criterion J_2 must be modified. Recall from Section 6-4 that a major problem in the design is to avoid saturation in physical devices. This constraint however is not reflected in J_2. Therefore a more realistic performance criterion should be

$$\text{minimize} \int_0^\infty [y(t) - y_d(t)]^2 \, dt \quad (8\text{-}3a)$$

subject to the constraint

$$\max |u(t)| \leq k \quad (8\text{-}3b)$$

---

[1] This includes the criteria, such as $\int_0^\infty e^{-at}[y_d(t) - y(t)]^2 \, dt$, that can be reduced to the form in Equation (8-2).

for some constant k. The constant k is determined by the linear range of the plant.

Clearly similar constraints should also be imposed on devices, such as electronic amplifiers and compensating networks, that are yet to be introduced. However this will make the design extremely complicated. Besides, electronic amplifiers are, compared with a plant, rather inexpensive, and hence, if saturated, they can be replaced by better ones. Therefore the saturation constraint is generally imposed only on a plant.

Although the criterion expressed in (8-3) can be used in the design, the resulting optimal system is not necessarily a linear system. In other words, in order to implement the optimal system, nonlinear and/or time-varying devices are required. In order to obtain a linear time-invariant optimal system, (8-3) has to be replaced by

$$J = \int_0^\infty \{q[y(t) - y_d(t)]^2 + u^2(t)\}\, dt \tag{8-4}$$

where q is a *positive* constant. It is called the *weighting factor*. We shall call Equation (8-4) a quadradic performance criterion. If q in (8-4) is negative, then the criterion is useless because a system with a large error and a large u may still have a small J. If q is a large positive number, more weight is imposed on the error. As q approaches infinity, the contribution of u in (8-4) becomes less significant, and at the extreme, (8-4) reduces to (8-2). In this case, since no penalty is imposed on the actuating signal u, its magnitude will be very large, possibly infinity if y_d is not continuous; hence the constraint of (8-3b) will be violated. If $q = 0$, then (8-4) reduces to

$$\int_0^\infty u^2(t)\, dt$$

and the optimal system that minimizes this expression is the one with $u = 0$. From these two extreme cases, we conclude that if q in (8-4) is properly chosen, then the constraint of (8-3b) will be satisfied. Hence although we are forced to use the quadratic performance criterion (8-4) in order to obtain a linear overall system, if q is properly chosen it is an acceptable substitution of (8-3).

Although the quadratic criterion (8-4) is an acceptable substitution of (8-3), it is unnecessarily stringent. We see that if $y(t) \neq y_d(t)$ or $u(t) \neq 0$ as t approaches infinity, then the performance index J will go to infinity. Hence an optimal system with a finite J must have $u(t) \to 0$ as $t \to \infty$. This is for some cases rather unnecessary, because we require only $|u(t)| \leq k$ for some constant k. However for control systems such as space vehicles that have a limited supply of energy, the requirement $u(t) \to 0$ make sense. In conclusion, although the quadratic performance criterion is not necessarily the most desirable one, it is acceptable in practice and it does yield a linear overall system. Hence we shall use it as a criterion in the design of optimal systems.

8-2 Optimal Control Systems: Transfer-Function Approach

Consider the plant shown in Figure 8-1. The plant is assumed to be completely characterized by the transfer function

$$\hat{g}(s) \triangleq \frac{N(s)}{D(s)} = \frac{\beta_0 + \beta_1 s + \cdots + \beta_n s^n}{\alpha_0 + \alpha_1 s + \cdots + \alpha_n s^n} \qquad \alpha_n > 0 \qquad (8\text{-}5)$$

where α_i and β_i are real numbers, not necessarily nonzero. It is also assumed that $N(s)$ and $D(s)$ have no common factor. The design problem is to find an overall transfer function that minimizes the quadratic performance criterion

$$J = \int_0^\infty \{q[y(t) - y_d(t)]^2 + u^2(t)\}\, dt \qquad (8\text{-}6)$$

where q is a positive constant, y_d is the command or desired signal, y is the output, and u is the actuating signal. Let $\hat{g}_f(s)$ be the optimal overall transfer function. In order to compute $\hat{g}_f(s)$, the concept of spectral factorization is needed.

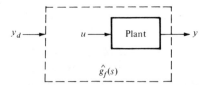

Figure 8-1
A design problem.

Consider the polynomial

$$\prod(s) \triangleq D(s)D(-s) + qN(s)N(-s) \qquad (8\text{-}7)$$

which is formed from the denominator and numerator of the plant transfer function and from the weighting factor q in the quadratic performance criterion. It is clear that $\prod(s) = \prod(-s)$; hence if s_1 is a root of $\prod(s) = 0$, so is $-s_1$. Since all the coefficients of $\prod(s)$ are real numbers by assumption, if s_1 is a root of $\prod(s) = 0$, so is its complex conjugate $\bar{s}_1$. Consequently all the roots of $\prod(s) = 0$ are symmetric with respect to the real axis, the imaginary axis, and the origin of the s-plane, as shown in Figure 8-2. We next show that $\prod(s)$ has no root on the imaginary axis. Consider

$$\begin{aligned}\prod(j\omega) &= D(j\omega)D(-j\omega) + qN(j\omega)N(-j\omega) \\ &= |D(j\omega)|^2 + q|N(j\omega)|^2\end{aligned} \qquad (8\text{-}8)$$

The assumptions that $q > 0$ and that $D(s)$ and $N(s)$ have no common factor imply that $\prod(j\omega) \neq 0$ for all ω. Hence the polynomial $\prod(s)$ has no roots on the imaginary axis. Consequently it is always possible to factor $\prod(s)$ into

$$D_f(s)D_f(-s) = D(s)D(-s) + qN(s)N(-s) \qquad (8\text{-}9)$$

where $D_f(s)$ consists of all the open left-half s-plane roots of $\prod(s)$; $D_f(-s)$ consists of all the open right-half s-plane roots of $\prod(s)$. Clearly $D_f(s)$ is a Hurwitz polynomial. The factorization in (8-9) is called a *spectral factorization*.

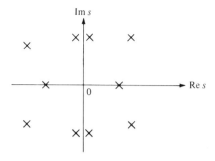

Figure 8-2
The roots of $\Pi(s) = 0$.

There are many ways to achieve spectral factorizations. If all the roots of $\Pi(s)$ are found, by grouping the left-half s-plane roots, the factorization is completed. There are subroutines available in the IBM Scientific Subroutine Package for computing the roots of a polynomial. These subroutines can be modified to take the advantage of the fact that the roots of $\Pi(s)$ are symmetric with respect to the origin of the s-plane. By so doing, the computing time can be substantially reduced. Spectral factorization can also be achieved as follows. Since $\Pi(s) = \Pi(-s)$ for all s, $\Pi(s)$ must be of the form

$$\Pi(s) \triangleq a_0 + a_2 s^2 + a_4 s^4 + \cdots + a_{2n} s^{2n} \qquad (8\text{-}10)$$

Let

$$D_f(s) = \gamma_0 + \gamma_1 s + \cdots + \gamma_n s^n \qquad (8\text{-}11)$$

By equating the coefficients of $D_f(s)D_f(-s) = \Pi(s)$, we obtain $\gamma_0^2 = a_0$, $\gamma_1^2 = -a_2 + 2\gamma_0\gamma_2, \ldots, \gamma_n^2 = (-1)^n a_{2n}$. These can be arranged as

$$\begin{aligned}
\gamma_0 &= \sqrt{a_0} \\
\gamma_1 &= \sqrt{-a_2 + 2\gamma_0\gamma_2} \\
\gamma_2 &= \sqrt{a_4 + 2\gamma_1\gamma_3 - 2\gamma_0\gamma_4} \\
&\vdots \\
\gamma_n &= \sqrt{(-1)^n a_{2n}}
\end{aligned} \qquad (8\text{-}12)$$

These equations can be solved by iteration on a digital computer. For a discussion of spectral factorization, see Reference [9].

With spectral factorization, we are ready to derive optimal overall transfer functions.

Problem

Consider a plant that is completely characterized by its irreducible proper transfer function $\hat{g}(s) = N(s)/D(s)$, as shown in Figure 8-1. The problem is to find an overall system with transfer function $\hat{g}_f(s)$ that minimizes the quadratic performance criterion

$$J = \int_0^\infty \{q[y(t) - y_d(t)]^2 + u^2(t)\}\, dt \qquad (8\text{-}13)$$

where q is a positive constant and y_d is the command signal. It is also required that the zeros of $\hat{g}_f(s)$ contain all the closed right-half s-plane zeros of $\hat{g}(s)$.

Solution

The first step in the solution is to perform the spectral factorization

$$D_f(s)D_f(-s) = D(s)D(-s) + qN(s)N(-s) \qquad (8\text{-}14)$$

where $D_f(s)$ is a Hurwitz polynomial. Then the optimal overall transfer function $\hat{g}_{f1}(s)$, when $y_d(t)$ is a step function, is given by

$$\hat{g}_{f1}(s) = \frac{qN(0)}{D_f(0)} \cdot \frac{N(s)}{D_f(s)} \qquad (8\text{-}15)$$

The optimal overall transfer function when $y_d(t)$ is a ramp function is given by

$$\hat{g}_{f2}(s) = q(k_1 + k_2 s)\frac{N(s)}{D_f(s)} = \left(1 + \frac{k_2}{k_1}s\right)\frac{qN(0)}{D_f(0)}\frac{N(s)}{D_f(s)} \qquad (8\text{-}16)$$

where

$$k_1 = \frac{N(0)}{D_f(0)} \qquad (8\text{-}17)$$

and

$$k_2 = \frac{d}{ds}\left[\frac{N(-s)}{D_f(-s)}\right]\bigg|_{s=0} \qquad (8\text{-}18)$$

or, if $N(s) = \beta_0 + \beta_1 s + \cdots + \beta_n s^n$ and $D_f(s) = \gamma_0 + \gamma_1 s + \cdots + \gamma_n s^n$, then

$$k_1 = \frac{\beta_0}{\gamma_0} \qquad (8\text{-}19)$$

$$k_2 = \frac{\beta_0 \gamma_1 - \gamma_0 \beta_1}{\gamma_0^2} \qquad (8\text{-}20)$$

∎

We first give examples to illustrate the application of this result.

Example 1

Consider a plant with transfer function

$$\hat{g}(s) = \frac{20}{s(s+2)}$$

The q in the performance criterion (8-13) is chosen, rather arbitrarily, as 1. Simple computation yields

$$D(s)D(-s) + qN(s)N(-s) = s^4 - 4s^2 + 400$$

Clearly $D_f(s)$ is of the form

$$D_f(s) = s^2 + \gamma_1 s + 20$$

8-2 OPTIMAL CONTROL SYSTEMS: TRANSFER-FUNCTION APPROACH

The first and last coefficients are obtained by inspection. By equating the coefficients of

$$(s^2 + \gamma_1 s + 20)(s^2 - \gamma_1 s + 20) = s^4 - 4s^2 + 400$$

the coefficient γ_1 can be computed as $\gamma_1 = \sqrt{44} = 6.63$. Hence the optimal transfer function when the command signal is a step function is, by using (8-15),

$$\hat{g}_{f1}(s) = \frac{20}{s^2 + 6.63s + 20} \tag{8-21}$$

The optimal transfer function when y_d is a ramp function is, by using (8-16),

$$\hat{g}_{f2}(s) = \left(1 + \frac{6.63}{20} s\right) \cdot \frac{20}{s^2 + 6.63s + 20} = \frac{6.63s + 20}{s^2 + 6.63s + 20} \tag{8-22}$$

Example 2

Consider the tracking antenna problem discussed in Section 6-1. For convenience, the plant chosen in Figure 6-1 is redrawn in Figure 8-3(a). Its block diagram is shown in Figure 8-3(b). The transfer functions in the block diagram are obtained from Equations (3-49) and (3-50) with proper modification. Let $N_m/N_L = \frac{1}{2}$, $J_L = 1.6$ Nm·rad^{-1}·s, $f = 0.04$ Nm·rad^{-1}·sec, $k_t = 1.2$ Nm/A, $k_b = 1.2$ V·rad^{-1}·s, and $k_g = 100$ V/A. It is assumed that the moments of

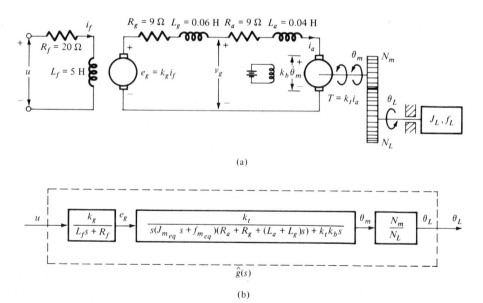

Figure 8-3 The plant of a tracking antenna control problem.

inertia of the motor and the gear train are negligible. Then the transfer function of the plant is equal to

$$\hat{g}(s) = \frac{\hat{\theta}_L(s)}{\hat{u}(s)} = \frac{1}{2} \cdot \frac{100}{5s + 20} \cdot \frac{1.2}{s(0.4s + 0.01)(18 + 0.1s) + 1.44s} \quad (8\text{-}23)$$

$$= \frac{300}{s^4 + 184s^3 + 760.5s^2 + 162s} \triangleq \frac{N(s)}{D(s)}$$

Suppose that it is required to design an overall system to minimize the performance criterion

$$\int_0^\infty \{25[y(t) - 1]^2 + u^2\} \, dt \quad (8\text{-}24)$$

that is, q is chosen as 25 and y_d as a unit step function. We first compute

$$D(s)D(-s) + qN(s)N(-s) = s^8 - 32{,}335s^6 + 518{,}744.25s^4 \\
- 26{,}244s^2 + 25 \times 90{,}000 \quad (8\text{-}25)$$

Hence D_f is of the form

$$D_f(s) = s^4 + \gamma_3 s^3 + \gamma_2 s^2 + \gamma_1 s + 1500 \quad (8\text{-}26)$$

Equating the coefficients of $D_f(s)D_f(-s) = D(s)D(-s) + qN(s)N(-s)$ yields

$$-\gamma_1^2 + 3000\gamma_2 = 26{,}244$$
$$\gamma_2^2 - 2\gamma_1\gamma_3 + 3000 = 518{,}744.25 \quad (8\text{-}27)$$
$$-\gamma_3^2 + 2\gamma_2 = -32{,}335$$

This is a set of nonlinear algebraic equations. They can be solved by iteration. First, pick an arbitrary γ_2, say $\gamma_2 = 1000$, and solve

$$\gamma_1^2 = 3000\gamma_2 - 26{,}244$$
$$\gamma_3^2 = 2\gamma_2 + 32{,}335$$

Then use the solved γ_1 and γ_3 to update γ_2 from

$$\gamma_2^2 = 515{,}744.25 + 2\gamma_1\gamma_3$$

We repeat this process a number of times (four times in this example), the γ_i, for $i = 1, 2, 3$, are computed as

$$\gamma_1 = 1800 \qquad \gamma_2 = 1088.4 \qquad \gamma_3 = 185.8 \quad (8\text{-}28)$$

For this example, it is possible to carry out the computation by hand; generally a digital computer is indispensable. From Equations (8-15), (8-26), and (8-28), the optimal overall transfer function is obtained as

$$\hat{g}_f(s) = \frac{25 \times 300}{1500} \cdot \frac{300}{s^4 + 185.8s^3 + 1088.4s^2 + 1800s + 1500} \quad (8\text{-}29)$$

$$= \frac{1500}{s^4 + 185.8s^3 + 1088.4s^2 + 1800s + 1500}$$

∎

We see from these two examples that it is rather straightforward to compute optimal transfer functions. How to implement these optimal transfer functions will be discussed later.

Remarks

1. A proof of (8-15) and (8-16) can be found in Reference [2]. It is proved by first transforming (8-13) into the frequency domain by using Parseval's theorem and then carrying out the minimization. The requirement that the zeros of $\hat{g}_f(s)$ contain the closed right-half s-plane zeros of $\hat{g}(s)$ is used in the derivations of (8-15) and (8-16); otherwise the results will be different.

2. Since $\delta N(s) \leq \delta D(s)$, the spectral factorization implies that $\delta D(s) = \delta D_f(s)$, where δ denotes the degree of a polynomial. Hence the optimal transfer function $\hat{g}_{f1}(s)$ in (8-15) satisfies the pole-zero excess inequality (7-3) and is synthesizable. The optimal transfer function $\hat{g}_{f2}(s)$ in Equation (8-16) however does not satisfy the pole-zero excess inequality (7-3); hence it cannot be implemented by using compensators exclusively with proper transfer functions. If a differentiator is permitted, $\hat{g}_{f2}(s)$ can be synthesized through $\hat{g}_{f1}(s)$, as shown in Figure 8-4. If no differentiator is permitted, this problem can be resolved as follows: In the design, we use a fictitious plant transfer function with one extra pole, say

$$\hat{g}'(s) = \frac{1}{1 + \tau s} \hat{g}(s)$$

The time constant τ is chosen to be very small or, equivalently, the magnitude of the negative pole $-1/\tau$ is chosen to be large, so that the response of $\hat{g}'(s)$ will closely approximate the response of $\hat{g}(s)$. Now if the design is carried out for $\hat{g}'(s)$, the pole-zero excess of the resulting $\hat{g}_f(s)$ will be equal to that of $\hat{g}(s)$. Consequently $\hat{g}_f(s)$ is now synthesizable.

3. The choice of the weighting factor q in the quadratic performance criterion (8-13) is quite a problem in this design. Clearly its value depends entirely on the linear range of the plant. In order to design a good tracking problem (that is, y approaches y_d quickly), it is suggested that q be chosen in the range from 25 to 100. It is possible to carry out the spectral factorization graphically, on the complex s-plane, as a function of q by the so-called root-square-locus method. We can then choose a proper q from the pole locations. The interested reader is referred to Reference [2].

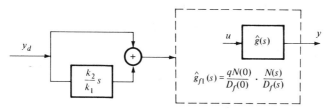

Figure 8-4 A synthesis of $\hat{g}_{f2}(s)$.

4. It is of interest to note that $\hat{g}_{f1}(s)$ and $\hat{g}_{f2}(s)$ in Equations (8-15) and (8-16) reduce to 1 if q increases to infinity. This confirms what we discussed in the previous section; that is, if (8-2) is used as a performance criterion, then, for any plant, the optimal overall transfer function is 1.

5. To be precise, in order for the design problem to be well defined, the optimal system must yield a finite J in (8-13). It can be verified that if y_d is a step function (a ramp function), the plant transfer function must have at least one pole (two poles) at the origin of the s-plane; otherwise no overall system will yield a finite J. Although it cannot be proved rigorously, Equations (8-15) and (8-16) might still yield the best design even for systems that are not so well defined. Hence (8-15) and (8-16) can be widely employed.

8-3 Optimal Control Systems: State-Variable Approach*

The problem posed and solved in Section 8-2 by the use of transfer functions will be restudied using dynamical equations. The plant shown in Figure 8-1 is assumed to be described by the controllable and observable dynamical equation

$$\dot{\mathbf{x}} = \mathbf{A}\mathbf{x} + \mathbf{b}u \qquad (8\text{-}30a)$$

$$y = \mathbf{c}\mathbf{x} \qquad (8\text{-}30b)$$

where $\mathbf{A}$, $\mathbf{b}$, and $\mathbf{c}$ are, respectively, $n \times n$, $n \times 1$, and $1 \times n$ real constant matrices. The problem is to design an overall system that minimizes the performance criterion

$$J = \lim_{T \to \infty} \int_0^T \{q[y(t) - y_d(t)]^2 + u^2(t)\}\, dt \qquad (8\text{-}31)$$

where q is a positive number. The solution of this problem in the transfer-function approach is in the form of an optimal overall transfer function. The solution in this section however will be in the form of an optimal actuating signal. This optimal actuating signal is often referred to as an *optimal control law* because it is expressed as a function of the state $\mathbf{x}$ and the desired signal y_d. To be more specific, the resulting system will minimize J in (8-31) if u is given by

$$u(t) = -\mathbf{b}'\mathbf{K}\mathbf{x}(t) + \mathbf{b}'\mathbf{g}(t) \qquad (8\text{-}32)$$

where $\mathbf{K}$ is the $n \times n$ symmetric matrix with all positive eigenvalues[2] satisfying the nonlinear algebraic matric equation

$$-\mathbf{K}\mathbf{A} - \mathbf{A}'\mathbf{K} + \mathbf{K}\mathbf{b}\mathbf{b}'\mathbf{K} - q\mathbf{c}'\mathbf{c} = 0 \qquad (8\text{-}33)$$

and the $n \times 1$ vector function $\mathbf{g}(t)$ is the solution of

$$\dot{\mathbf{g}}(t) = -(\mathbf{A} - \mathbf{b}\mathbf{b}'\mathbf{K})'\mathbf{g}(t) - q\mathbf{c}'y_d(t) \qquad (8\text{-}34)$$

---

* May be omitted without loss of continuity.
[2] Or equivalently, $\mathbf{K}$ is a symmetric positive definite matrix. See Reference [1] or [3].

with the boundary condition $g(T) = 0$, where T is defined as in (8-31) and the "prime" denotes the transpose. If K and $g(t)$ are computed and if all the state variables x are available (see Section 8-4), then the optimal system can be built as shown in Figure 8-5.

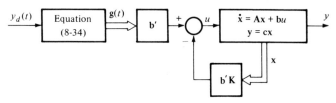

Figure 8-5 Optimal system.

We mention a special case where y_d is a step function. In this case the $g(t)$ in Equation (8-32) can be replaced by

$$g(t) = -[(A - bb'K)']^{-1}cqy_d \qquad (8\text{-}35)$$

A derivation of the results in this section can be found in Reference [1].

Example

Consider the system shown in Figure 8-1. It is assumed that the plant is described by

$$\begin{bmatrix} \dot{x}_1 \\ \dot{x}_2 \end{bmatrix} = \begin{bmatrix} 0 & 1 \\ 0 & -2 \end{bmatrix} x + \begin{bmatrix} 0 \\ 20 \end{bmatrix} u \qquad (8\text{-}36a)$$

$$y = \begin{bmatrix} 1 & 0 \end{bmatrix} x \qquad (8\text{-}36b)$$

Find the optimal control law that minimizes

$$\int_0^\infty \{[y(t) - 1]^2 + u^2(t)\}\, dt \qquad (8\text{-}37)$$

For this problem we have

$$A = \begin{bmatrix} 0 & 1 \\ 0 & -2 \end{bmatrix} \quad b = \begin{bmatrix} 0 \\ 20 \end{bmatrix} \quad c = \begin{bmatrix} 1 & 0 \end{bmatrix} \quad q = 1 \quad y_d(t) = 1 \quad \text{for } t \geq 0$$

Let the symmetric matrix K be of the form

$$K = \begin{bmatrix} k_{11} & k_{12} \\ k_{12} & k_{22} \end{bmatrix}$$

Then Equation (8-33) becomes

$$-\begin{bmatrix} k_{11} & k_{12} \\ k_{12} & k_{22} \end{bmatrix} \begin{bmatrix} 0 & 1 \\ 0 & -2 \end{bmatrix} - \begin{bmatrix} 0 & 0 \\ 1 & -2 \end{bmatrix} \begin{bmatrix} k_{11} & k_{12} \\ k_{12} & k_{22} \end{bmatrix}$$

$$+ \begin{bmatrix} k_{11} & k_{12} \\ k_{12} & k_{22} \end{bmatrix} \begin{bmatrix} 0 \\ 20 \end{bmatrix} \begin{bmatrix} 0 & 20 \end{bmatrix} \begin{bmatrix} k_{11} & k_{12} \\ k_{12} & k_{22} \end{bmatrix} - \begin{bmatrix} 1 \\ 0 \end{bmatrix} \begin{bmatrix} 1 & 0 \end{bmatrix} = \begin{bmatrix} 0 & 0 \\ 0 & 0 \end{bmatrix}$$

which becomes, after simplification,
$$400k_{12}^2 - 1 = 0$$
$$400k_{12}k_{22} + 2k_{12} - k_{11} = 0$$
$$400k_{22}^2 + 4k_{22} - 2k_{12} = 0$$

From these equations a **K** with all positive eigenvalues can be found as

$$\mathbf{K} = \begin{bmatrix} \dfrac{8.63}{20} & 0.05 \\ 0.05 & \dfrac{4.63}{400} \end{bmatrix} \tag{8-38}$$

The function $\mathbf{g}(t)$ in (8-35) can be computed as

$$\mathbf{g}(t) = -\left[\left(\begin{bmatrix} 0 & 1 \\ 0 & -2 \end{bmatrix} - \begin{bmatrix} 0 \\ 20 \end{bmatrix} \begin{bmatrix} 0 & 20 \end{bmatrix} \begin{bmatrix} \dfrac{8.63}{20} & 0.05 \\ 0.05 & \dfrac{4.63}{400} \end{bmatrix}\right)'\right]^{-1} \begin{bmatrix} 1 \\ 0 \end{bmatrix} y_d \tag{8-39}$$

$$= -\begin{bmatrix} 0 & -20 \\ 1 & -6.63 \end{bmatrix}^{-1} \begin{bmatrix} 1 \\ 0 \end{bmatrix} y_d = \begin{bmatrix} \dfrac{6.63}{20} \\ \dfrac{1}{20} \end{bmatrix} y_d$$

Hence the optimal control law is, by using (8-32),

$$u(t) = -\begin{bmatrix} 0 & 20 \end{bmatrix} \begin{bmatrix} \dfrac{8.63}{20} & 0.05 \\ 0.05 & \dfrac{4.63}{400} \end{bmatrix} \mathbf{x}(t) + \begin{bmatrix} 0 & 20 \end{bmatrix} \begin{bmatrix} \dfrac{6.63}{20} \\ \dfrac{1}{20} \end{bmatrix} y_d \tag{8-40}$$

$$= -\begin{bmatrix} 1 & \dfrac{4.63}{20} \end{bmatrix} \mathbf{x}(t) + y_d$$

The basic block diagram of the plant is shown in Figure 8-6. An implementation of the control law is also shown. We note that the transfer function of the plant in this example is the same as the one in Example 1 of Section 8-2. Their performance criteria are also the same. Hence the results should be identical. Indeed the transfer function from y_d to y in Figure 8-6 is, from Mason's formula,

$$\frac{20 \dfrac{1}{s^2}}{1 - \left(-\dfrac{2}{s} - \dfrac{4.63}{s} - \dfrac{20}{s^2}\right)} = \frac{20}{s^2 + 6.63s + 20}$$

which is the same as Equation (8-21). ∎

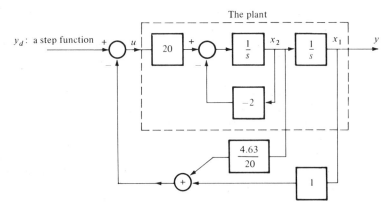

Figure 8-6 Implementation of the control law in (8-40).

The purpose of this section is to acquaint the reader with the state-variable approach; therefore its introduction is rather brief. It is now pertinent to compare this approach with the transfer-function approach in solving the optimal problem. First, in the transfer-function approach, the initial state must be assumed to be zero. The initial state in the state-variable approach is entirely arbitrary; however the result turns out to be independent of the initial state. Second, for the single-variable systems, the transfer-function approach is simpler, as demonstrated in the examples of this and the previous section, than the state-variable approach. Both approaches can be extended to the multivariable case. It seems however that, in the multivariable case, the state-variable approach is simpler than the transfer-function approach. Furthermore there are many digital computer programs, such as GASP and ASP, available in the state-variable approach. See References [7] and [8]. Third, in the transfer-function approach, it is possible to obtain some intuitive feeling of the system from the poles and zeros of $\hat{g}_f(s)$ and from the coefficients of $\hat{g}_f(s)$. For example, from the constant terms of the numerator and denominator of $\hat{g}_f(s)$, we know immediately the steady-state error of the optimal system. This might not be possible in the state-variable approach. Finally, the state-variable approach is applicable to time-varying and nonlinear systems. Its performance criterion can also be more general. The transfer-function approach is however restricted to linear time-invariant systems with a quadratic performance criterion. We see from the above discussion that each approach has its own merits. Hence a designer will benefit by familiarizing himself with both approaches.

8-4 The Regulator Problem*

The control problem studied in the previous section is often referred to as a *tracking problem* or *servomechanism*, for the output of the plant $y(t)$ is designed

---

* May be omitted without loss of continuity.

to follow the command, or desired, signal $y_d(t)$. In this section we introduce a special case of the tracking problem in which $y_d \equiv 0$. This special case is often called the *output regulator problem*. Since linearized models of many control systems reduce directly to the regulator problem, it is justifiable to have a section on this class of problems.

One may ask: If the command signal y_d is identically zero, will the actual output also be identically zero? This is true if all the initial conditions are zero. Hence the responses in regulator problems are due to nonzero initial states that, in turn, are caused by disturbances. To illustrate this point, consider an aircraft cruising at a fixed altitude, say h_0. Its altitude may change because of air pockets or for other reasons. Hence if h_0 is considered as the reference attitude, then maintaining the aircraft at height 0 (with respect to h_0) is a regulator problem.

A control system may be called to perform the jobs of tracking and regulating. A radiotelescope pointing to a star is a regulator problem. If the same radiotelescope is used to track a satellite, then it is a tracking problem. As we shall see immediately, a system designed for tracking can always perform the regulating job as well.

The responses of regulating systems are due to nonzero initial states; hence the problem must be formulated in terms of dynamical equations. Consider the system shown in Figure 8-1. The plant is assumed to be described by the controllable and observable dynamical equation

$$\dot{\mathbf{x}} = \mathbf{A}\mathbf{x} + \mathbf{b}u \qquad (8\text{-}41\text{a})$$

$$y = \mathbf{c}\mathbf{x} \qquad (8\text{-}41\text{b})$$

where $\mathbf{A}$, $\mathbf{b}$, and $\mathbf{c}$ are $n \times n$, $n \times 1$, and $1 \times n$ real constant matrices. The problem is to design a control system so that for any nonzero initial state, the resulting system minimizes

$$\int_0^\infty [qy^2(t) + u^2(t)]\, dt \qquad (8\text{-}42)$$

where q is a positive constant. We see that Equation (8-39) reduces to (8-42) if y_d is set to zero. The solution of this problem is that if u is given by

$$u(t) = -\mathbf{b}'\mathbf{K}\mathbf{x}(t) \qquad (8\text{-}43)$$

where $\mathbf{K}$ is the $n \times n$ symmetric matrix with all positive eigenvalues satisfying

$$-\mathbf{K}\mathbf{A} - \mathbf{A}'\mathbf{K} + \mathbf{K}\mathbf{b}\mathbf{b}'\mathbf{K} - q\mathbf{c}'\mathbf{c} = \mathbf{0} \qquad (8\text{-}44)$$

then the resulting system will be the optimal system. The structure of the optimal system is shown in Figure 8-7. By comparing Figures 8-5 and 8-7, we see why an optimal tracking system will perform the regulating task optimally.

Although the regulator problem is formulated in dynamical equations, it can also be solved by using transfer functions. Consider the system shown in

8-5 ARE ALL STATE VARIABLES AVAILABLE FOR FEEDBACK?

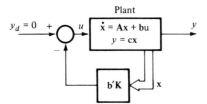

Figure 8-7
Optimal regulator.

Figure 8-7. The plant is assumed to be completely characterized by its transfer function

$$\hat{g}(s) = \frac{N(s)}{D(s)} \qquad (8\text{-}45)$$

where $N(s)$ and $D(s)$ have no common factors. If the transfer function from y_d to y is designed to be equal to

$$\hat{g}_f(s) = \frac{N(s)}{D_f(s)} \qquad (8\text{-}46)$$

where

$$D_f(s)D_f(-s) = D(s)D(-s) + qN(s)N(-s) \qquad (8\text{-}47)$$

then for any nonzero initial state of the plant, the overall system minimizes the performance criterion

$$\int_0^\infty [qy^2(t) + u^2(t)]\, dt \qquad (8\text{-}48)$$

This result can be derived from (8-42) through (8-44). Its derivation is beyond the scope of this text and is omitted. As with the tracking problem, for the single-variable plants it is easier to solve the regulator problem by using transfer functions.

8-5 Are All State Variables Available for Feedback?

We have discussed in the previous sections the determination of optimal control systems. Starting with this section, the implementations of these optimal control systems will be studied. As discussed in Section 7-4, closed-loop, or feedback, configurations of compensators are generally superior to open-loop configurations. Hence we use exclusively feedback compensations in the design of compensators.

As we see in (8-32) and (8-43), optimal actuating signals require feedback from all state variables. Hence we must discuss whether or not all state variables are available in feedback.

The state variables of a plant are not necessarily of the same type. They might be mechanical variables, electrical variables, temperature, pressure, or chemical ingredients. Hence if the actuating signal is a voltage, then transducers must be used to transform all state variables into a voltage. This however may sometimes be impossible, or very costly, as will be explained in the following example.

Consider the tracking antenna problem studied in Example 2 of Section 8-2. Its plant is shown in Figure 8-3. As discussed in Chapter 3, the state variables of the plant can be chosen as the motor shaft position θ_m and velocity $\dot{\theta}_m$, the armature current i_a, and the field circuit current i_f. If all state variables are used in feedback, then four voltages proportional, respectively, to these four state variables must be generated. As discussed in Chapter 3, the generation of voltages proportional to θ_m and $\dot{\theta}_m$ can be achieved by the use of a potentiometer and a tachometer. Hence their discussions will not be repeated here.

We discuss now the generation of a voltage proportional to the armature current i_a. By examining Figure 8-3, one may suggest that this can be easily achieved by branching out the voltage across R_g or R_a. This is not possible however because the circuit in Figure 8-3 is just a model; in reality R_g, or R_a, is not lumped as shown but rather distributed across the entire circuit. Hence a voltage proportional to i_a must be generated by some other method. We discuss, in the following, two possible methods of generating a voltage proportional to i_a. The first method is to insert a resistor R_1 between the generator and the motor, as shown in Figure 8-8. Then a voltage proportional to i_a can be generated. However this is achieved not without a price. If R_1 is very small, the voltage across R_1 may consist largely of noise. If R_1 is large, the voltage is more accurate, but a considerable amount of power is wasted in R_1, and the efficiency of the system is reduced. Clearly after the employment of R_1, the mathematical description of the plant has to be properly modified. Another possible way of generating a voltage proportional to i_a is as follows: Since L_a is usually very small, the voltage v_g across the motor terminals is equal to

$$v_g = i_a R_a + k_b \dot{\theta}_m \qquad (8\text{-}49)$$

Hence if a tachometer is used to generate a voltage proportional to $\dot{\theta}_m$, say $v = k\dot{\theta}_m$, and if R_a can be measured accurately, then a voltage proportional to i_a can be obtained as

$$i_a = \frac{1}{R_a}\left(v_g - \frac{k_b}{k} v\right) \qquad (8\text{-}50)$$

This voltage can then be used in feedback. This arrangement requires the employment of a tachometer, exact measurement of R_a, and the assumption that $L_a = 0$; hence its employment is not as easy as the method of inserting R_1. Note that if both L_a and L_g are neglected, then i_a is no longer a state variable, and the voltage v_g will be a linear combination of the state variables i_f and $\dot{\theta}_m$.

If the voltage $k_g i_f$ can be picked up from the generator, then it can be used directly in feedback. Unfortunately the voltage across the generator terminals is v_g, which is quite different from the voltage $k_g i_f$, and hence a different method of generating a voltage proportional to i_f is needed. This can be achieved by inserting a resistor R_2, as shown in Figure 8-8. Although the power consumption

8-5 ARE ALL STATE VARIABLES AVAILABLE FOR FEEDBACK? 201

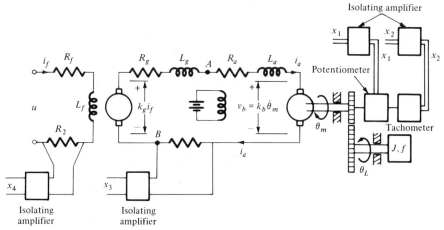

Figure 8-8 Generation of state variables.

of R_2 is comparatively small, because of its low power level, the effect of R_2 on the transfer function cannot be neglected. In using R_1 or R_2, its loading on other parts of the system should be avoided. This can be achieved by using isolating amplifiers, as shown in Figure 8-8.

From the foregoing discussion we see that it is possible, at least for the system in Figure 8-8, to introduce feedback from all state variables. It is certainly permissible to introduce feedback just from the output or just from part of the state variables. Hence in synthesizing an optimal system, there are many possible configurations of compensators: feedback from all state variables, feedback from part of the state variables, and feedback from just the output.

In order to compare these possibilities the degree of compensators must be discussed. Let $\hat{g}(s) = N(s)/D(s)$ be the transfer function of a plant, and let $\hat{g}_f(s) = kN(s)/D_f(s)$ be an overall system to be synthesized. Let the degree of $D(s)$ be n. As will be shown later, if feedback is introduced from only the output, then the degree of compensators required in synthesizing $\hat{g}_f(s)$ will generally be $(n - 1)$. However if feedback is introduced from all the state variables, then the degree of compensators required will be zero. That is, if all state variables are available, then $\hat{g}_f(s)$ can be synthesized by using constant gains alone. If some but not all of the state variables are used in feedback, then the degree of compensators required in synthesizing $\hat{g}_f(s)$ will be a number between zero and $(n - 1)$. Hence we conclude that the more state variables are used in feedback, the smaller the degree of compensators.

We summarize what we have discussed in this section. If the optimal overall transfer function of a given plant is found and is known to be synthesizable, then there are many different ways to synthesize it. If all state variables are used in feedback, the degree of compensators used will be the smallest. However in order to generate voltages proportional to state variables, transducers or other devices have to be used. This not only is costly but also may reduce the efficiency of the system. If only the output is used in feedback, the degree of compensators required will be the largest. However in this case, only one transducer—for example, a potentiometer—is needed to transform the output into a voltage signal. Since compensating networks are rather inexpensive compared with transducers, although the degree of compensators is larger in output feedback, it is not clear that the state feedback is definitely better than the output feedback. A similar remark also holds for feedback that uses part of the state variables. Hence in the design of compensators a designer has ample room for manipulation.

8-6 Design of Compensators: Output Feedback

If the design of optimal systems with respect to quadratic performance criteria is carried out by using transfer functions, then the solutions are in the form of optimal overall transfer functions. If the design is carried out by means of dynamical equations, then the solutions yield the optimal actuating signals, as shown in Figure 8-5. Clearly once an optimal actuating signal is found, the overall transfer function from y_d to y can be computed. In this section, no matter which method is used, the overall transfer function is assumed to be known.

Given a plant with a transfer function $\hat{g}(s)$ of degree n and a synthesizable $\hat{g}_f(s)$, if feedback is introduced from the output, then the degree of the compensator is generally of degree $(n - 1)$. Now if the degree of $\hat{g}_f(s)$ is less than $(2n - 1)$, then pole-zero cancellation, as discussed in Section 4-5, must occur inside the system. These canceled or missing poles cannot be neglected in the design.

Example 1

Given a plant with the transfer function $\hat{g}(s) = 20/s(s + 2)$. The overall transfer function that minimizes

$$\int_0^\infty \{[y(t) - 1]^2 + u^2\} \, dt$$

is found in (8-21) as $\hat{g}_f(s) = 20/(s^2 + 6.63s + 20)$. If the configuration of the compensator is chosen as shown in Figure 8-9, then the compensator $C(s)$ can be solved from

$$\hat{g}_f(s) = \frac{C(s)\hat{g}(s)}{1 + C(s)\hat{g}(s)}$$

as

$$C(s) = \frac{\hat{g}_f(s)}{\hat{g}(s)[1 - \hat{g}_f(s)]} = \frac{s + 2}{s + 6.63}$$

For this problem the canceled pole $(s + 2)$ is a stable pole. The pole will not cause any serious effect on the response of the system. Hence this configuration of compensator is acceptable in practice.

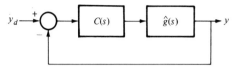

Figure 8-9
Compensator configuration.

Example 2

Consider a plant with the transfer function $1/(s + 1)(s - 2)$. Design an optimal system to minimize

$$J = \int_0^\infty \{100[y(t) - 1]^2 + u^2(t)\} \, dt \qquad (8\text{-}51)$$

First, we perform the following spectral factorization:

$$\begin{aligned} D(s)D(-s) + qN(s)N(-s) \\ = (s + 1)(s - 2)(-s + 1)(-s - 2) + 100 \\ = s^4 - 5s^2 + 104 = (s^2 + 5.04s + 10.2)(s^2 - 5.04s + 10.2) \\ \triangleq D_f(s)D_f(-s) \end{aligned}$$

Hence the optimal overall transfer function is, by using (8-15),

$$\hat{g}_f(s) = \frac{qN(0)}{D_f(0)} \cdot \frac{N(s)}{D_f(s)} = \frac{100}{10.2} \cdot \frac{1}{s^2 + 5.04s + 10.2} = \frac{9.8}{s^2 + 5.04s + 10.2}$$

Note that the steady-state error due to a step function of the system is not zero, and hence this optimal system has an infinite J in (8-51). For a discussion of this difficulty, see the remarks at the end of Section 8-2.

If the configuration of the compensator in Figure 8-9 is chosen for this system, then the compensator required can be computed as

$$C(s) = \frac{\hat{g}_f(s)}{\hat{g}(s)[1 - \hat{g}_f(s)]} = \frac{9.8(s + 1)(s - 2)}{s^2 + 5.04s + 10.2}$$

This compensator and the plant both have degree 2. Since the overall transfer function $\hat{g}_f(s)$ has degree 2, there are two poles missing in $\hat{g}_f(s)$. In other words, there are two pairs of pole-zero cancellations inside the system. The canceled poles are $(s + 1)$ and $(s - 2)$; one of them is an unstable pole. Hence we conclude that the configuration of the compensator chosen in Figure 8-9 is not suitable for this problem.

If the configuration of the compensator is chosen as shown in Figure 8-10, then it can be easily verified that $C_1(s)$ is equal to $(-0.898s^2 + 1.514s + 3.04)$, which is not a proper rational function. Hence Figure 8-10 is again not acceptable for this problem. ∎

One may start to wonder whether it is possible at all to synthesize the optimal system in Example 2 by using output feedback alone without using any state variables. The answer is affirmative if we choose the configuration of compensators shown in Figure 8-11. If this configuration is used, not only can the

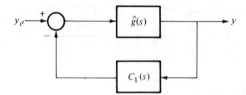

Figure 8-10
Compensator configuration.

canceled poles but also the poles of the compensators be arbitrarily chosen by the designer.

Given a plant with the transfer function $\hat{g}(s) = N(s)/D(s)$, we introduce in the following a procedure of computing $C_0(s)$ and $C_1(s)$ so that the resulting system has $N(s)/D_f(s)$ as its transfer function. We note that this part of the design is needed in the synthesis of optimal systems with respect to quadratic performance criteria. After the synthesis of $N(s)/D_f(s)$, if the block denoted as M in Figure 8-11 is 1, then the system in Figure 8-11 is an optimal regulating system; if M is $qN(0)/D_f(0)$, then it is an optimal system for a step command signal; if M is $q(k_1 + k_2 s)$, then it is an optimal system for a ramp command signal. Hence if we succeed in synthesizing $N(s)/D_f(s)$, then the design of optimal systems studied in this chapter can be completed.

Let
$$\hat{g}(s) = \frac{N(s)}{D(s)}$$

$$C_0(s) = \frac{N_0(s)}{D_c(s)}$$

and
$$C_1(s) = \frac{N_1(s)}{D_c(s)}$$

Note that the poles of $C_0(s)$ and $C_1(s)$ are chosen to be the same. The overall transfer function of the system in Figure 8-11 is, by Mason's formula,

$$\hat{g}_f(s) = \frac{\hat{g}(s)}{1 + C_0(s) + \hat{g}(s)C_1(s)} \quad (8\text{-}52)$$

$$= \frac{N(s)D_c(s)}{D(s)[D_c(s) + N_0(s)] + N(s)N_1(s)}$$

If $\hat{g}_f(s)$ is required to be $N(s)/D_f(s)$, then Equation (8-52) implies that

$$D_c(s)D_f(s) = D(s)[D_c(s) + N_0(s)] + N(s)N_1(s) \quad (8\text{-}53)$$

or

$$D_c(s)[D_f(s) - D(s)] = D(s)N_0(s) + N(s)N_1(s) \quad (8\text{-}54)$$

In this equation $N(s)$, $D(s)$, and $D_f(s)$ are known polynomials; $D_c(s)$, $N_0(s)$, and $N_1(s)$ are unknown polynomials to be solved. We see from Equations (8-52) and (8-53) that in this design the polynomial $D_c(s)$ is canceled from the

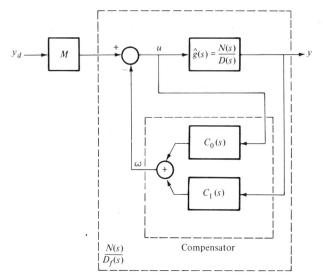

Figure 8-11 Compensator configuration.

denominator and numerator of (8-52). Hence the roots of $D_c(s)$ are the missing poles of $\hat{g}_f(s)$. As discussed earlier, these poles should be stable poles; otherwise the system can never function properly. Hence in (8-54), we like to have the freedom of choosing $D_c(s)$. Consequently the design problem becomes: Given $N(s)$, $D(s)$, $D_f(s)$, and an arbitrarily chosen $D_c(s)$, find $N_0(s)$ and $N_1(s)$ to satisfy Equation (8-54). Furthermore the degrees of $N_0(s)$ and $N_1(s)$ are required to be equal to or smaller than that of $D_c(s)$ in order for $C_0(s)$ and $C_1(s)$ to be proper. It is also required that the degree of $D_c(s)$ be as small as possible for economical reasons.

It turns out that if $N(s)$ and $D(s)$ have no common factor and if $D(s)$ has a degree n, then by choosing the degree of $D_c(s)$ to be $(n-1)$, it is always possible to solve $N_0(s)$ and $N_1(s)$, both of degree $(n-1)$ or less, to satisfy (8-54) for any $D_f(s)$ and any $D_c(s)$. This fact will be established in the following. Let

$$\hat{g}(s) = \frac{N(s)}{D(s)} = \frac{\beta_0 + \beta_1 s + \cdots + \beta_n s^n}{\alpha_0 + \alpha_1 s + \cdots + \alpha_n s^n} \qquad \alpha_n > 0 \qquad (8\text{-}55)$$

and let

$$\mathbf{C}(s) = [C_0(s) \quad C_1(s)] = \left[\frac{N_0(s)}{D_c(s)} \quad \frac{N_1(s)}{D_c(s)}\right] \qquad (8\text{-}56)$$

where

$$D_c(s) = \gamma_0 + \gamma_1(s) + \cdots + \gamma_{n-1} s^{n-1} \qquad (8\text{-}57)$$

$$N_0(s) = c_{00} + c_{01} s + \cdots + c_{0(n-1)} s^{n-1} \qquad (8\text{-}58)$$

$$N_1(s) = c_{10} + c_{11} s + \cdots + c_{1(n-1)} s^{n-1} \qquad (8\text{-}59)$$

If $D(s)$ is of degree n, then $D_f(s)$ will also be of degree n if the optimal design discussed in Sections 8-2 and 8-3 is used. After $D_f(s)$ and $D_c(s)$ are known [that is, $D_f(s)$ is computed from the spectral factorization and $D_c(s)$ is chosen by the designer], we may compute

$$D_c(s)[D_f(s) - D(s)] \triangleq d_0 + d_1 s + \cdots + d_{2n-1} s^{2n-1} \qquad (8\text{-}60)$$

Note that this is a polynomial of degree $(2n - 1)$ or less. Now using Equations (8-55), (8-58), (8-59), and (8-60) and equating the coefficients of the same power of s in (8-54), we obtain the following linear algebraic equation:

$$\mathbf{\Lambda c} \triangleq \begin{bmatrix} \alpha_0 & 0 & \cdots & 0 & \vdots & \beta_0 & 0 & \cdots & 0 \\ \alpha_1 & \alpha_0 & \cdots & 0 & \vdots & \beta_1 & \beta_0 & \cdots & 0 \\ \vdots & \ddots & & \vdots & \vdots & \vdots & & & \vdots \\ & & & \alpha_0 & \vdots & & & & \beta_0 \\ \alpha_n & \alpha_{n-1} & \cdots & \alpha_1 & \vdots & \beta_n & \beta_{n-1} & \cdots & \beta_1 \\ 0 & \alpha_n & \cdots & \alpha_2 & \vdots & 0 & \beta_n & \cdots & \beta_2 \\ \vdots & & & \vdots & \vdots & \vdots & & & \vdots \\ 0 & 0 & \cdots & \alpha_{n-1} & \vdots & 0 & 0 & \cdots & \beta_{n-1} \\ 0 & 0 & \cdots & \alpha_n & \vdots & 0 & 0 & \cdots & \beta_n \end{bmatrix} \begin{bmatrix} c_{00} \\ c_{01} \\ \vdots \\ c_{0(n-1)} \\ \hdashline c_{10} \\ c_{11} \\ \vdots \\ c_{1(n-1)} \end{bmatrix} = \begin{bmatrix} d_0 \\ d_1 \\ \vdots \\ \vdots \\ \vdots \\ \vdots \\ d_{(2n-2)} \\ d_{(2n-1)} \end{bmatrix} \triangleq \mathbf{d}$$

$\underbrace{\qquad\qquad}_{n \text{ columns}} \underbrace{\qquad\qquad}_{n \text{ columns}}$ $(8\text{-}61)$

We see that $\mathbf{\Lambda}$, as defined in (8-61), is a $2n \times 2n$ square matrix. It can be read out directly from the coefficients of $\hat{g}(s)$. In (8-61) $\mathbf{\Lambda}$ and $\mathbf{d}$ are known, and $\mathbf{c}$ is unknown. It is a well-known fact in linear algebra that, for any $\mathbf{d}$, there exists a vector $\mathbf{c}$ satisfying (8-61) if and only if $\mathbf{\Lambda}$ is a nonsingular matrix. It is proved in Reference [3] that if $N(s)$ and $D(s)$ have no common factor, then $\mathbf{\Lambda}$ is nonsingular. This establishes the assertion that for any $\hat{g}(s) = N(s)/D(s)$ of degree n, a compensator $\mathbf{C}(s) = [C_0(s) \; C_1(s)]$ of degree $(n - 1)$ and of arbitrarily chosen poles can be found to synthesize any $N(s)/D_f(s)$. We summarize the procedure for computing $\mathbf{C}(s)$ in the following. Let $\mathbf{C}(s)$ be of the form given in Equations (8-56) through (8-59). Choose the poles of $\mathbf{C}(s)$. Compute Equation (8-60), and then form the linear algebraic equation (8-61). The solution of the equation immediately yields the compensator.

Example 2 (continued)

Given the plant transfer function

$$\hat{g}(s) = \frac{1}{(s+1)(s-2)} = \frac{1}{-2 - s + s^2}$$

8-6 DESIGN OF COMPENSATORS: OUTPUT FEEDBACK

find $C_0(s)$ and $C_1(s)$ and M in Figure 8-11 so that the overall transfer function is equal to

$$\hat{g}_f(s) = 9.8 \times \frac{1}{s^2 + 5.04s + 10.2}$$

After assigning gain 9.8 to the block denoted as M, we need to synthesize $1/(s^2 + 5.04s + 10.2)$. Since the plant is of degree 2, the compensators can be chosen as

$$C(s) = \frac{1}{2 + s} [c_{00} + c_{01}s \mid c_{10} + c_{11}s] \qquad (8\text{-}62)$$

The pole $(s + 2)$ is chosen arbitrarily. Now we compute

$$D_c(s)[D_f(s) - D(s)] = (s + 2)(s^2 + 5.04s + 10.2 - s^2 + s + 2)$$
$$= 24.4 + 24.28s + 6.04s^2 \qquad (8\text{-}63)$$

From the coefficients of $\hat{g}(s)$ and (8-63), we form the equation

$$\begin{bmatrix} -2 & 0 & 1 & 0 \\ -1 & -2 & 0 & 1 \\ 1 & -1 & 0 & 0 \\ 0 & 1 & 0 & 0 \end{bmatrix} \begin{bmatrix} c_{00} \\ c_{01} \\ c_{10} \\ c_{11} \end{bmatrix} = \begin{bmatrix} 24.4 \\ 24.28 \\ 6.04 \\ 0 \end{bmatrix}$$

Its solution is $c_{00} = 6.04$, $c_{01} = 0$, $c_{10} = 36.48$, and $c_{11} = 30.32$. If $C(s)$ is expressed as

$$C(s) = \frac{1}{s + 2} [6.04 \mid 30.32s + 36.48] = [0 \mid 30.32] + \frac{1}{s + 2} [6.04 \mid -24.16]$$

a basic block diagram of $C(s)$ can be found as shown in Figure 8-12. It is obtained from the observable-form realization of $C(s)$. (See Section 5-5.) ∎

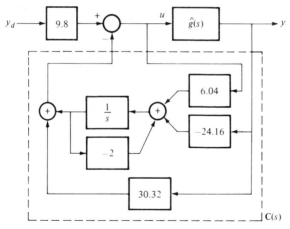

Figure 8-12 Complete design of the system in Example 2, Section 8-6.

Let us recapitulate what we have discussed in this section. In the design of compensators, by using output feedback, we may first try the configurations shown in Figures 8-9 and 8-10. In these configurations the compensators can be very easily computed. Since the design always involves pole-zero cancellation, the canceled poles must be examined. If the canceled poles are stable and not very close to the imaginary axis, then the configuration in Figure 8-9 or 8-10 can be used, and the design is completed. If the canceled poles are unstable or very close to the imaginary axis of the s-plane, then they will cause saturation or too much oscillation. In this case the configurations in Figure 8-9 and 8-10 cannot be used, and a different one must be chosen. If we use the configuration in Figure 8-11, then we have complete freedom in choosing the canceled poles. The compensators in this configuration can be obtained by solving a set of linear algebraic equations. The procedure can be easily programmed on a digital computer.

A remark is in order concerning the poles of the compensators in Figure 8-11. Clearly these poles should be chosen to be stable. It is however unnecessary for them to have very large negative real parts. Poles with large negative real parts might make the compensating network susceptible to saturation. As a rule, the poles of compensators may be chosen to be comparable with the poles of their overall transfer functions.

8-7 Design of Compensators: State Feedback

The design of optimal systems can be carried out by using either dynamical equations or transfer functions. The solution in the state-variable approach is in the form of state feedback, as given in (8-32). Consequently if all state variables are available for feedback, the optimal system can be easily implemented as shown in Figure 8-6. Hence in this section we discuss only the transfer function approach. To be more specific, we study the synthesis of $N(s)/D_f(s)$ from the plant transfer function $\hat{g}(s) = N(s)/D(s)$ by using feedback from all the state variables of the plant. This problem is often referred to as *pole placement by state feedback*, for it uses state-variable feedback to force the plant to have a desired $D_f(s)$ or, equivalently, a set of desired poles. Before presenting a general design procedure, we use an example to illustrate the underlying idea.

Example 1

Consider a plant with transfer function $20/s(s + 2)$. This is a simplified model of an armature-controlled dc motor (by assuming $L_a = 0$). The state variables are θ_m, the motor shaft position, and $\dot{\theta}_m$, the motor shaft angular velocity. These two state variables can be generated by using a potentiometer and a tachometer in the feedback path, as shown in Figure 8-13. We first show that, by adjusting k_1 and k_2, the poles of the overall system can be arbitrarily assigned. Indeed, the overall transfer function is, by applying Mason's formula,

$$\hat{g}_f(s) = \frac{\dfrac{20}{s(s+2)}}{1 + \dfrac{20k_2}{(s+2)} + \dfrac{20k_1}{s(s+2)}} = \frac{20}{s^2 + (2 + 20k_2)s + 20k_1} \quad (8\text{-}64)$$

Clearly by the adjustment of k_1 and k_2, the poles of $\hat{g}_f(s)$ can be arbitrarily chosen. For example, if it is desired to have $\hat{g}_f(s) = 20/(s^2 + 6.63s + 20)$, k_1 and k_2 can be chosen as $k_1 = 1$, $k_2 = 0.2315$. ∎

In this example, the pole placement is achieved by using constant gains in the feedback; no dynamics are introduced. We explain why this is possible. For a system of degree n, there are n state variables. In state feedback we have n parameters to be adjusted; hence we shall be able to control the n poles of the system without introducing additional dynamics. In constant-gain state feedback, since the degree of compensators is zero, if the degree of an overall transfer function is equal to that of a plant, then there is no pole-zero cancellation problem in the design.

In Example 1 we were able to obtain the feedback gains by simple manipulation; this approach may be difficult for more complicated systems. Hence a systematic design procedure is desirable.

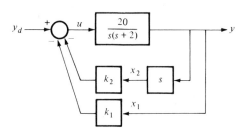

Figure 8-13 State feedback.

Problem

Given a plant with input u, output y, and state $\mathbf{x}$. The plant is assumed to be completely characterized by the irreducible proper rational function

$$\hat{g}(s) = \frac{\hat{y}(s)}{\hat{u}(s)} = \frac{N(s)}{D(s)} = \frac{N(s)}{\alpha_0 + \alpha_1 s + \alpha_2 s^2 + \cdots + \alpha_n s^n} \qquad \alpha_n > 0 \quad (8\text{-}65)$$

Find the feedback gains so that the resulting system has

$$\frac{N(s)}{\bar{\alpha}_0 + \bar{\alpha}_1 s + \bar{\alpha}_2 s^2 + \cdots + \bar{\alpha}_n s^n} \quad (8\text{-}66)$$

with $\bar{\alpha}_n = \alpha_n$ as its transfer function.

Algorithm

1. Find the transfer-function matrix from u to $\mathbf{x}$; say,

$$\begin{bmatrix} \hat{x}_1(s)/\hat{u}(s) \\ \hat{x}_2(s)/\hat{u}(s) \\ \vdots \\ \hat{x}_n(s)/\hat{u}(s) \end{bmatrix} = \begin{bmatrix} M_1'(s)/D_1(s) \\ M_2'(s)/D_2(s) \\ \vdots \\ M_n'(s)/D_n(s) \end{bmatrix} = \frac{1}{D(s)} \begin{bmatrix} M_1(s) \\ M_2(s) \\ \vdots \\ M_n(s) \end{bmatrix}$$

$$= \frac{1}{D(s)} \begin{bmatrix} \beta_{10} + \beta_{11}s + \cdots + \beta_{1(n-1)}s^{n-1} \\ \beta_{20} + \beta_{21}s + \cdots + \beta_{2(n-1)}s^{n-1} \\ \vdots \\ \beta_{n0} + \beta_{n1}s + \cdots + \beta_{n(n-1)}s^{n-1} \end{bmatrix} \tag{8-67}$$

Note that it can be shown that $D(s)$ is the least common multiplier of $\{D_1(s), D_2(s), \ldots, D_n(s)\}$. Since the transfer function from u to $\mathbf{x}$ is strictly proper, the numerator $M_i(s)$ is of degree $(n-1)$ or less.

2. Form the algebraic equation

$$\Lambda \mathbf{k} = \begin{bmatrix} \beta_{10} & \beta_{20} & \cdots & \beta_{n0} \\ \beta_{11} & \beta_{21} & \cdots & \beta_{n1} \\ \vdots & \vdots & & \vdots \\ \beta_{1(n-1)} & \beta_{2(n-1)} & \cdots & \beta_{n(n-1)} \end{bmatrix} \begin{bmatrix} k_1 \\ k_2 \\ \vdots \\ k_n \end{bmatrix} = \begin{bmatrix} \bar{\alpha}_0 - \alpha_0 \\ \bar{\alpha}_1 - \alpha_1 \\ \vdots \\ \bar{\alpha}_{n-1} - \alpha_{n-1} \end{bmatrix} \tag{8-68}$$

Note that Λ is an $n \times n$ matrix. It can be deduced from Theorem 5-7 of Reference [3] that Λ is nonsingular. Hence for any α_i and $\bar{\alpha}_i$, a solution $\mathbf{k}$ exists in Equation (8-68). The solution k_i is the gain associated with the ith state variable x_i. ∎

Note that this algorithm is applicable only if the numerator of (8-66) is equal to that of (8-65). Hence in the synthesis of $\hat{g}_f(s) = N_f(s)/D_f(s)$, if $N_f(s)$ is different from $N(s)$, the numerator of the plant, then $\hat{g}_f(s)$ must be arranged into the form $\hat{g}_f(s) = MN(s)/D_f(s)$, and then the algorithm can be applied.

Example 2

Consider again the problem in Example 1. The transfer function from u to $\mathbf{x}$ can be obtained from Figure 8-13 as

$$\begin{bmatrix} \hat{x}_1(s)/\hat{u}(s) \\ \hat{x}_2(s)/\hat{u}(s) \end{bmatrix} = \begin{bmatrix} \dfrac{20}{s(s+2)} \\ \dfrac{20}{s+2} \end{bmatrix} = \frac{1}{s^2 + 2s} \begin{bmatrix} 20 \\ 0 + 20s \end{bmatrix}$$

Since $\hat{g}(s) = 20/(s^2 + 2s)$, $\hat{g}_f(s) = 20/(s^2 + 6.63s + 20)$, we have

$$\begin{bmatrix} 20 & 0 \\ 0 & 20 \end{bmatrix} \begin{bmatrix} k_1 \\ k_2 \end{bmatrix} = \begin{bmatrix} 20 - 0 \\ 6.63 - 2 \end{bmatrix}$$

Its solution is $k_1 = 1$, $k_2 = 0.2315$.

Example 3

Consider the tracking antenna problem studied in Example 2 of Section 8-2. The plant of the system is shown in Figure 8-3. As discussed in Section 8-5, in order to introduce feedback from all the state variables, the plant has to be modified to the one in Figure 8-8. It is assumed that the moments of inertia of the potentiometer and tachometer are, compared with the load, negligible. The resistors R_1 and R_2 in Figure 8-8 are chosen to be 1 ohm. Because of the loading of R_1 and R_2, the transfer function computed in (8-23) cannot be used. It should be modified to

$$\hat{g}(s) = \frac{\hat{\theta}_L(s)}{\hat{u}(s)} = \frac{1}{2} \cdot \frac{100}{5s + 21} \cdot \frac{1.2}{s(0.4s + 0.01)(19 + 0.1s) + 1.44s}$$

$$= \frac{300}{s^4 + 194.2s^3 + 838.9s^2 + 171s} \triangleq \frac{N(s)}{D(s)} \qquad (8\text{-}69)$$

This is obtained by adding R_2 to R_f and adding R_1 to $(R_g + R_a)$; see Figure 8-3(b). Because of the modification of $\hat{g}(s)$, the optimal overall transfer function computed in (8-29) cannot be used. Hence we have to compute the new optimal transfer function that minimizes

$$\int_0^\infty \{25[y(t) - 1]^2 + u^2\}\, dt$$

We first perform the spectral factorization

$$D(s)D(-s) + qN(s)N(-s) = s^8 - 36{,}035.84s^6 + 637{,}336.81s^4 - 29{,}241s^2$$
$$+ (25 \times 90{,}000) = D_f(s)D_f(-s)$$
$$D_f(s) = s^4 + 195.9s^3 + 1169s^2 + 1865s + 1500$$

This is obtained by iteration, as in Example 2 of Section 8-2. Hence the optimal overall transfer function is, by using (8-15),

$$\hat{g}_f(s) = 5 \cdot \frac{300}{s^4 + 195.9s^3 + 1169s^2 + 1865s + 1500} = 5 \cdot \frac{N(s)}{D_f(s)} \qquad (8\text{-}70)$$

We discuss now the implementation of $\hat{g}_f(s)$ by using state feedback. A schematic diagram of the overall system is shown in Figure 8-14. In order to determine the feedback gains k_i, for $i = 1, 2, 3, 4$, we first compute the transfer function matrix from u to $\mathbf{x}$. Recall that the state variables are $x_1 = \theta_m$, $x_2 = \dot{\theta}_m$, $x_3 = i_a$, and $x_4 = i_f$. The transfer functions from u to x_i, $i = 1, 2,$ and 4, can be easily computed as

$$\frac{\hat{x}_4(s)}{\hat{u}(s)} = \frac{1}{5s + 21}$$

$$\frac{\hat{x}_1(s)}{\hat{u}(s)} = \frac{N_L}{N_m} \hat{g}(s) = \frac{600}{s^4 + 194.2s^3 + 838.9s^2 + 171s}$$

$$\frac{\hat{x}_2(s)}{\hat{u}(s)} = s \cdot \frac{\hat{x}_1(s)}{\hat{u}(s)} = \frac{600}{s^3 + 194.2s^2 + 838.9s + 171}$$

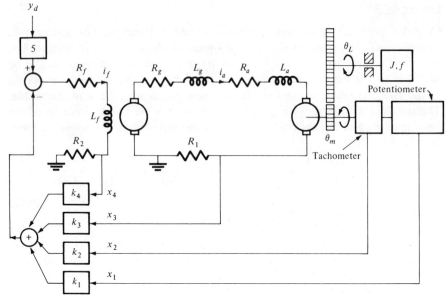

Figure 8-14 An implementation of $\hat{g}_f(s)$ by state feedback.

Using $\hat{T}(s) = k_t \hat{i}_a(s) = (J_{m_{eq}} s^2 + f_{m_{eq}} s)\hat{\theta}_m(s)$, and expressing $\hat{\theta}_m(s)$ in terms of $\hat{u}(s)$ from Figure 8-3(b), the transfer function from u to x_3 can be computed as

$$\frac{\hat{x}_3(s)}{\hat{u}(s)} = \frac{\hat{i}_a(s)}{\hat{u}(s)}$$

$$= \frac{k_g(J_{m_{eq}} s + f_{m_{eq}})}{(L_f s + R_f + 1)[(J_{m_{eq}} s + f_{m_{eq}})(R_a + R_q + 1 + L_a s + L_g s) + k_t k_b]}$$

$$= \frac{100}{5s + 21} \cdot \frac{0.4s + 0.01}{(0.4s + 0.01)(19 + 0.1s) + 1.44}$$

$$= \frac{200s + 5}{s^3 + 194.2s^2 + 838.9s + 171}$$

These transfer functions can be arranged as

$$\begin{bmatrix} \hat{x}_1(s)/\hat{u}(s) \\ \hat{x}_2(s)/\hat{u}(s) \\ \hat{x}_3(s)/\hat{u}(s) \\ \hat{x}_4(s)/\hat{u}(s) \end{bmatrix} = \frac{1}{s(s^3 + 194.2s^2 + 838.9s + 171)} \begin{bmatrix} 600 \\ 600s \\ 200s^2 + 5s \\ 0.2s^3 + 38s^2 + 8.15s \end{bmatrix}$$

(8-71)

From the coefficients of $D(s)$, $D_f(s)$, and (8-71), we form the equation

$$\begin{bmatrix} 600 & 0 & 0 & 0 \\ 0 & 600 & 5 & 8.15 \\ 0 & 0 & 200 & 38 \\ 0 & 0 & 0 & 0.2 \end{bmatrix} \begin{bmatrix} k_1 \\ k_2 \\ k_3 \\ k_4 \end{bmatrix} = \begin{bmatrix} 1500 - 0 \\ 1865 - 171 \\ 1169 - 838.9 \\ 195.9 - 194.2 \end{bmatrix}$$

Its solution is $k_1 = 2.5$, $k_2 = 2.7$, $k_3 = 0.035$, and $k_4 = 8.5$. ∎

The computation of state feedback gains can also be carried out by using dynamical equations. In this case it often requires the employment of equivalence transformations; the interested reader is referred to Reference [3].

A remark is in order concerning the plant in Example 2 of this section. As mentioned in Section 3-3, the inductance of the armature circuit is often very small and can be neglected in the design. If this is done, the transfer function of the plant reduces to degree 3, and the design can be considerably simplified.

8-8 Design of Compensators: Partial State Feedback*

In this section the design of compensators by using only part of the state variables in feedback will be discussed. We present first the problem and the algorithm.

Problem

Given a plant with the irreducible proper transfer function

$$\hat{g}(s) = \frac{N(s)}{D(s)} = \frac{N(s)}{\alpha_0 + \alpha_1 s + \cdots + \alpha_n s^n} \qquad \alpha_n > 0 \qquad (8\text{-}72)$$

It is assumed that feedback is introduced from $z_1, z_2, \ldots, z_p$, where $p < n$, and the z_i's are the state variables or the output of the plant. It is also assumed that the configuration of compensators is chosen as shown in Figure 8-15. Find the compensators $C_i(s)$, for $i = 0, 1, \ldots, p$, so that the resulting system has

$$\frac{N(s)}{D_f(s)} = \frac{N(s)}{\bar{\alpha}_0 + \bar{\alpha}_1 s + \cdots + \bar{\alpha}_n s^n} \qquad (8\text{-}73)$$

as its transfer function.

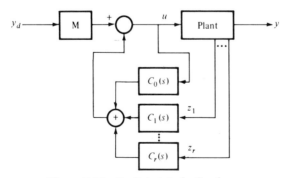

Figure 8-15 Partial state feedback.

---

* May be omitted in the first reading.

Algorithm

1. Find the transfer-function matrix from u to $z_1, z_2, \ldots, z_p$, say

$$\begin{bmatrix} \hat{z}_1(s)/\hat{u}(s) \\ \hat{z}_2(s)/\hat{u}(s) \\ \vdots \\ \hat{z}_p(s)/\hat{u}(s) \end{bmatrix} = \begin{bmatrix} M_1'(s)/D_1(s) \\ M_2'(s)/D_2(s) \\ \vdots \\ M_p'(s)/D_p(s) \end{bmatrix} = \frac{1}{D(s)} \begin{bmatrix} M_1(s) \\ M_2(s) \\ \vdots \\ M_p(s) \end{bmatrix}$$

$$= \frac{1}{D(s)} \begin{bmatrix} \beta_{10} + \beta_{11}s + \cdots + \beta_{1n}s^n \\ \beta_{20} + \beta_{21}s + \cdots + \beta_{2n}s^n \\ \vdots \\ \beta_{p0} + \beta_{p1}s + \cdots + \beta_{pn}s^n \end{bmatrix} \quad (8\text{-}74)$$

2. Form the $(n + m + 1) \times (q + 1)(m + 1)$ matrix of rank $(n + m + 1)$:

$$\bar{\Lambda}(m) = \begin{bmatrix} \alpha_0 & 0 & \cdots & 0 & \vline & \beta_{10} & 0 & \cdots & 0 & \vline & \cdots \\ \alpha_1 & \alpha_0 & \cdots & 0 & \vline & \beta_{11} & \beta_{10} & \cdots & 0 & \vline & \\ \vdots & \vdots & & \vdots & \vline & \vdots & \vdots & & \vdots & \vline & \\ \alpha_n & \alpha_{n-1} & \cdots & \alpha_0 & \vline & \beta_{1n} & \beta_{1(n-1)} & \cdots & \beta_{10} & \vline & \\ 0 & \alpha_n & \cdots & \cdot & \vline & 0 & \beta_{1n} & \cdots & \cdot & \vline & \\ 0 & 0 & \cdots & \cdot & \vline & 0 & 0 & \cdots & \cdot & \vline & \\ \vdots & \vdots & & \vdots & \vline & \vdots & \vdots & & \vdots & \vline & \\ 0 & 0 & \cdots & \alpha_{n-1} & \vline & 0 & 0 & \cdots & \beta_{1(n-1)} & \vline & \\ 0 & 0 & \cdots & \alpha_n & \vline & 0 & 0 & \cdots & \beta_{1n} & \vline & \end{bmatrix}$$

$\underbrace{}_{m+1 \text{ columns}}$ $\underbrace{}_{m+1 \text{ columns}}$

with upper arrows: $2p \downarrow$, $p \downarrow$, $(p+2) \downarrow$, $2 \downarrow$

$$\begin{bmatrix} \beta_{p0} & 0 & \cdots & 0 \\ \beta_{p1} & \beta_{p0} & \cdots & 0 \\ \vdots & \vdots & & \vdots \\ \beta_{pn} & \beta_{p(n-1)} & \cdots & \beta_{p0} \\ 0 & \beta_{pn} & \cdots & \cdot \\ 0 & 0 & \cdots & \cdot \\ \vdots & \vdots & & \vdots \\ 0 & 0 & \cdots & \beta_{p(n-1)} \\ 0 & 0 & \cdots & \beta_{pn} \end{bmatrix} \quad n + m + 1 \text{ rows}$$

with upper arrows: $(p+1) \downarrow$, $1 \downarrow$

$\underbrace{}_{m+1 \text{ columns}}$

(8-75)

8-8 DESIGN OF COMPENSATORS: PARTIAL STATE FEEDBACK

It can be shown that the integer m is in the range

$$(n - p)/p \leq m \leq n - p \tag{8-76}$$

The formation of $\bar{\Lambda}(m)$ can be achieved constructively by starting from the least integer larger than $(n - p)/p$ and then increasing m repeatedly by one until $\bar{\Lambda}(m)$ is of rank $n + m + 1$.

3. Reduce $\bar{\Lambda}(m)$ to a nonsingular square matrix of order $(n + m + 1)$ by deleting the linearly dependent columns of $\bar{\Lambda}(m)$ in the order as indicated on the top of $\bar{\Lambda}(m)$. Let the reduced matrix be

$$\Lambda \triangleq \begin{bmatrix}
\alpha_0 & \cdots & 0 & \beta_{10} & \cdots & 0 & & \beta_{p0} & \cdots & 0 \\
\alpha_1 & & \vdots & \beta_{11} & & \vdots & & \beta_{p1} & \cdots & \vdots \\
\vdots & & 0 & \vdots & & \beta_{10} & & \vdots & & \beta_{p0} \\
\alpha_n & & \alpha_0 & \beta_{1n} & & \beta_{11} & \cdots & \beta_{pn} & & \beta_{p1} \\
0 & & \alpha_1 & 0 & & \vdots & & 0 & & \vdots \\
\vdots & & \vdots & \vdots & & \beta_{1n} & & \vdots & & \beta_{pn} \\
0 & \cdots & \alpha_n & 0 & \cdots & 0 & & 0 & \cdots & 0
\end{bmatrix} \begin{array}{c} \\ \\ \\ n + m + 1 \text{ rows} \\ \\ \\ \end{array}$$

$$\underbrace{}_{\substack{m + 1 \\ \text{columns}}} \underbrace{}_{\substack{m_1 + 1 \\ \text{columns}}} \underbrace{}_{\substack{m_p + 1 \\ \text{columns}}} \tag{8-77}$$

where $m_i \leq m$, for $i = 1, 2, \ldots, p$, and $(m_1 + 1) + \cdots + (m_p + 1) = n$.

4. Let

$$\mathbf{C}(s) = \begin{bmatrix} \dfrac{N_0(s)}{D_c(s)} & \dfrac{N_1(s)}{D_c(s)} & \cdots & \dfrac{N_p(s)}{D_c(s)} \end{bmatrix} \tag{8-78}$$

where

$$D_c(s) = \gamma_0 + \gamma_1 s + \cdots + \gamma_m s^m \tag{8-79}$$

$$N_i(s) = c_{i0} + c_{i1} s + \cdots + c_{im_i} s^{m_i} \qquad i = 0, 1, 2, \ldots, p \tag{8-80}$$

and $m_0 \triangleq m$.

5. Choose the poles of $\mathbf{C}(s)$ or, equivalently, the roots of $D_c(s)$, and then compute

$$D_c(s)[D_f(s) - D(s)] \stackrel{\text{(say)}}{=} d_0 + d_1 s + \cdots + d_{n+m} s^{n+m} \tag{8-81}$$

Note that the degree of (8-81) is, at most, $n + m$.

6. Solve the linear algebraic equation

$$\Lambda \begin{bmatrix} c_{00} \\ c_{01} \\ \vdots \\ c_{0m} \\ \cdots \\ c_{10} \\ \vdots \\ c_{1m_1} \\ \cdots \\ \vdots \\ \cdots \\ c_{p0} \\ \vdots \\ c_{pm_p} \end{bmatrix} = \begin{bmatrix} d_0 \\ d_1 \\ \vdots \\ \cdots \\ \vdots \\ \vdots \\ \vdots \\ d_{n+m} \end{bmatrix} \qquad (8\text{-}82)$$

The solution gives immediately the compensator in Equation (8-78). ∎

Example 1

Design compensators for the control system studied in Example 2 of Section 8-2 by using potentiometer and tachometer feedbacks from the motor shaft. It is assumed that the moments of inertia of the potentiometer and tachometer used can be neglected.

Because the moments of inertia of the transducers are negligible, the plant transfer function computed in (8-23) need not be modified, as we did in Example 2 of Section 8-7. Hence the problem is: Given

$$\hat{g}(s) = \frac{300}{s^4 + 184s^3 + 760.5s^2 + 162s}$$

find compensators so that the resulting system has

$$\hat{g}_f(s) = 5 \cdot \frac{300}{s^4 + 185.8s^3 + 1088.4s^2 + 1800s + 1500}$$

as its transfer function.

The configuration of compensators is chosen as shown in Figure 8-16. It is clear that

$$\begin{bmatrix} \hat{z}_1(s)/\hat{u}(s) \\ \hat{z}_2(s)/\hat{u}(s) \end{bmatrix} = \frac{1}{s^4 + 184s^3 + 760.5s^2 + 162s} \begin{bmatrix} 600 \\ 600s \end{bmatrix}$$

For this problem, the m in (8-76) is bounded by

$$\frac{4-2}{2} \le m \le 4 - 2$$

or

$$1 \le m \le 2$$

8-8 DESIGN OF COMPENSATORS: PARTIAL STATE FEEDBACK

We form the matrix $\Lambda(1)$ as follows:

$$\bar{\Lambda}(1) = \begin{bmatrix} 0 & 0 & \vdots & 600 & 0 & \vdots & 0 & 0 \\ 162 & 0 & \vdots & 0 & 600 & \vdots & 600 & 0 \\ 760.5 & 162 & \vdots & 0 & 0 & \vdots & 0 & 600 \\ 184 & 760.5 & \vdots & 0 & 0 & \vdots & 0 & 0 \\ 1 & 184 & \vdots & 0 & 0 & \vdots & 0 & 0 \\ 0 & 1 & \vdots & 0 & 0 & \vdots & 0 & 0 \end{bmatrix}$$

The matrix $\bar{\Lambda}(1)$ has rank 5 that is smaller than $(n + m + 1) = 6$. Hence we proceed to $\bar{\Lambda}(2)$:

$$\bar{\Lambda}(2) = \begin{bmatrix} & & & 6 & 4 & 2 & & 5 & 3 & 1 \\ & & & \downarrow & \downarrow & \downarrow & & \downarrow & \downarrow & \downarrow \\ 0 & 0 & 0 & 600 & 0 & 0 & & 0 & 0 & 0 \\ 162 & 0 & 0 & 0 & 600 & 0 & & 600 & 0 & 0 \\ 760.5 & 162 & 0 & 0 & 0 & 600 & & 0 & 600 & 0 \\ 184 & 760.5 & 162 & 0 & 0 & 0 & & 0 & 0 & 600 \\ 1 & 184 & 760.5 & 0 & 0 & 0 & & 0 & 0 & 0 \\ 0 & 1 & 184 & 0 & 0 & 0 & & 0 & 0 & 0 \\ 0 & 0 & 1 & 0 & 0 & 0 & & 0 & 0 & 0 \end{bmatrix}$$

which is of rank $n + m + 1 = 7$. Since $\bar{\Lambda}(2)$ is a 7×9 matrix, there are two linearly dependent columns in $\bar{\Lambda}(2)$. We delete them in the order indicated on the top of $\bar{\Lambda}(2)$. Column 2 depends on column 3, and hence it should be deleted. After deleting column 2, the next linearly dependent column is column 4; hence we obtain

$$\Lambda = \begin{bmatrix} 0 & 0 & 0 & \vdots & 600 & \vdots & 0 & 0 & 0 \\ 162 & 0 & 0 & \vdots & 0 & \vdots & 600 & 0 & 0 \\ 760.5 & 162 & 0 & \vdots & 0 & \vdots & 0 & 600 & 0 \\ 184 & 760.5 & 162 & \vdots & 0 & \vdots & 0 & 0 & 600 \\ 1 & 184 & 760.5 & \vdots & 0 & \vdots & 0 & 0 & 0 \\ 0 & 1 & 184 & \vdots & 0 & \vdots & 0 & 0 & 0 \\ 0 & 0 & 1 & \vdots & 0 & \vdots & 0 & 0 & 0 \end{bmatrix} \quad \text{(8-83)}$$

$$\underbrace{\qquad\qquad\qquad}_{m+1}\ \underbrace{\quad}_{m_1+1}\ \underbrace{\qquad\qquad}_{m_2+1}$$

Clearly we have $m = 2$, $m_1 = 0$, and $m_2 = 2$. Hence the compensator may be chosen as

$$C(s) = \frac{1}{s^2 + \gamma_1 s + \gamma_0} [c_{00} + c_{01}s + c_{02}s^2 \;\vdots\; c_{10} \;\vdots\; c_{20} + c_{21}s + c_{22}s^2]$$

Rather arbitrarily, $(s^2 + \gamma_1 s + \gamma_0)$ is chosen as $(s^2 + 4s + 4)$. We compute

$$D_c(s)[D_f(s) - D(s)] = (s^2 + 4s + 4)(1.8s^3 + 327.9s^2 + 1638s + 1500)$$
$$= 1.8s^5 + 335.1s^4 + 2956.8s^3 + 9363.6s^2 + 12{,}552s + 6000$$

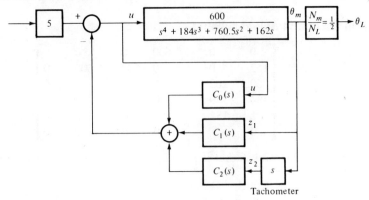

Figure 8-16 Partial state feedback.

and form

$$\begin{bmatrix} 0 & 0 & 0 & | & 600 & | & 0 & 0 & 0 \\ 162 & 0 & 0 & | & 0 & | & 600 & 0 & 0 \\ 760.5 & 162 & 0 & | & 0 & | & 0 & 600 & 0 \\ 184 & 760.5 & 162 & | & 0 & | & 0 & 0 & 600 \\ 1 & 184 & 760.5 & | & 0 & | & 0 & 0 & 0 \\ 0 & 1 & 184 & | & 0 & | & 0 & 0 & 0 \\ 0 & 0 & 1 & | & 0 & | & 0 & 0 & 0 \end{bmatrix} \begin{bmatrix} c_{00} \\ c_{01} \\ c_{02} \\ c_{10} \\ c_{20} \\ c_{21} \\ c_{22} \end{bmatrix} = \begin{bmatrix} 6000 \\ 12{,}552 \\ 9363.6 \\ 2956.8 \\ 335.1 \\ 1.8 \\ 0 \end{bmatrix}$$

Its solution is $c_{00} = 3.9$, $c_{01} = 1.8$, $c_{02} = 0$, $c_{10} = 10$, $c_{20} = 19.87$, $c_{21} = 10.18$, and $c_{22} = 1.45$. Hence the compensator is

$$C(s) = \frac{1}{s^2 + 4s + 4} [1.8s + 3.9 \ \vdots \ 10 \ \vdots \ 1.45s^2 + 10.18s + 19.87]$$

By using the observable form realization of $C(s)$, an operational amplifier circuit can be built for $C(s)$. (See Section 5-5.) ∎

We give now some remarks concerning the algorithm.

Remarks

1. The configuration of compensators chosen in Figure 8-15 is only one of many possible configurations. The chosen configuration however has the distinct property that the designer has the freedom of choosing the poles of the

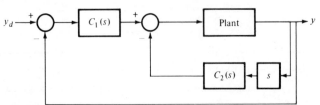

Figure 8-17 A different configuration of compensators.

compensators that turn out to be the canceled poles. Not every other configuration, such as the one in Figure 8-17, has this kind of property.

2. If $p = 1$, then (8-76) becomes $m = n - 1$, and the algorithm developed in this section reduces to the one in Section 8-6. If $p = n$, then (8-76) becomes $m = 0$. In this case, if $\beta_{in} = 0$ and $\alpha_n = \bar{\alpha}_n$, then $C_0(s) \equiv 0$, and the algorithm in this section reduces to the one in Section 8-7.

3. The derivation of this algorithm is similar to the one in Section 8-6. The existence of the compensators is proved in Reference [5].

8-9 Remarks and Review Questions

The design of optimal systems with respect to quadratic performance criteria consists of two independent parts: the search of optimal overall systems and the design of compensators. The search of optimal systems can be carried out by using either transfer functions or state variable equations. For single-variable systems, the transfer function approach is simpler; for multivariable systems, the state variable approach is simpler. The state variable approach can also be extended to time-varying and nonlinear systems.

Given a plant $\hat{g}(s)$ with input $u(t)$ and output $y(t)$, what is the optimal transfer function that minimizes

$$\int_0^\infty [y(t) - y_d(t)]^2 \, dt$$

where $y_d(t)$ is a desired signal? Is it always possible to design such an optimal system in practice?

What is the reason for including $u^2(t)$ in the performance criterion (8-4)?

In replacing constraint (8-3b) by (8-4), one may obtain the impression that if

$$\int_0^\infty u_1^2(t) \, dt < \int_0^\infty u_2^2(t) \, dt$$

then

$$\max_t |u_1(t)| < \max_t |u_2(t)|$$

Is this impression necessarily correct?

Why should the weighting factor q in (8-4) be positive? Is there any systematic way of choosing q?

How many configurations of compensators can you use in a design?

The configuration of compensators shown in Figure 8-9 is often called *series compensation*. The configuration of the type shown in Figure 8-17 is often called *feedback compensation*. These two compensations are most often used in control texts. The configuration shown in Figure 8-11 appeared only after the introduction of the so-called state observer in the late 1960s. The major advantage of the configuration in Figure 8-11 is that the poles of the compensators and the canceled poles can be arbitrarily chosen. This is possible because of the introduction of additional compensation from u.

What are the degrees of compensators needed in state feedback, partial state feedback, and output feedback?

Is it possible to state which configuration of compensators is the best to use in the design of a control system?

What is the procedure for computing feedback gains in the state feedback?

The quadratic performance criterion is chosen mainly for mathematical convenience, although it also yields practically acceptable systems. In employing the criterion, the choice of the weighting factor q in (8-4) has to be chosen by trial and error. From these two drawbacks and the fact that the linear time-invariant equation descriptions of physical systems are often obtained by approximation, one may wonder at the practical significance of the term "optimal." Indeed, we shall view the optimal design as an approach to the design, rather than just striving for the optimality. As we will see later, this approach may be the most systematic and the most satisfactory method in the design of control systems.

The design procedure developed in this chapter can be programmed on a digital computer. In developing such a program, the existing subroutines, such as the ones of checking linear independence of vectors and of solving linear algebraic equations in IBM Scientific Subroutine Package, should be utilized. The input of the program may be the transfer function of a plant, the performance criterion, and the configuration of compensators; its output is the required compensators.

In addition to the quadratic performance criterion, the criterion

$$\int_0^\infty t|y(t) - y_d(t)| \, dt$$

is also studied in the literature. This criterion is called the *integral of time multiplied by the absolute error*, abbreviated as ITAE. This criterion has a better selectivity than the quadratic criterion. Unfortunately this criterion cannot be studied analytically, though some limited results are obtained by using computer simulations. There are some other performance criteria, but no analytical results are available. See Reference [6].

Every optimal overall system is BIBO stable, because its denominator, $D_f(s)$, is chosen as a Hurwitz polynomial in the spectral factorization. In the design of compensators, if care is taken to ensure that the missing poles are stable, then the optimal overall system will be totally stable.

References

[1] Athans, M., and P. L. Falb, *Optimal Control.* New York: McGraw-Hill, 1966.
[2] Chang, S. S. L., *Synthesis of Optimal Control Systems.* New York: McGraw-Hill, 1961.
[3] Chen, C. T., *Introduction to Linear System Theory.* New York: Holt, Rinehart and Winston, 1970.
[4] Chen, C. T., "A new look at transfer function design," *Proc. IEEE,* vol. 59, no. 9, pp. 1580–1585, 1971.
[5] Chen, C. T., and C. H. Hsu, "Design of dynamic compensators for multivariable systems," *Preprints 1971 JACC,* pp. 893–900.
[6] Dorf, R. C., *Modern Control Systems.* Reading, Mass.: Addison-Wesley, 1967.
[7] Freested, W. C., R. E. Webber, and R. W. Bass, "The 'GASP' computer program: An integrated tool for optimal control and filter design," *Preprints 1968 JACC,* pp. 198–202.
[8] Kalman, R. E., and T. S. Engler, *A User's Manual for Automatic Synthesis Programs.* Washington, D.C.: National Aeronautics and Space Administration, 1966.
[9] Riddle, A. C., and B. D. Anderson, "Spectral factorization: Computational aspects," *IEEE Trans. Automatic Control,* vol. AC-11, pp. 764–765, 1966.

Problems

8-1 Determine, if you can, which complex number, $2 + 3j$ or $3 + 2j$, is smaller. Do the same with the two vectors

$$\begin{bmatrix} 1 \\ 2 \end{bmatrix} \quad \text{and} \quad \begin{bmatrix} 2 \\ 1 \end{bmatrix}$$

8-2 Is the function

$$J = \int_0^\infty \{q[y(t) - y_d(t)] + u(t)\} \, dt$$

a good performance criterion?

8-3 Consider the tracking antenna problem studied in Example 2 of Section 8-2. It is now assumed that the inductances of the motor and generator are negligible; that is, $L_a = 0$ and $L_g = 0$. Find the optimal system that minimizes

$$J = \int_0^\infty \{q[y(t) - y_d(t)]^2 + u^2(t)\} \, dt$$

where $q = 25$, and y_d is a unit step function. Find the necessary compensators if feedback is introduced from θ_L.

8-4 Design compensators for Problem 8-3 by using state feedback.

8-5 Design compensators for Problem 8-3 by using feedback from θ_L and $\dot{\theta}_L$. The moments of inertia of the required transducers may be neglected.

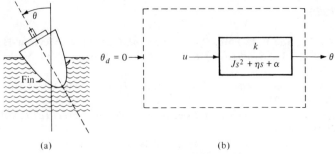

Figure P8-1

8-6 One way to stabilize an ocean liner, for passengers' comfort, is to use a pair of fins, as shown in Figure P8-1(a). The fins are controlled by an actuator, which is itself a feedback system consisting of a hydraulic motor. The transfer function of the actuator, compared with the dynamics of the liner, may be simplified as a constant k. The equation governing the roll motion of the liner is

$$J\ddot{\theta}(t) + \eta\dot{\theta}(t) + \alpha\theta(t) = ku(t)$$

where θ is the roll angle, and $ku(t)$ is the roll moment generated by the fins. The block diagram of the liner and actuator is shown in Figure P8-1(b). It is assumed that $\alpha/J = 0.3$, $\eta/2\sqrt{\alpha J} = 0.1$, and $k/\alpha = 0.05$. Design an overall system that minimizes the performance index

$$J = \int_0^\infty \{100[\theta(t) - \theta_d(t)]^2 + u^2(t)\}\, dt$$

where the desired roll angle, θ_d, is clearly identically zero (a regulator problem). It is assumed that the roll angle θ can be measured and used in feedback.

8-7 The transfer function from the actuating signal u to the actual heading y of an idealized ship-steering system can be written as $4/s^2$. Design compensators so that the resulting system minimizes

$$\int_0^\infty \{[y(t) - y_d(t)]^2 + u^2\}\, dt$$

where y_d may be assumed to be a step function.

8-8 A highly simplified model for controlling the yaw of an aircraft is shown in Figure P8-2, where θ is the yaw error, ϕ is the rudder deflection. The rudder is controlled by an actuator whose transfer function can be approximated as a constant k. Let J be the moment of inertia of the aircraft with respect to the yaw axis. For simplicity, it is assumed that the restoring torque is proportional to the rudder deflection $\phi(t)$; that is,

$$J\ddot{\theta}(t) = -k_1\phi(t)$$

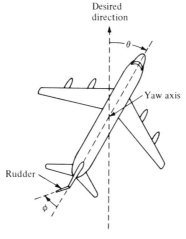

Figure P8-2

Suppose both yaw angle θ and its rate of change $\dot{\theta}$ are available for feedback, design an overall system to minimize the performance index

$$J = \int_0^\infty [100\theta^2(t) + \phi^2(t)]\, dt$$

Design the system by using the transfer-function approach *and* the state-variable approach.

8-9 Show that the coefficients of $\prod(s)$ and $D_f(s)$ in (8-10) and (8-11) are related by

$$a_0 = r_0^2$$
$$a_2 = 2r_0r_2 = r_1^2$$
$$a_4 = 2r_0r_4 - 2r_1r_3 + r_2^2$$
$$a_6 = 2r_0r_6 - 2r_1r_5 + 2r_2r_4 - r_3^2$$
$$\vdots$$
$$a_{2n} = 2r_0r_{2n} - 2r_1r_{2n-1} + 2r_2r_{2n-2} - \cdots + (-1)^n r_n^2$$

where $r_i = 0$ for $i > n$.

8-10 The depth of a submarine can be maintained automatically by a control system. The transfer function of a submarine from the stern plane angle θ to the actual depth y can be approximated as

$$\frac{10(s+2)^2}{(s+10)(s^2+0.1)}$$

Design a control system to minimize the performance criterion

$$\int_0^\infty \{[y(t) - y_d]^2 + \theta^2\}\, dt$$

where y_d is the desired depth. It is assumed that the actual depth can be measured by a pressure transducer and used in the feedback.

CHAPTER

9

The Root-Locus Method

The design procedure introduced in Chapter 8 proceeds from the determination of an overall transfer function and then the computation of the required compensators. The design techniques that will be introduced in this and following chapters take an entirely different approach: We first choose the configuration and type of compensators and then adjust the parameters of the compensators so that the resulting system will satisfy the specifications. The technique that will be introduced in this chapter is called the *root-locus method*.

9-1 Problem Formulation: One-Parameter Variation

In designing by means of the root-locus method, we first choose the configuration of compensators. Since the method is essentially a searching process by trial and error, we like to proceed from the simplest possible compensator and then to a more complicated one. For example, we may first use a pair of synchros with an adjustable gain. If, after trying all the gains, we still cannot obtain a satisfactory system, then we may use an additional tachometer feedback or compensating network, or both. Clearly the more compensators used, the more parameters we can adjust, and consequently we have a better chance of obtaining a satisfactory system. Unfortunately we are able, in practice, to handle

9-1 PROBLEM FORMULATION: ONE-PARAMETER VARIATION

the variation of one parameter only, or more precisely, one parameter at a time; hence the compensator introduced in this design cannot be too complicated.

We state formally the type of control problems that can be designed by using the root-locus method. Consider a plant with transfer function $\hat{g}(s)$. After the configuration of compensators and the type of compensators are chosen, if the overall transfer function can be expressed as

$$\hat{g}_f(s) = \frac{N_f(s, k)}{p(s) + kq(s)} \tag{9-1}$$

where $p(s)$ and $q(s)$ are two polynomials independent of k, and k is a real parameter to be adjusted, then the root-locus method can be applied.

Example 1

Consider a plant with the transfer function $20/s(s + 2)$. If the configuration of compensator is chosen as shown in Figure 9-1, then

$$\hat{g}_f(s) = \frac{20k}{s(s + 2) + k \cdot 20}$$

This is in the form of (9-1); hence the root-locus method can be applied.

Example 2

Consider a plant with the transfer function $10/s(s^2 + 2s + 2)$. Suppose the adjustment of a gain alone as in Figure 9-1 cannot yield a satisfactory system, we then try the compensators shown in Figure 9-2. Its overall transfer function is

$$\hat{g}_f(s) = \frac{10k_1(s + k_2)}{s(s + 2)(s^2 + 2s + 2) + 10k_1(s + k_2)}$$

After k_1 is chosen, say 5, $\hat{g}_f(s)$ becomes

$$\hat{g}_f(s) = \frac{50(s + k_2)}{[s(s + 2)(s^2 + 2s + 2) + 50s] + k_2 \cdot 50}$$

This is in the form of (9-1); hence the root-locus method can be applied.

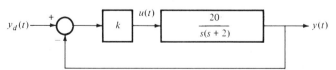

Figure 9-1 A control system.

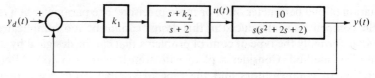

Figure 9-2 A control system.

Example 3

Consider again the plant in Example 2. If the configuration of compensator is chosen as shown in Figure 9-3, then the overall transfer function is

$$\hat{g}_f(s) = \frac{10k_1}{s(s^2 + 2s + 2) + 10k_3 s + 10k_1}$$

If either k_1 or k_3 is chosen, then $\hat{g}_f(s)$ is in the form of (9-1), and the root-locus technique can be applied. ∎

From these examples, we see that the root-locus method is widely applicable.

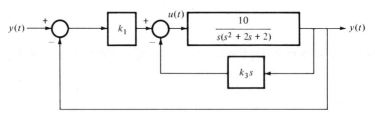

Figure 9-3 A control system.

9-2 Desired Pole Locations

In addition to being totally stable, every control system is designed to satisfy a set of specifications. As discussed in Section 6-2, the specifications can be divided into the steady-state performance and the transient performance. Once the overall transfer function $\hat{g}_f(s)$ is obtained, the range of k that meets the steady-state performance specification can be easily obtained from the coefficients of $\hat{g}_f(s)$ associated with s^0 and s^1, as discussed in Section 6-2. The range of k in which the system is BIBO stable can also be obtained by the application of the Routh-Hurwitz criterion. Hence the design problem in the root-locus method is to determine, if possible, a k that meets the transient performance from the range in which the system is stable and satisfies the steady-state performance.

In the design of a control system by using the root-locus method, we plot the poles of the overall system as a function of k, and then try to pick a k from the plot of these pole trajectories. Therefore it is essential to establish some relationship between the pole locations and the specifications on the transient performance. The determination of desired pole locations from the specifications on overshoot, rise time, and settling time is very difficult, because the response depends not only on the poles but also on the zeros. Therefore for the same set of specifications, we may have different desired pole locations for transfer functions with the same denominator but different numerators. Hence a general discussion of this problem is not possible; we discuss in the following only a very special case.

Consider a control system with transfer function

$$\hat{g}_f(s) = \frac{\omega_n^2}{s^2 + 2\delta\omega_n s + \omega_n^2} \tag{9-2}$$

where δ is called the *damping ratio*. The poles of $\hat{g}_f(s)$ in (9-2) are

$$\delta\omega_n \pm \omega_n\sqrt{\delta^2 - 1} \tag{9-3}$$

If $0 < \delta < 1$, the poles are a pair of complex conjugate poles, and the system is said to be *underdamped*. If $\delta = 1$, the system has two poles at $-\omega_n$ and is said to be *critically damped*. If $\delta > 1$, the system has two distinct negative real poles and is said to be *overdamped*. The response of the system due to a unit step function can be computed, for $\delta < 1$, as

$$\begin{aligned}
y(t) &= \mathscr{L}^{-1}\left[\frac{\omega_n^2}{s^2 + 2\delta\omega_n s + \omega_n^2} \cdot \frac{1}{s}\right] \\
&= 1 + \frac{e^{-\delta\omega_n t}}{\sqrt{1-\delta^2}}\sin\left[\omega_n\sqrt{1-\delta^2}\,t - \tan^{-1}\frac{\sqrt{1-\delta^2}}{-\delta}\right]
\end{aligned} \tag{9-4}$$

By using this equation, it is straightforward, though tedious, to show that the percentage overshoot depends *only* on the damping ratio δ. The relationship between the overshoot and the damping ratio is a nonlinear function and is as shown in Figure 9-4(b). Since $\delta = \cos\theta$, where θ is defined as in Figure 9-4(a),

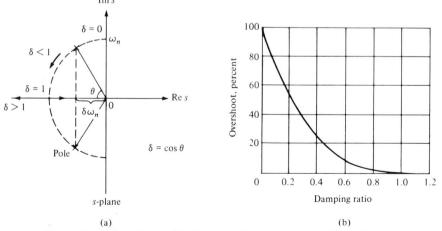

Figure 9-4 The relationship between damping ratio and overshoot.

from the curve in Figure 9-4 and the specification on the overshoot we can determine immediately the region in which the poles should locate. For example, if the overshoot is required to be less than 10 percent, then Figure 9-4 implies that $\delta > 0.6$, which in turn implies, by using $\delta = \cos\theta$, that $\theta < 53°$. If the overshoot is required to be less than 5 percent, then $\delta > 0.707$, and $\theta < 45°$. The regions in which the poles of $\hat{g}_f(s)$ should locate in order to have less than 5 percent overshoot is shown in the shaded area in Figure 9-5.

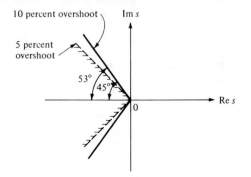

Figure 9-5
Translation of the specification on overshoot into pole location.

The settling time, as defined in Section 6-2, is the time required for the step response of the system to settle within 2 percent of the steady-state value. From (9-4), we see that the deviation of the response from the steady-state output is bounded by

$$D \triangleq \left| \frac{e^{-\delta\omega_n t}}{\sqrt{1-\delta^2}} \sin\left[\omega_n\sqrt{1-\delta^2}\, t - \tan^{-1}\frac{\sqrt{1-\delta^2}}{-\delta}\right]\right| \leq \frac{e^{-\delta\omega_n t}}{\sqrt{1-\delta^2}}$$

(9-5)

for $\delta < 1$. Note that $\delta\omega_n$ is, as shown in Figure 9-4, the magnitude of the real part of the poles. Now if $\delta < 0.8$, then (9-5) becomes

$$D \leq \frac{e^{-\delta\omega_n t}}{\sqrt{1-\delta^2}} \leq 1.7 e^{-\delta\omega_n t} \leq 0.02 \quad \text{for } t \geq \frac{4.5}{\delta\omega_n}$$

Hence given the settling time, if $\delta\omega_n \geq 4.5/t_s$ or, equivalently,

$$-(\text{real parts of the poles}) \geq \frac{4.5}{t_s} \quad (9\text{-}6)$$

then the specification on the settling time can be met. In other words, for the second-order system shown in (9-2), if the damping factor δ is smaller than 0.8, by locating the poles on the left-hand side of the vertical line shown in Figure 9-6, the specification on the settling time can be met. By combining the specifications on the overshoot and settling time, we obtain in Figure 9-6 an allowable region for the poles.

The *rise time* is, by definition, the time required for the step response of a system to rise from 0 to 90 percent of its steady-state value. The translation of

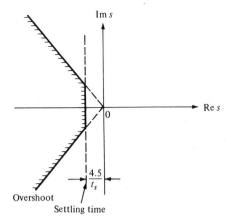

Figure 9-6 Translation of the specifications on overshoot and settling time into pole location.

the rise time into the pole location cannot be done precisely as in the cases of overshoot and settling time. All we can say is that, generally, *the farther away the closest pole from the origin of the s-plane, the smaller the rise time.*[1] This can be verified by observing the responses shown in Figure 9-7. We show in Figure 9-7(a) the responses of two systems whose poles have the same real part but different imaginary parts. Note that the real part of a pole governs the envelope of the oscillation, and the imaginary part governs the frequency of oscillation. Hence from Figure 9-7(a), we conclude that for complex poles with the same real parts, the farther away a pole from the origin, the smaller the rise time. The same statement also holds for poles with zero imaginary parts as shown in Figure 9-7(b). This statement cannot be established rigorously and, strictly speaking, is not necessarily correct for all cases. Since there is no other information available concerning the rise time and pole locations, we shall still use it in the design of control systems.

We now summarize what we have discussed up to this point. For a system with a transfer function of the form in (9-2), from the specifications on overshoot and settling time, we can find from Figure 9-4 and (9-6) a region, as shown

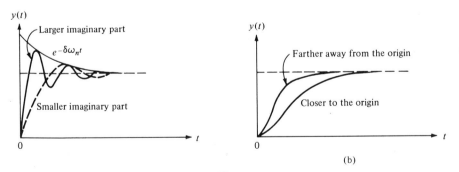

Figure 9-7 Time responses.

---

[1] A system may have many poles. The closest pole is the one that is closest to the origin.

in Figure 9-6, in which the poles should be located. As far as the rise time is concerned, all we can say is that the farther away the closest pole from the origin, the smaller the rise time. It is not easy to specify it more quantitatively, such as the rise time is less than, say, 10 seconds for certain pole locations.

With the allowable region for poles shown in Figure 9-6, we still have ample room for manipulation. We can locate the poles any place in the region and the system will satisfy all the specifications. Clearly a system with all real poles located deep inside the allowable region will be better than a system with poles located on the right-hand-side boundary of the region in terms of rise time, settling time, and overshoot. Therefore one may suggest that the poles should be located deep inside the region. This however should never be done in practice. Recall that the specifications introduced are for output signals only; nothing has been said about the saturation of actuating signals. Recall also that linear plants are usually valid only for certain input ranges. Hence if the magnitude of an input signal is very large, then the plant will saturate and will not function as expected. Now if the poles of an overall system are located far left deep inside the allowable region, then the system will respond faster. This is usually achieved with a large actuating signal. Hence the deeper the location of the poles inside the region, the larger the magnitude of the actuating signal; and consequently the system has more chance to saturate. Therefore the saturation problem will govern the final choice of the pole locations. Because of saturation and because of only one adjustable parameter,[2] the closest poles of the resulting systems in most designs will be located on the right-hand-side boundary of the allowable region.

The region in Figure 9-6 is developed for a second-order transfer function with a constant numerator. Is it possible to extend it to the general case? Unfortunately the answer is negative. The only exception is the class of systems that can very well be approximated by transfer functions of the form in (9-2). This class of systems is said to have a pair of *dominant* poles, that is, a pair of poles close to the $j\omega$-axis (the rest of the poles are very far away from the $j\omega$-axis). Although the region in Figure 9-6 is not necessarily valid for transfer functions of forms other than (9-2), since there is no other information available for general transfer functions, we shall still use the region in the design. Therefore, in the design by using the root-locus method, a final check of the resulting system by computer simulations is advisable.

We shall now present an example to illustrate the ideas involved in this section.

Example

Given a plant with the transfer function $1/s(s+2)$. Design an overall system that satisfies the following specifications: steady-state error due to a step input

---

[2] If we have a sufficient number of adjustable parameters, then we can locate all poles arbitrarily. See Chapter 8.

<10 percent; overshoot ≤5 percent; settling time ≤9 seconds; rise time as small as possible.

The configuration of compensator for this system is chosen as shown in Figure 9-8. This can be accomplished by using a pair of potentiometers or using a synchro with a demodulator (see Chapter 3). The overall transfer function is

$$\hat{g}_f(s) = \frac{k}{s^2 + 2s + k} \qquad (9\text{-}7)$$

The overall system is totally stable for any $k > 0$ and satisfies the steady-state error specification for any $k > 0$. Hence in the following, we shall find a k, with $k > 0$, to meet the specifications on the transient performance. The specification on overshoot requires that the poles of $\hat{g}_f(s)$ be located inside the sector bounded by $\theta = 45°$, as shown in Figure 9-9. The settling time requires the poles to be located on the left-hand side of the vertical line passing through $-4.5/t_s = -0.5$. Hence if all the poles of $\hat{g}_f(s)$ are located inside the shaded area of Figure 9-9, then the overall system will satisfy the specifications on the overshoot and settling time.

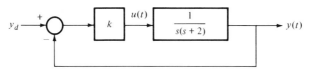

Figure 9-8 A control system.

The poles of the overall system for $k = 0.36, 0.75, 1, 2,$ and 5 are plotted in Figure 9-9. Note that there are two poles for each k. For $k = 0.36$, although one pole is inside the region, the other is not. Hence if we choose $k = 0.36$, the system satisfies the specification on the overshoot but not that of the settling time. If we choose $k = 5$, then the system satisfies the specification on the settling time but not that on the overshoot. However for $k = 0.75, 1,$ and 2, all the poles are inside the allowable region, and the system satisfies both the overshoot and settling-time specifications. Hence if the rise time (the speed of response) is of no concern, then k can be chosen as $k = 0.75, 1,$ or 2, and the design is completed. For this system, as k increases, the closest pole of the system will move away from the origin; hence the larger the gain k, the smaller the rise time. However as k increases, the magnitude of the actuating signal will be larger, and the system may saturate. Hence in choosing k, the saturation problem has to be considered. There is no simple analytical method for determining whether or not a system will saturate. It may be easiest to check it by computer (analog or digital) simulation. For this problem, if saturation will not occur, then we may choose $k = 2$, and the design is completed.

A side bonus for choosing large k for this problem is that the steady-state error due to a ramp input will be the smallest. The steady-state error due to a ramp input is, from (6-8), $(2 - 0)/k$; hence the larger the k, the smaller the error.

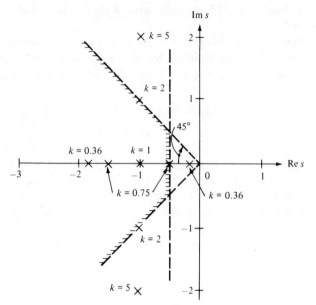

Figure 9-9 Pole locations of $s^2 + 2s + k$ for various values of k.

For this example we are able to find a gain k to meet all the specifications. However we may not be able to do so for all cases. For example, if the settling time in this example is required to be less than 4 seconds, then no k will meet the specification. In this case a different configuration of compensators or more complicated compensators have to be used. ∎

9-3 The Plot of Root Loci

Consider a control system with the overall transfer function

$$\hat{g}_f(s) = \frac{N_f(s, k)}{p(s) + kq(s)} \tag{9-8}$$

In the design of the system, by using the root-locus method, we try to pick a k, if it exists, so that the poles of $\hat{g}_f(s)$ or, equivalently, the roots of $p(s) + kq(s)$ are in the desired locations. Therefore if the plot of the roots of $p(s) + kq(s)$ as a function of k is available, then the existence of a k that will give a satisfactory response can be determined immediately. The plots of the roots of $p(s) + kq(s)$ as a function of the real variable k are called the *root loci*. The root loci of $p(s) + kq(s)$ can be obtained by solving for the roots numerically. They can also be obtained graphically. In this section we introduce the graphical method of plotting the root loci of $p(s) + kq(s)$.

In order to plot the root loci of $p(s) + kq(s)$, we first factor $p(s)$ and $q(s)$ as

$$p(s) = k_p(s + p_1)(s + p_2) \cdots (s + p_m) \triangleq k_p \bar{p}(s) \tag{9-9}$$

$$q(s) = k_q(s + q_1)(s + q_2) \cdots (s + q_l) \triangleq k_q \bar{q}(s) \tag{9-10}$$

where k_p and k_q are real constants, and p_i and q_i are, respectively, the roots of $p(s)$ and $q(s)$. Note that $p(s)$ and $q(s)$ are assumed to have real coefficients; hence complex conjugate roots appear in pairs. Equation $p(s) + kq(s) = 0$ can be written[3] as

$$\frac{(s + q_1)(s + q_2) \cdots (s + q_l)}{(s + p_1)(s + p_2) \cdots (s + p_m)} = -\frac{k_p}{k_q}\frac{1}{k} \triangleq -\frac{1}{\bar{k}} \tag{9-11}$$

Then the root loci consist of all s satisfying Equation (9-11) for all $\bar{k}$. Note that for each s, say s_1, each factor in the left-hand side of (9-11) is a vector as shown in Figure 9-10. If we use

$$s + q_i = |s + q_i|e^{j\theta_i} \tag{9-12}$$

and

$$s + p_i = |s + p_i|e^{j\phi_i} \tag{9-13}$$

where θ_i and ϕ_i are the angles, as shown in Figure 9-10, then (9-11) becomes

$$\frac{|(s + q_1)(s + q_2) \cdots (s + q_l)|e^{j(\theta_1 + \theta_2 + \cdots + \theta_l)}}{|(s + p_1)(s + p_2) \cdots (s + p_m)|e^{j(\phi_1 + \phi_2 + \cdots + \phi_m)}} = -\frac{1}{\bar{k}} \tag{9-14}$$

This actually consists of two equations—namely,

$$\sum_1^l \theta_i - \sum_1^m \phi_i = \text{the phase of } \frac{-1}{\bar{k}} = (2n + 1)\pi \text{ for } \bar{k} > 0 \tag{9-15}$$

$$= 2n\pi \quad \text{for } \bar{k} < 0$$

for $n = 0, \pm 1, \pm 2, \ldots$, and

$$\left|\frac{(s + q_1)(s + q_2) \cdots (s + q_l)}{(s + p_1)(s + p_2) \cdots (s + p_m)}\right| = \left|\frac{1}{\bar{k}}\right| \tag{9-16}$$

We note that (9-15) depends only on the sign of $\bar{k}$. Clearly any point on the s-plane that satisfies (9-15) for some n is a root of $p(s) + kq(s) = 0$. Hence in the first stage of plotting the root loci, we use only (9-15); Equation (9-16) is used in the determination of the gain of the root loci in the second stage.

We shall now discuss some general properties of the root loci of $p(s) + kq(s) = 0$. With these properties, the plotting of the root loci can be facilitated.

---

[3] The equation $p(s) + kq(s) = 0$ can also be written as $\bar{p}(s)/\bar{q}(s) = -(k_q/k_p)k$, and the subsequent discussion is still directly applicable. The reason for choosing (9-11) is that the p_i's and q_i's are, in many cases, just the poles and zeros of a plant.

234 THE ROOT-LOCUS METHOD

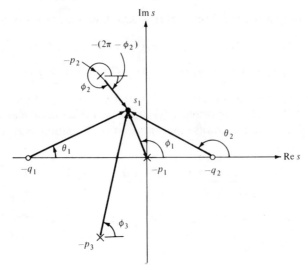

Figure 9-10 Vectors from the roots of $p(s)$ and $q(s)$. Roots of $p(s)$ are denoted by crosses; roots of $q(s)$, by open circles.

Properties of root loci. Consider the equation $p(s) + kq(s) = 0$. It is written as

$$\frac{(s + q_1)(s + q_2) \cdots (s + q_l)}{(s + p_1)(s + p_2) \cdots (s + p_m)} = -\frac{k_p}{k_q}\frac{1}{k} \triangleq \frac{-1}{\bar{k}} \quad (9\text{-}17)$$

Its root loci have the following properties:

1. Let the degree of the polynomial $p(s) + kq(s)$ be n. That is, $n = \max[\delta p(s), \delta q(s)]$, where δ denotes the degree. Then for each $k \neq 0$, there are n roots. Therefore as k or, equivalently, $\bar{k}$ varies, there are n trajectories of the roots. It can be shown that the roots of a polynomial are continuous functions of its coefficients. Hence *as* k *varies continuously, the roots form* n *continuous trajectories.*

2. If $\bar{k} = 0$ or, equivalently, $k = 0$, the roots of $p(s) + kq(s)$ are just the roots of $p(s)$. If $\bar{k} = \pm\infty$, the roots of $p(s) + kq(s) = 0$ are just the roots of $q(s)$. Hence *as* $\bar{k}$ *(or* k*) varies from zero to positive or negative infinity, the root loci move continuously from the roots of* p(s) *to the roots of* q(s). Note that if $\delta q(s) < n$ [or $\delta p(s) < n$], then $q(s)[p(s)]$ is considered to have $n - \delta q(s)$ roots [$n - \delta p(s)$ roots] at infinity.

3. *The root loci are symmetric with respect to the real axis of the* s-*plane*. This follows from the fact that the coefficients of $p(s)$ and $q(s)$ and k are all real; hence the complex roots of $p(s) + kq(s) = 0$ appear in pairs.

4. In checking whether or not a point on the real axis of the s-plane satisfies (9-15), the complex roots of $p(s)$ and $q(s)$ need not be considered, because the net angle due to each pair of complex conjugate roots is 2π. With this fact, it

can be easily verified that, *if $\bar{k} > 0$, the part of the real axis whose right-hand side has odd number of roots of* p(s) *and* q(s) *constitutes a part of the root loci. If $\bar{k} < 0$, the part of the real axis whose right-hand side has even number of roots of* p(s) *and* q(s) *constitutes a part of the root loci.*

5. Let $m \triangleq \delta p(s)$, $l \triangleq \delta q(s)$. If $m \neq l$, we draw $|m - l|$ number of straight lines emitting from the point

$$\left(-\frac{(\sum_i^m p_i) - (\sum_1^l q_i)}{m - l}, 0\right) \quad (9\text{-}18)$$

on the s-plane, with angles

$$\frac{(2n + 1)\pi}{|m - l|} \quad n = 0, 1, \ldots, |m - l| - 1 \quad \text{for } \bar{k} > 0 \quad (9\text{-}19a)$$

or

$$\frac{2n\pi}{|m - l|} \quad n = 0, 1, \ldots, |m - l| - 1 \quad \text{for } \bar{k} < 0 \quad (9\text{-}19b)$$

Then, for very large s, the root loci will approach these straight lines. We call these lines *asymptotes*.

We prove this assertion for the case $m > l$. For extremely large s, Equation (9-17) can be approximated by

$$\frac{-1}{\bar{k}} = \frac{(s + q_1)(s + q_2)\cdots(s + q_l)}{(s + p_1)(s + p_2)\cdots(s + p_m)} = \frac{s^l + (\Sigma q_i)s^{l-1} + \cdots}{s^m + (\Sigma p_i)s^{m-1} + \cdots}$$

$$= \frac{1}{s^{m-l} + (\Sigma p_i - \Sigma q_i)s^{m-l-1} + \cdots} \doteq \frac{1}{\left(s + \dfrac{\Sigma p_i - \Sigma q_i}{m - l}\right)^{m-l}}$$

This can be looked upon as having $(m - l)$ repeated roots at (9-18). Hence, for $\bar{k} > 0$, if each root contributes a degree $(2n + 1)\pi/(m - l)$, then $(m - l)$ roots will contribute a degree $(2n + 1)\pi$. This proves the assertion. The case for $\bar{k} < 0$ and $l > m$ can be similarly proved.

With these properties, we may now discuss the plotting of the root loci. We do it by examples. Consider the polynomial $2s(s + 2) + 4k(s + 4)$. It can be written as

$$\frac{s + 4}{s(s + 2)} = -\frac{2}{4k} = -\frac{1}{2k}$$

where $\bar{p}(s) = s(s + 2)$, and $\bar{q}(s) = s + 4$. Their roots are plotted in Figure 9-11. Since the degree of the polynomial is 2, there are two roots for each k. We plot in Figure 9-11(a) the root loci for $k > 0$. The intervals $(-\infty, -4]$ and $[-2, 0]$ on the real axis are clearly parts of the root loci, because their right-hand sides have odd numbers of roots. Property 5 implies that there is one asymptote that happens to be the negative real axis. The rest of the root

236 THE ROOT-LOCUS METHOD

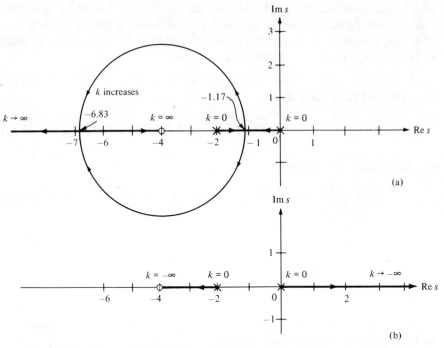

Figure 9-11 The root loci of $2s(s + 2) + 4k(s + 4)$. (a) $k > 0$. (b) $k < 0$.

loci is obtained graphically as follows: Pick an arbitrary point, and measure the angles. If $\theta_1 - (\phi_1 + \phi_2)$ is equal to $(2n + 1)\pi$ for some integer n, then it is a point on the root loci. Otherwise we try another point. The process is carried out by trial and error. There is a tool called the *Spirule* that is especially designed for plotting the root loci. By use of a Spirule, the condition $\theta_1 - (\phi_1 + \phi_2) = (2n + 1)\pi$ can be easily checked. After the root loci on the upper-half s-plane are obtained, those in the lower half can be obtained immediately because of property 3. In Figure 9-11(b) we show the root loci of $2s(s + 2) + 4k(s + 4)$ for $k < 0$.

Although the root loci are obtained by trial and error, the approximate shape of the curve can often be obtained by an intelligent guess before any angles are measured. For example, consider the root loci in Figure 9-11(a). As k increases from 0, the two roots move away from 0 and -2, and finally they will coincide. As k continues to increase, the two roots must break away from the real axis. As k approaches infinity, the root loci are again on the real axis; one approaches the root -4, and one approaches $-\infty$ (see property 2). Since the root loci are continuous, after breaking away from the interval $[-2, 0]$, they must come back to a point somewhere on the interval $(-\infty, -4]$. From these arguments a rough shape of the root loci can be obtained.

It is appropriate at this point to discuss the breakaway points shown in Figure 9-11(a). They can be obtained by either direct measurements or by an analytical method. We note that a breakaway point has at least two roots in common. Therefore the point should satisfy the equations

$$p(s) + kq(s) = 0 \qquad (9\text{-}20a)$$

and

$$\frac{d}{ds}[p(s) + kq(s)] = 0 \qquad (9\text{-}20b)$$

Since Equation (9-20b) contains parameter k, it is easier to work on the equation

$$0 = \frac{d}{ds}\left\{q(s)\left[\frac{p(s)}{q(s)} + k\right]\right\} = \left[\frac{d}{ds}q(s)\right]\left[\frac{p(s)}{q(s)} + k\right] + q(s)\frac{d}{ds}\left[\frac{p(s)}{q(s)}\right]$$

$$= q(s)\frac{d}{ds}\left[\frac{p(s)}{q(s)}\right]$$

Since $q(s) \neq 0$ at a breakaway point, hence a breakaway point must satisfy

$$\frac{d}{ds}\left[\frac{p(s)}{q(s)}\right] = 0 \qquad (9\text{-}21)$$

If a solution of Equation (9-21), say s_0, also satisfies (9-20a) for some real k, say k_0, then s_0 is a breakaway point, and k_0 gives the gain at that breakaway point.

For example, for the breakaway points in Figure 9-11, we solve

$$\frac{d}{ds}\left[\frac{s(s+2)}{s+4}\right] = \frac{(s+4)(2s+2) - s(s+2)}{(s+4)^2} = 0$$

Its solutions are $s_1 = -6.83$ and $s_2 = -1.17$. They also satisfy $2s(s+2) + 4k(s+4) = 0$ for some real k, and hence they are breakaway points.

Although breakaway points can be solved analytically, it is not recommended, except for simple cases, to do so for the following reasons: First, the root-locus technique is essentially a graphical method. A breakaway point can be obtained with other points by direct measurements. Second, the exact location of a breakaway point is often not important in the design; and finally, the analytical method may have to solve a polynomial of degree 3 or higher.

We give one more example to illustrate the plot of root loci. Consider the equation

$$s(s+4)(s+2+2j)(s+2-2j) + 2k(s+1) = 0 \qquad (9\text{-}22)$$

or

$$\frac{s+1}{s(s+4)(s+2+2j)(s+2-2j)} = -\frac{1}{2k}$$

Its root loci for $k > 0$ are plotted in Figure 9.12 by using the solid lines. Property 5 implies that there are three asymptotes intersecting at

$$\left(\frac{(0+4+2+2j+2-2j)-1}{3}, 0\right) = (-2.33, 0)$$

with angles 60°, 180°, and 240°. For very large s, the root loci approach these asymptotes. The exact loci are obtained by point-by-point measurements.

The angle at which a root locus departed from (arrived at) a root is called the *angle of departure (arrival)*. This angle can be obtained as follows: Consider the root $-p_3$ in Figure 9-12. Let s_0 be a point very close to the root. If s_0 is a point on the root locus, then

$$\theta_1 - (\phi_1 + \phi_2 + \phi_3 + \phi_4) = (2n + 1)\pi$$

for some integer n. In measuring the angles θ_1 and ϕ_i, for $i = 1, 2$, and 4, the point s_0 can be looked upon as coinciding with $-p_3$. Hence the angle of departure ϕ_3 of the root $-p_3$, choosing $n = -1$, is

$$\phi_3 = \theta_1 - (\phi_1 + \phi_2 + \phi_4) - (2n + 1)\pi$$
$$= 116.5° - (135° + 45° + 90°) + 180° = 26.5°$$

The angle of departure or arrival at other roots can be similarly obtained. The root loci of (9-22) for $k < 0$ are shown in Figure 9-12 by using the dotted lines.

We recapitulate what we have discussed up to this point. In the plot of root loci, we first obtain the root loci on the real axis and the asymptotes by using (9-18) and (9-20). If necessary, we then compute the angle of departure or

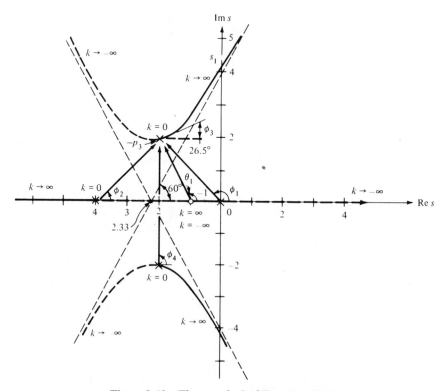

Figure 9-12 The root loci of Equation (9-22).

arrival for each complex root. With this information we are often able to predict the rough shape of root loci. The exact root loci however will have to be plotted point by point by checking the angle condition. This is purely a trial-and-error process.

Every point on the root loci of $p(s) + kq(s) = 0$ is a root for some real k_1. This k_1 can be obtained by using Equation (9-16). For example, the k_1 corresponding to s_1 shown in Figure 9-12 can be obtained from

$$\left|\frac{1}{2k_1}\right| = \left|\frac{s_1 + 1}{s_1(s_1 + 4)(s_1 + 2 + 2j)(s_1 + 2 - 2j)}\right|$$

$$= \left|\frac{4.4}{4.25 \times 5.9 \times 3 \times 6.6}\right| = \left|\frac{1}{113}\right|$$

It is obtained by measuring the distance between s_1 and each root. From this equation, k_1 can be computed as 56.5. Note that if the polynomial $p(s) + kq(s)$ has degree n, then there are n points on the root loci with the same k.

If the roots of a polynomial are all in the left-half s-plane, then the polynomial is a Hurwitz polynomial. Clearly from the root loci, the range of k for a polynomial to be Hurwitz can be easily obtained. For example, from Figure 9-11 we know immediately that the polynomial $2s(s + 2) + 4k(s + 4)$ is Hurwitz for $k > 0$. From Figure 9-12 we know that the polynomial $s(s + 4) \times (s^2 + 4s + 8) + 2k(s + 1)$ is Hurwitz for $0 < k < 56.5$. Certainly these ranges of k can also be obtained by applying the Routh-Hurwitz criterion.

9-4 Design by the Root-Locus Technique

In this section the design of control systems by means of the root-locus technique will be studied.

Example 1

Consider again the design problem studied in the Example in Section 9-2. The overall transfer function is $\hat{g}_f(s) = k/(s^2 + 2s + k)$. The root loci of $(s^2 + 2s + k) = 0$ or of

$$\frac{1}{s(s + 2)} = -\frac{1}{k}$$

for $k > 0$ are shown in Figure 9-13. There are two asymptotes emanating from the point $(-(2 + 0)/2, 0) = (-1, 0)$, with degrees $90°$ and $270°$. For this example, it happens that the root loci coincide with the asymptotes. Suppose it is decided to locate the poles of $\hat{g}_f(s)$ at s_1 and $\bar{s}_1$. Then the required k can be computed from

$$\left|\frac{1}{k}\right| = \left|\frac{1}{s_1(s_1 + 2)}\right| = \frac{1}{1.4 \times 1.4} = \frac{1}{1.96}$$

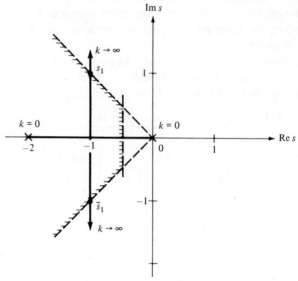

Figure 9-13 The root loci of $s^2 + 2s + k = 0$, for $k \geq 0$.

as $k = 1.96$. Note that $|s_1| = 1.4$ and $|s_1 + 2| = 1.4$ are obtained by direct measurement.

Example 2

Consider a plant with the transfer function

$$\frac{2}{s(s + 1)(s + 5)}$$

Design an overall system to satisfy the following specifications: The steady-state error due to a ramp function should be as small as possible; the overshoot should be less than 5 percent; the settling time should be less than 5 seconds; and the rise time should be as small as possible.

First, we choose the configuration of compensator as shown in Figure 9-14. The overall transfer function is $\hat{g}_f(s) = 2k/(s^3 + 6s^2 + 5s + 2k)$. The BIBO stability requires that $k > 0$. Hence in plotting the root loci of $s^3 + 6s^2 + 5s + 2k = 0$, or of

$$\frac{1}{s(s^2 + 6s + 5)} = \frac{-1}{2k}$$

we need to plot only for $k > 0$. The root loci are shown in Figure 9-15. It has three asymptotes emanating from $(-(0 + 1 + 5)/3, 0) = (-2, 0)$, with angles

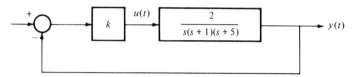

Figure 9-14 Configuration of a compensator.

60°, 180°, and 300°. The exact root loci are obtained by direct measurements. The breakaway point can also be computed analytically by solving

$$0 = \frac{d}{ds}\left[\frac{s(s^2 + 6s + 5)}{1}\right] = 3s^2 + 12s + 5$$

Its solutions are -0.46 and -3.5. Clearly -0.46 is a breakaway point but not -3.5.[4]

In order for the resulting system to have a settling time of less than 5 seconds, all the poles of $\hat{g}_f(s)$ should be located on the left-hand side of the vertical line passing through the point $-4.5/t_s = -0.9$. From the root loci in Figure 9-15 we see that this is not possible for any $k > 0$. Hence the configuration chosen in Figure 9-14 cannot meet the specifications.

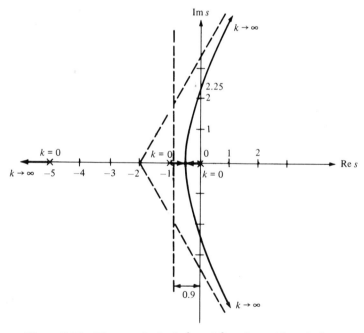

Figure 9-15 The root loci of $s^3 + 6s^2 + 5s + 2k = 0$, for $k \geq 0$.

---

[4] It is a breakaway point of the root loci for $k < 0$.

242 THE ROOT-LOCUS METHOD

As a next try, we introduce the compensating network shown in Figure 9-16. Rather arbitrarily, we choose $a = 1$. We note that there is a pole-zero cancellation for this a. However the canceled pole is located on the left-hand side of the vertical line shown in Figure 9-15, and hence this cancellation is permitted in the design. The advantage of having this cancellation is that it will simplify the plotting of the root loci.

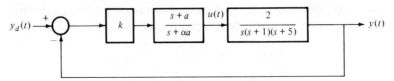

Figure 9-16 Configuration of compensators.

The overall transfer function of the system in Figure 9-16 with $a = 1$ is equal to

$$\hat{g}_f(s) = \frac{2k}{s(s + \alpha)(s + 5) + 2k} \qquad (9\text{-}23)$$

If k is chosen as $k = 5$, then Equation (9-23) becomes

$$\hat{g}_f(s) = \frac{10}{s^3 + (5 + \alpha)s^2 + 5\alpha s + 10} = \frac{10}{(s^3 + 5s^2 + 10) + \alpha s(s + 5)} \qquad (9\text{-}24)$$

The root loci of $(s^3 + 5s^2 + 10) + \alpha s(s + 5) = 0$, or of

$$\frac{-1}{\alpha} = \frac{s(s + 5)}{s^3 + 5s^2 + 10} = \frac{s(s + 5)}{(s + 5.35)(s - 0.18 + j1.36)(s - 0.18 - j1.36)} \qquad (9\text{-}25)$$

are plotted in Figure 9-17.

Now the specification on overshoot requires that the roots be inside the sector bounded by 45° as shown. The settling time requires that the roots be on the left-hand side of the vertical line passing through the point $(-0.9, 0)$. We see from Figure 9-17 that if $\alpha_1 \leq \alpha \leq \alpha_3$, then these two specifications are satisfied. In order to have the rise time as small as possible, the closest root to the origin should be as far away from the origin as possible. Again from Figure 9-17, we rule out the range from α_2 to α_3. The roots corresponding to any α in (α_1, α_2) have approximately the same distance from the origin; hence the rise time corresponding to any α in (α_1, α_2) is roughly the same. The last specification is that the steady-state error due to a ramp function be as small as possible. From (9-24) and (6-8), we know that the steady-state error is equal to

$$\left| \frac{0 - 5\alpha}{10} \right| = \left| \frac{\alpha}{2} \right| \times 100 \text{ percent}$$

9-4 DESIGN BY THE ROOT-LOCUS TECHNIQUE   243

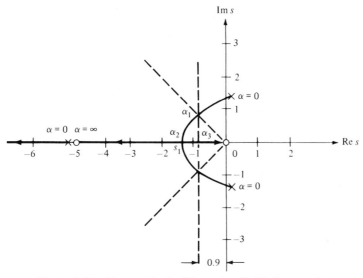

Figure 9-17 The root loci of Equation (9-25) for $\alpha > 0$.

Hence in order to have the smallest possible error, we choose α to be α_1. The parameter α_1 can be obtained from (9-25) as

$$\left|\frac{1}{\alpha_1}\right| = \left|\frac{s_1(s_1 + 5)}{(s_1 + 5.35)(s_1 - 0.18 + j1.36)(s_1 - 0.18 - j1.36)}\right|$$

$$= \frac{1.4 \times 4.1}{4.45 \times 1.2 \times 2.7} = \frac{1}{2.52}$$

or $\alpha_1 = 2.52$. Hence by choosing $k = 5$, $a = 1$, and $\alpha = 2.52$, the system in Figure 9-17 satisfies the design specifications, and the design is completed.

We see that this design is carried out entirely by trial and error. In fact, we have tried $k = 1$ without success. Clearly we may also choose an α, and then plot the root loci with respect to k. We may also choose an entirely different configuration, such as the one in Figure 9-18. The overall transfer function of the system in Figure 9-18 is, for $k = 5$,

$$\hat{g}_f(s) = \frac{10}{s^3 + 6s^2 + 5s + 10 + 2k_1 s}$$

The root loci of $(s^3 + 6s^2 + 5s + 10) + 2k_1 s$, or of

$$\frac{-1}{2k_1} = \frac{s}{s^3 + 6s^2 + 5s + 10} = \frac{s}{(s + 5.42)(s + 0.3 + j1.33)(s + 0.3 - j1.33)}$$

(9-26)

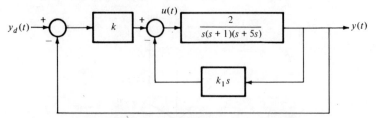

Figure 9-18 Configuration of compensators.

is plotted in Figure 9-19. It is obtained with the aid of a digital computer by using the POLRT subroutine in the IBM System/360 Scientific Subroutine Package. We see that if $3 \leq k_1 \leq 5.5$, then the system satisfies the specifications on the overshoot and settling time. Since the steady-state error due to a ramp function is given by $(2k_1 + 5)/10$, if k_1 is chosen as 3, then the system has the smallest steady-state error. If k_1 is chosen as 4, then the closest pole to the origin has a greater distance than the one corresponding to the other k_1 in $3 \leq k_1 \leq 5.5$. Hence if $k_1 = 4$, then the system has the smallest rise time.

The unit step responses of the systems in Figure 9-16 and 9-18 are shown in Figure 9-20. They are obtained by the employment of System/360 CSMP on a digital computer. The three designs are all satisfactory.

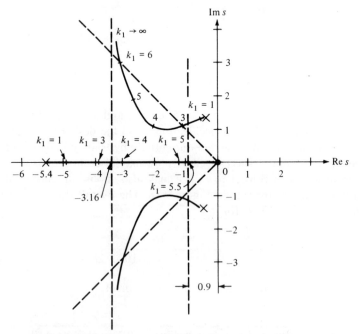

Figure 9-19 The root loci of Equation (9-26) for $k_1 \geq 0$.

9-4 DESIGN BY THE ROOT-LOCUS TECHNIQUE

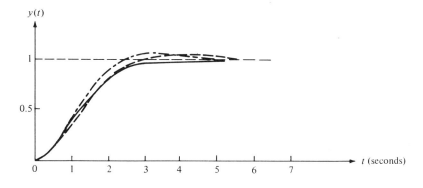

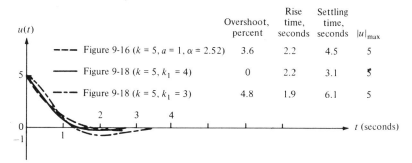

	Overshoot, percent	Rise time, seconds	Settling time, seconds	$\|u\|_{max}$
--- Figure 9-16 ($k = 5, a = 1, \alpha = 2.52$)	3.6	2.2	4.5	5
—— Figure 9-18 ($k = 5, k_1 = 4$)	0	2.2	3.1	5
—·— Figure 9-18 ($k = 5, k_1 = 3$)	4.8	1.9	6.1	5

Figure 9-20 The unit step responses of the systems in Figures 9-16 and 9-18.

Example 3

Consider the tracking antenna problem studied in Example 2 of Section 8-2. The transfer function of the plant, as computed in (8-23), is

$$\hat{g}(s) = \frac{300}{s^4 + 184s^3 + 760.5s^2 + 162s} = \frac{300}{s(s + 3.997)(s + 0.225)(s + 179.77)}$$

Design an overall system that has a satisfactory step response.

As a first try, the configuration of compensator is chosen as shown in Figure 9-21. Note that there is a pole-zero cancellation. The canceled pole is however a stable pole and has a large negative real part; hence it will not seriously affect the response of the system. The transfer function of the overall system is

$$\hat{g}_f(s) = \frac{300k_1}{s^4 + (179.995 + k_2)s^3 + (40.448 + 179.995k_2)s^2 + 40.448k_2s + 300k_1}$$

The gain k_1 is chosen rather arbitrarily as $k_1 = 10$. The root loci of

$$s^4 + (179.995 + k_2)s^3 + (40.448 + 179.995k_2)s^2 + 40.448k_2s + 300k_1$$

246   THE ROOT-LOCUS METHOD

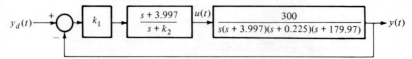

Figure 9-21   Configuration of compensators.

with $k_1 = 10$ or, equivalently, of

$$\frac{-1}{k_2} = \frac{s^3 + 179.995s^2 + 40.448s}{s^4 + 179.995s^3 + 40.448s^2 + 3000}$$

$$= \frac{s(s + 0.225)(s + 179.76947)}{(s + 179.76999)(s + 2.65)(s - 1.21 + j2.2)(s - 1.21 - j2.2)}$$

(9-27)

are shown in Figure 9-22. For k_2 larger than 9, the overall system is totally stable. However two of the poles of the overall system are very close to the imaginary axis; hence the response may be very oscillatory and the percentage overshoot may be very large. Therefore we conclude that the system is not satisfactory for any k_2, and a different configuration of compensators has to be chosen.

Consider the configuration of compensators shown in Figure 9-23. The overall transfer function is

$$\hat{g}_f(s) = \frac{300k_1}{s^4 + 184s^3 + 760.5s^2 + (162 + 300k_2)s + 300k_1}$$

Since the constant terms of the denominator and numerator of $\hat{g}_f(s)$ are both equal to $300k_1$, the steady-state error due to a step function of the overall

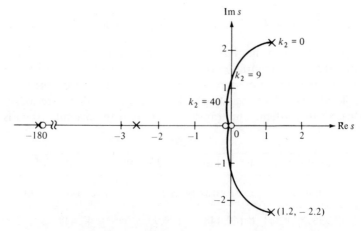

Figure 9-22   The root loci of Equation (9-27) for $k_2 \geq 0$.

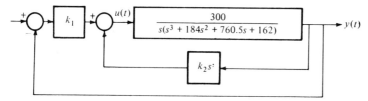

Figure 9-23 Configuration of compensators.

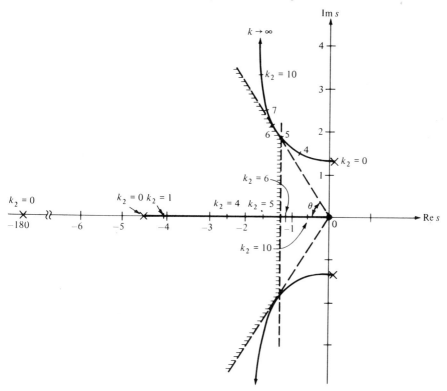

Figure 9-24 The root loci of Equation (9-28) for $k_2 \geq 0$.

system is equal to zero for any k_1. The gain k_1 is chosen rather arbitrarily as $k_1 = 5$. The root loci of

$$s^4 + 184s^3 + 760.5s^2 + (162 + 300k_2)s + 300k_1$$

with $k_1 = 5$ or, equivalently, of

$$\frac{-1}{300k_2} = \frac{s}{s^4 + 184s^3 + 760.5s^2 + 162s + 1500}$$

$$= \frac{s}{(s + 4.45)(s + 179.77)(s - 0.11 + j1.36)(s - 0.11 - j1.36)}$$

(9-28)

are shown in Figure 9-24. From the root loci, we see that if k_2 is chosen as 5, then all the poles of the overall system will be located in the region shown. Hence by choosing $k_1 = 5$ and $k_2 = 5$, the design is completed.

The step response of the resulting system is shown in Figure 12-5 of the Epilogue. The overshoot is 2 percent, the rise time is 1.75 seconds, and the settling time is 4.9 seconds. The step response is indeed very good. ∎

If Equation (9-6) and Figure 9-4 are used to estimate the overshoot and the settling time of the system in this example, then the overshoot should be 18 percent and the settling time should be 3.6 seconds. These data are quite different from the measured ones. This is conceivable because Equation (9-6) and Figure 9-4 are derived from a second-order transfer function with a constant numerator. Therefore there is no reason for the measured data to be close to those obtained from Equation (9-6) and Figure 9-4. However the guidelines— namely, to keep all poles far away from the imaginary axis and from the origin and to have all poles bounded by a small sector—seem applicable to any case.

9-5 Remarks and Review Questions

What is the main difference in approach between the root-locus method and the optimal-design method?

In applying the root-locus method, the overall transfer function must be first transformed into the form shown in Equation (9-1). Why should $p(s)$ and $q(s)$ in (9-1) be independent of k?

Why should the root loci of a real polynomial be symmetric with respect to the real axis?

As k increases from zero to infinity, how will the roots of $p(s) + kq(s)$ migrate?

Why, in the determination of the root loci of $p(s) + kq(s)$ on the real axis, need the complex conjugate roots of $p(s)$ and $g(s)$ not be considered?

What is an asymptote? How do you find asymptotes?

What is the angle of departure (arrival)? How do you compute it?

Can we use the root-locus method to determine the range of a parameter of a polynomial for the polynomial to be Hurwitz?

How do you translate the specification on the overshoot and settling time into the pole location? Is the translation applicable to any transfer function? If not, what do you do in the design of control systems?

If all the poles of a system lie on the negative real axis with very large magnitude, then the system will have a very small rise time, settling time, and overshoot. Why not design a system with this kind of pole distribution?

The root-locus method introduced in this chapter has, compared with the optimal design, the following disadvantages: First, it is a trial-and-error method. Second, if there are two or more parameters to be adjusted, unless we plot the entire family of root loci it is not possible to obtain the best design. Third, except for second-order transfer functions with constant numerators, it is not easy to obtain the desired pole locations from the specifications. Fourth, the saturation problem is, strictly speaking, not considered in the design. Finally, although the response depends on the poles and zeros of an overall transfer function, the zeros are essentially not considered in the root-locus method.

Is there any advantage at all in using the root-locus method over the optimal design technique? The answer is affirmative. First, the degree of compensators used in the root-locus method is always smaller than that used in the optimal design. In the optimal design, if only output feedback is used, the degree of compensators required is one less than the degree of the plant. In the root-locus method, the degree of compensators used is generally either one or zero. The second advantage of the root-locus method is that, from the root loci and the zeros of the overall system, we can visualize an approximate time response of the resulting system. This is however not possible in the optimal design.

In the root-locus method, if the compensator used has a degree one less than that of the plant, then the plotting of the root loci is entirely unnecessary. We can locate the poles of the overall system at any desired locations by the methods discussed in Chapter 8 or by direct manipulation.

The root-locus method involves essentially the adjustment of parameters to obtain a good system. Hence this can also be achieved, maybe more easily, by analog or digital computer simulations. However a rough sketch of root loci will immediately tell us whether the design is possible or a different configuration must be chosen.

The root loci of the polynomial $p(s) + kq(s)$ can also be obtained on a digital computer by using the existing subroutines, such as the ones in the IBM System/360 Scientific Subroutine Package. In fact, the root loci in Figures 9-19, 9-21, and 9-23 are obtained with the aid of a digital computer.

References

[1] D'Azzo, J. J., and C. H. Houpis, *Feedback Control System Analysis and Synthesis*, 2d ed. New York: McGraw-Hill, 1965.
[2] Evans, W. R., "Graphical analysis of control systems," *Trans. AIEE*, vol. 67, pp. 547–551, 1948.
[3] Kuo, B. C., *Automatic Control Systems*, 2d ed. Englewood Cliffs, N.J.: Prentice-Hall, 1967.
[4] Moore, A. W., "Phase-locked loops for motor-speed control," *IEEE Spectrum*, vol. 10, pp. 61–67, 1973.
[5] Truxall, J. G., *Automatic Feedback Control System Synthesis*. New York: McGraw-Hill, 1955.

Problems

9-1 The root loci of the system in Figure P9-1(a) are assumed to be of the form shown in Figure P9-1(b). Do all graphical work directly on Figure P9-1(b) to answer the following:

a. Find the stability range of k.

b. If $k = 2$, where are the three poles of the closed-loop system?

c. Where is the real pole that has the same value of k as the pair of pure imaginary poles?

d. If it is required to have overshoot less than 20 percent, settling time smaller than 10 seconds, and the steady-state error due to a step input as small as possible, what k will you choose?

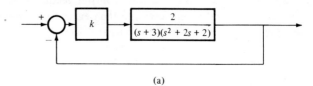

(a)

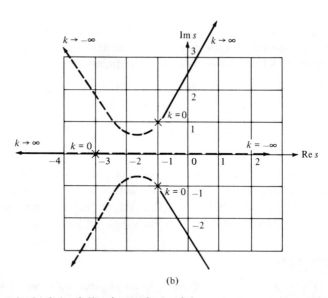

Figure P9-1

(b)

9-2 Find the root loci of the following polynomials:

a. $s^3 + 2s^2 + 3s + ks + 2k$

b. $s^3(1 + 0.001s)(1 + 0.002s) + k(1 + 0.1s)(1 + 0.25s)$

9-3 The transfer function from the thrust deflection angle u to the pitch angle θ of a guided missile is found as

$$\hat{g}(s) = \frac{4(s + 0.05)}{s(s + 2)(s - 1.2)}$$

The configuration of compensator is chosen as shown in Figure P9-2. The actuator is designed to have transfer function $\hat{g}_1(s) = 1/(s + 6.1)$. If $k_1 = 2k_2$, determine whether it is possible to design the system so that the steady-state error due to a step input is less than 10 percent, the overshoot is less than 15 percent, and the settling time is less than 10 seconds.

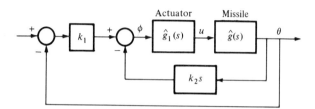

Figure P9-2

9-4 Consider the problem of controlling the yaw of an airplane discussed in Problem 8-8. The configurations of compensator are chosen as shown in Figure P9-3, where $\hat{g}(s) = -k/Js^2 = -2/s^2$. It is required to design an overall system so that the steady-state error due to a ramp input is smaller than 10 percent, the overshoot is smaller than 10 percent, the settling time is smaller than 5 seconds, and the rise time is as small as possible. Is it possible to achieve this design by using configuration (a)? How about configurations (b) and (c)? In using (b) and (c), do you have to plot the root loci? (Answer to the last question: No.)

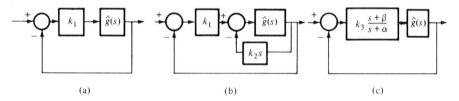

Figure P9-3

9-5 Consider the tracking antenna problem studied in Problem 8-3. Design an overall system by using the configuration shown in Figure 9-23 with $k_1 = 5$ so that the system has a satisfactory step response. Compare your result with the one in Example 3 of Section 9-4.

9-6 Consider the ship stabilization problem discussed in Problem 8-6. The configuration of compensators is chosen as shown in Figure P9-4. Is it possible

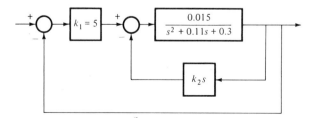

Figure P9-4

to choose a k_2 so that, in the overall system, the steady-state error due to a step input is smaller than 15 percent, the overshoot is smaller than 5 percent, and the settling time is smaller than 30 seconds? (The rise time is of no concern here.) If no such k_2 exists, choose a different k_1, and repeat the design.

9-7 A machine tool can be automatically controlled by a punched tape, as shown in Figure P9-5(a). This class of control systems is often referred to as *numerical control*. By neglecting the so-called quantization problem, the system can be represented by the block diagram shown in Figure P9-5(b). Find a gain k so that the system has a zero steady-state error due to a step function and has a settling time as small as possible. (The overshoot and rise time are of no concern in this problem.)

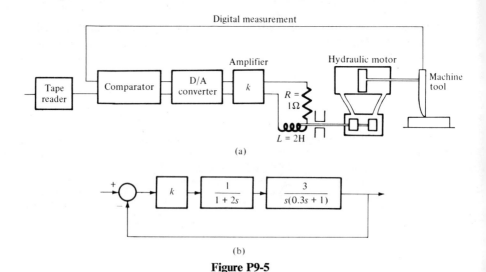

Figure P9-5

9-8 Consider the feedback system shown in Figure P9-6. It is assumed that a is non-negative and is to be designed. Find the range of a in which the feedback system is stable by using the Routh-Hurwitz criterion. Verify this by using the root-locus technique and find the value of a at which the system has the smallest settling time and overshoot.

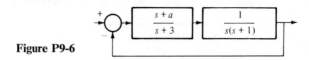

Figure P9-6

9-9 The depth of a submarine can be maintained automatically by the control system shown in Figure P9-7. The transfer function from the stern plane angle θ

to the actual depth y of the submarine can be approximated by $10(s+2)^2/(s+10)(s^2+0.1)$. The actual depth of the submarine is measured by a pressure transducer, which is assumed to have a transfer function of 1. Find the smallest k so that the steady-state error of the system due to a step function is less than 5 percent, the settling time is less than 10 seconds, and the overshoot is less than 2 percent.

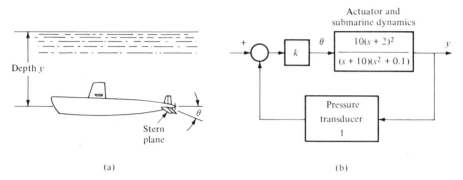

Figure P9-7

9-10 An analog computer generally consists of many control systems. They are used to position potentiometers, induction resolvers, dials, and other devices. Such a control system is shown in Figure P9-8. Find k_1 and k_2 so that the steady-state error of the system due to a step function is zero, the settling time is less than 1 second, and the overshoot is less than 5 percent.

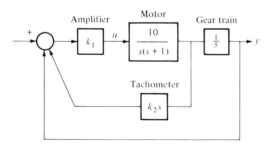

Figure P9-8

9-11 The speed of a motor shaft can be controlled very accurately by using a phase-locked loop [3]. The schematic diagram of such a system and its block diagram are shown in Figure P9-9. The desired speed is transformed into a pulse sequence with a fixed frequency. The encoder at the motor shaft generates a pulse stream whose frequency is proportional to the motor speed. The phase comparator generates a voltage proportional to the difference in phase and, consequently, in frequency. Draw the root loci of the system. Does there exist a k so that the settling time of the system is smaller than 1 second and the overshoot is smaller than 10 percent?

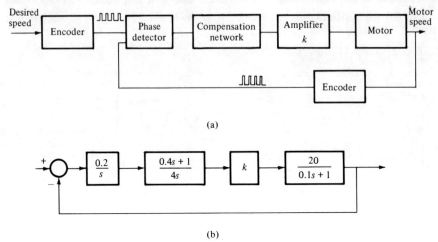

Figure P9-9

CHAPTER

10
FREQUENCY-DOMAIN TECHNIQUE

The design techniques introduced in the preceding chapters use either transfer functions $\hat{g}(s)$ or dynamical equations. The design technique that will be introduced in this chapter uses only the information of $\hat{g}(s)$ on the imaginary axis—that is, $\hat{g}(j\omega)$, for $0 \leq \omega < \infty$—and is called the *frequency-domain technique*.

Chronologically the frequency-domain technique was the first method developed to design control systems. There are several reasons. First, the method does not need precise mathematical descriptions of systems; it uses only $\hat{g}(j\omega)$, which can be obtained by direct measurement [see the remarks following Equation (4-39) and Problem 4-16]. Second, the method is independent of the complexity of systems and is applicable to systems containing time-delay elements. Finally, the method requires only simple calculation.

This chapter is organized as follows: The specifications in the time domain of control systems are translated into the frequency domain in Section 10-1. Three frequency-domain plots of the same transfer function are introduced in Section 10-2. They are the polar plot, log magnitude-phase plot, and Bode plot. These three plots are actually the same plot but with different sets of coordinates. In Section 10-3 the Nyquist stability criterion is introduced. It can be used to check BIBO stability of unit feedback systems. In Section 10-4 the specifications in the frequency domain of overall systems are translated into those of open-loop systems. To do so, the concepts of gain margin and phase margin are introduced.[1] The design method in the frequency domain is then discussed in Section 10-5.

---

[1] The reader is urged to keep Table 10-1 on page 294 in mind, for it provides a bird's-eye view of what will be discussed in the early part of this chapter.

10-1 Frequency-Domain Specifications

Given a plant with the transfer function $\hat{g}(s)$, the design problem is to find an overall system with the transfer function $\hat{g}_f(s)$ that satisfies a certain set of specifications. The specifications may involve rise time, settling time, overshoot, and steady-state errors. Since we shall use only $\hat{g}_f(j\omega)$ in the design, the correlation between these specifications and $\hat{g}_f(j\omega)$ should be established.

Although $\hat{g}_f(s)$ is a rational function with real coefficients, $\hat{g}_f(j\omega)$ is generally a complex-valued function. Let us write $\hat{g}_f(j\omega)$ as

$$\hat{g}_f(j\omega) = |\hat{g}_f(j\omega)|e^{j\theta(\omega)}$$

where $\theta = \tan^{-1}\left[\operatorname{Im}\hat{g}_f(j\omega)/\operatorname{Re}\hat{g}_f(j\omega)\right]$. Typical plots of $\hat{g}_f(j\omega)$ and $\theta(\omega)$ with respect to ω of a control system are shown in Figure 10-1. The specifications on the gain plot $[|\hat{g}_f(j\omega)|$ versus $\omega]$ can be stated in three frequency ranges: low, middle, and high. In the low frequency range, $\hat{g}_f(0)$ and its derivatives are the key information. In the middle frequency range, the gain plot is specified by the peak resonance M_p and the bandwidth ω_b, as shown in Figure 10-1. The *peak resonance* M_p is defined as

$$M_p \triangleq \max_{0 \leq \omega < \infty} |\hat{g}_f(j\omega)|$$

The *bandwidth* ω_b is defined by the equation $|\hat{g}_f(j\omega_b)| = 0.707|\hat{g}_f(0)|$. The gain plot in the high-frequency range is specified as $|\hat{g}_f(j\omega)| < \varepsilon$ for all $\omega > \omega_d$. In the following the relationships between these specifications in the frequency-domain and the time-domain specifications will be discussed.

Let

$$\hat{g}_f(s) = \frac{\beta_0 + \beta_1 s + \cdots + \beta_m s^m}{\alpha_0 + \alpha_1 s + \cdots + \alpha_n s^n} \qquad n \geq m \qquad (10\text{-}1)$$

be an overall transfer function yet to be designed. It is assumed that $\hat{g}_f(s)$ is BIBO stable. The steady-state error due to a step function is equal to, as derived in (6-5),

$$\left|1 - \frac{\beta_0}{\alpha_0}\right| = |1 - \hat{g}_f(0)| \times 100 \text{ percent} \qquad (10\text{-}2)$$

Hence from $|\hat{g}_f(0)|$ and $\theta(0)$, the steady-state error due to a step function can be immediately determined. For example, if $\theta(0) = 0$, and if $|\hat{g}_f(0)| = 1$ [that is, $\hat{g}_f(0) = 1$], then the error is zero; if $\hat{g}_f(0) = 0.95$, then the error is 5 percent. The steady-state error due to a ramp function, as derived in (6-7), is equal to

$$\left|\left(1 - \frac{\beta_0}{\alpha_0}\right)t - \frac{\alpha_0\beta_1 - \beta_0\alpha_1}{\alpha_0^2}\right| = |[1 - \hat{g}_f(0)]t - \hat{g}_f'(0)| \times 100 \text{ percent} \qquad (10\text{-}3)$$

Hence in order to have a finite steady-state error due to a ramp function, it is required that $\hat{g}_f(0) = 1$, and then the error is equal to the slope of $\hat{g}_f(j\omega)$ at $\omega = 0$. Hence from the gain plot at $\hat{g}_f(0)$, the steady-state performance can be easily determined.

10-1 FREQUENCY-DOMAIN SPECIFICATIONS   257

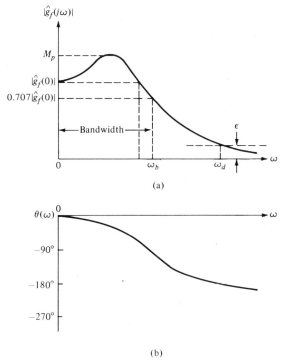

Figure 10-1 (a) Gain and (b) phase plots of $\hat{g}_f(j\omega)$.

Although the steady-state performance can be easily translated into the plot of $\hat{g}_f(j\omega)$, it is not so for the transient performance specifications. A general and rigorous discussion of this problem is not possible; therefore we shall discuss the following very special case: Consider

$$\hat{g}_f(s) = \frac{\omega_n^2}{s^2 + 2\delta\omega_n s + \omega_n^2} \quad (10\text{-}4)$$

or

$$|\hat{g}_f(j\omega)| = \frac{\omega_n^2}{[(\omega_n^2 - \omega^2)^2 + (2\delta\omega_n\omega)^2]^{1/2}} \quad (10\text{-}5)$$

By algebraic manipulation, it can be shown that

$$M_p = \max_\omega |\hat{g}_f(j\omega)| = \frac{1}{2\delta\sqrt{1-\delta^2}} \quad \text{for } \delta \leq 0.707$$

$$= 1 \quad \text{for } \delta \geq 0.707 \quad (10\text{-}6)$$

We see that the peak resonance M_p depends only on the damping ratio δ; their relationship is shown in Figure 10-2, in which the relationship between the overshoot and the damping ratio is also plotted (see Figure 9-4). We see from

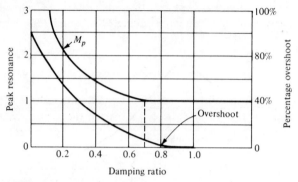

Figure 10-2 Relationship between peak resonance and overshoot.

Figure 10-2 that the larger the overshoot is, the larger the peak resonance. It is also possible to specify the range of M_p for a given overshoot. For example, if the percentage overshoot is required to be less than 10 percent, then the peak resonance M_p should be less than 1.05; if the overshoot is less than 20 percent, then M_p should be less than 1.25. We note that this relation is derived from the transfer function given in Equation (10-4). Hence it is not necessarily valid for other transfer functions. However since there is no other information available, we shall use Figure 10-2 for the general case. The situation here is quite similar to the one in Section 9-2.

The bandwidth of $\hat{g}_f(s)$ as defined in Figure 10-1(a) is closely related to the transient response of the system. Unfortunately a precise statement is not possible; all we can say is that

$$\text{response time} \sim \frac{1}{\text{bandwidth}} \qquad (10\text{-}7)$$

where the response time indicates either the rise time or the settling time. In words, the response time is inversely proportional to the bandwidth; that is, the larger the bandwidth, the smaller the response time. We give in the following two intuitive arguments to support the relation (10-7): A large bandwidth indicates that higher-frequency signals can pass through the system to the outputs (see Section 4-1). In other words, the system can respond swiftly to fast-changing signals; hence the system should have a smaller rise time. The relation (10-7) can also be explained by using the Laplace transform. The Laplace-transform pair is

$$g_f(t) = \frac{1}{2\pi j} \int_{c-j\infty}^{c+j\infty} \hat{g}_f(s) e^{st}\, ds \qquad (10\text{-}8)$$

and

$$\hat{g}_f(s) = \int_0^\infty g_f(t) e^{-st}\, dt \qquad (10\text{-}9)$$

10-1 FREQUENCY-DOMAIN SPECIFICATIONS

Since $\hat{g}_f(s)$ is assumed to be BIBO stable, the c in Equation (10-8) can be taken as $c = 0$. Substituting $s = 0$ and multiplying (10-8) and (10-9) together, we obtain

$$\frac{\int_0^\infty g_f(t)\, dt}{g_f(0)} \cdot \frac{\int_{-\infty}^\infty \hat{g}_f(j\omega)\, d\omega}{2\hat{g}_f(0)} = \pi \qquad (10\text{-}10)$$

The two factors can be viewed as the equivalent bandwidth of $\hat{g}_f(s)$ and the equivalent duration of its impulse response. Hence the larger the bandwidth, the narrower the impulse response. If u is a step function then the response of the system is given by

$$y(t) = \int_0^t g_f(\tau) u(t - \tau)\, d\tau = \int_0^t g_f(\tau)\, d\tau \qquad (10\text{-}11)$$

This equation implies that the output $y(t)$ will reach the steady state in the duration of the impulse response. Hence we conclude that the larger the bandwidth, the smaller the settling time.

The speed of response depends not only on the bandwidth but also on the shape of the gain plot. Clearly it is highly impossible for two systems with exactly the same bandwidth but different gain plots as shown in Figure 10-3 to have the same speed of response. Hence in addition to the bandwidth, the cutoff rate at frequency ω_b should also be discussed. One way to insure a proper cutoff rate is to require that the gain $|\hat{g}_f(j\omega)|$ be less than, say, ε for all $\omega > \omega_d$. The implication of this high-frequency specification on the time response is not clear. However it has a very simple interpretation on noise rejection: Noise or unwanted signals with frequency spectra lying in the range $\omega > \omega_d$ will be greatly suppressed or eliminated.

The ability of the actual output $y(t)$ of a control system to follow a desired signal $y_d(t)$ depends not only on the gain plot but also on the phase plot of $\hat{g}_f(s)$. If the phase plot is not a linear function of ω, then distortion will occur. Distortion of signals in control systems is not as important or critical as in communication systems; hence in the design of control systems, generally no specification is imposed on the phase plot of $\hat{g}_f(s)$.

We summarize what we have discussed in this section. The steady-state performance criteria impose constraints on the gain plot of $\hat{g}_f(j\omega)$ at $\omega = 0$.

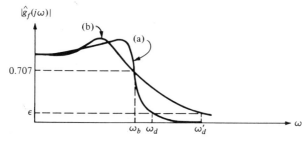

Figure 10-3 Two gain plots with the same bandwidth but different cutoff rates.

If $\hat{g}_f(0) = 1$, then the steady-state error due to a step function is zero. If in addition, the slope of $\hat{g}_f(j\omega)$ at $\omega = 0$ is zero, then the steady-state error due to a ramp function is zero. From the specification on the overshoot a constraint on the peak resonance can be obtained from Figure 10-2. Although the figure is derived from a second-order transfer function with a constant numerator, because of the lack of other information we shall use it for any transfer function. In order to have a fast response, the overall system should have a large bandwidth. However because high-frequency noise is always present in control systems, it is not desirable to have an unnecessarily large bandwidth. High-frequency noise can be eliminated by imposing $|\hat{g}_f(j\omega)| < \varepsilon$ at high frequencies.

The frequency-domain specifications introduced up to this point are for overall transfer functions. The design technique to be introduced starts from the given plant transfer function and then searches for compensators, so that the overall transfer function will satisfy the specifications. Therefore if it is possible to translate the specifications from an overall system to a plant and compensator, then the design can be greatly simplified. Unfortunately this is not easy, except for the configuration of compensators shown in Figure 10-4. Therefore in the remainder of this chapter we study exclusively the unit feedback system shown. Before proceeding we introduce three different frequency plots of $\hat{g}_c(s) = \hat{g}(s)C(s)$.

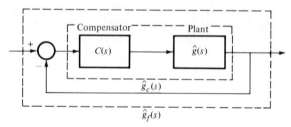

Figure 10-4 A unit feedback system.

10-2 Frequency-Domain Plots: Polar Plot, Bode Plot, and Log Magnitude-Phase Plot

We introduce in this section the plots of transfer function $\hat{g}_c(s)$ along the positive $j\omega$-axis [that is, $\hat{g}_c(j\omega)$ for $0 \le \omega < \infty$] with respect to three different sets of coordinates.

The plots of $\hat{g}_c(j\omega)$ with three different sets of coordinates are shown in Figure 10-5. The one in Figure 10-5(a) is called the *polar plot*. It is a plot of $\hat{g}_c(j\omega)$ on the $\hat{g}_c(j\omega)$-plane as shown. It is obtained point by point by computing $\hat{g}_c(j\omega) = |\hat{g}_c(j\omega)|e^{j\theta} = \text{Re } \hat{g}_c(j\omega) + j \text{ Im } \hat{g}_c(j\omega)$ at each ω. The behavior of $\hat{g}_c(j\omega)$ as $\omega \to 0$ and $\omega \to \infty$ can be estimated by using the approximated equations of $\hat{g}_c(j\omega)$. This is illustrated in the following example.

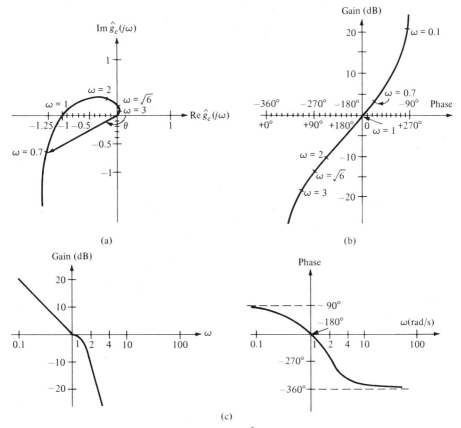

Figure 10-5 The plots of $\hat{g}_c(s) = (2 - s)/s(s^2 + 2s + 2)$ with three different coordinates. (a) Polar plot. (b) Log magnitude-phase plot. (c) Bode plot.

Example 1

Plot the polar plot of $\hat{g}_c(s) = (2 - s)/s(s^2 + 2s + 2)$.

We first compute

$$\hat{g}_c(j\omega) = \frac{2 - j\omega}{j\omega(-\omega^2 + 2j\omega + 2)} = -\frac{6 - \omega^2}{4 + \omega^4} - j\frac{4(1 - \omega^2)}{\omega(4 + \omega^4)}$$

Using this equation, we compute $\hat{g}_c(j0.1) = -1.5 - j10$, $\hat{g}_c(j0.7) = -1.3 - j0.69$, $\hat{g}_c(j1) = -1$, $\hat{g}_c(j2) = -0.17 + j0.3$, $\hat{g}_c(j\sqrt{6}) = j0.2$, and $\hat{g}_c(j3) = 0.035 + j0.13$. These points are plotted in Figure 10-5(a), as shown.

We discuss now the behavior of $\hat{g}_c(j\omega)$ as $\omega \to 0$ and $\omega \to \infty$. As ω becomes very large, we have $2 - j\omega \doteq -j\omega$, and $-\omega^2 + 2j\omega + 2 \doteq -\omega^2$; hence $\hat{g}_c(j\omega)$ can be approximated by

$$\hat{g}_c(j\omega) \doteq \frac{-j\omega}{j\omega(-\omega^2)} = \frac{1}{\omega^2} \quad \text{as } \omega \to \infty$$

262 FREQUENCY-DOMAIN TECHNIQUE

This implies that $\hat{g}_c(j\omega)$ approaches zero with phase $0°$ as ω approaches infinity. As ω approaches 0, the real part of $\hat{g}_c(j\omega)$ approaches $-6/4 = -1.5$, and the imaginary part becomes $-\infty$. With this information the polar plot can be completed as shown. ∎

The plot in Figure 10-5(b) is called the *log magnitude-phase plot*. It is a plot of gain versus phase on rectangular coordinates as shown. The gain is expressed in decibels (dB), defined as

$$\text{dB} = 20 \log |\hat{g}_c(j\omega)| \qquad (10\text{-}12)$$

This plot can be obtained by computing $\hat{g}_c(j\omega) = |g_c(j\omega)|e^{j\theta}$ or obtained from the polar plot. For example, the point at $\omega = 0.7$ can be measured on Figure 10-5(a) to have magnitude 1.5 or $20 \log 1.5 = 3.52$ dB and phase $-153°$. This is plotted on Figure 10-5(b). The other points can be similarily obtained.

The plot in Figure 10-5(c) is called the *Bode plot*. It actually consists of two plots: the gain plot versus frequency and the phase plot versus frequency. The gain is expressed in decibels, the phase in degrees, and the frequency in logarithmic units, as shown in Figure 10-5(c). Clearly the Bode plot of $\hat{g}_c(s)$ can be obtained from the polar plot, the log magnitude-phase plot, or directly from $\hat{g}_c(j\omega)$ point by point.

The polar plot, the log magnitude-phase plot, and the Bode plot are actually the same plot except for different coordinates. Hence if any one of them is known, the other two can be obtained merely by coordinate transformations. The Bode plot, which we shall now discuss, is the easiest of these three plots to obtain. Consider $\hat{g}_c(s) = N_c(s)/D_c(s)$. It is first written as

$$\hat{g}_c(s) = \frac{k \prod_q (1 + \tau_q s) \prod_j \left[1 + (2\delta_j/\omega_{n_j})s + s^2/\omega_{n_j}^2\right]}{s^l \prod_p (1 + \tau_p s) \prod_i \left[1 + (2\delta_i/\omega_{n_i})s + s^2/\omega_{n_i}^2\right]} \qquad (10\text{-}13)$$

or

$$\hat{g}_c(j\omega) = \frac{k \prod_q (1 + j\tau_q \omega) \prod_j \left[1 + j(2\delta_j/\omega_{n_j})\omega - \omega^2/\omega_{n_j}^2\right]}{(j\omega)^l \prod_p (1 + j\tau_p \omega) \prod_i \left[1 + j(2\delta_i/\omega_{n_i})\omega - \omega^2/\omega_{n_i}^2\right]} \qquad (10\text{-}14)$$

where the quadratic factors contain complex conjugate poles. The gain of $g_c(j\omega)$ in decibels is given by

$$20 \log |\hat{g}_c(j\omega)| = 20 \log |k| + 20 \sum_q \log |1 + j\tau_q \omega|$$

$$+ 20 \sum_j \log |1 + j(2\delta_j/\omega_{n_j})\omega - (\omega/\omega_{n_i})^2|$$

$$- 20 \log |j\omega|^l - 20 \sum_p \log |1 + j\tau_p \omega|$$

$$- 20 \sum_i \log |1 + j(2\delta_i/\omega_{n_i})\omega - (\omega/\omega_{n_i})^2|$$

$$(10\text{-}15)$$

The phase of $\hat{g}_c(j\omega)$ is given by

$$\arg \hat{g}_c(j\omega) = \arg k + \sum_q \arg(1 + j\tau_q\omega) \quad (10\text{-}16)$$

$$+ \sum_j \arg[1 + j(2\delta_j/\omega_{n_j})\omega - (\omega/\omega_{n_j})^2] - \arg(j\omega)^l$$

$$- \sum_p \arg[1 + j\tau_p\omega] - \sum_i \arg(1 + j(2\delta_i/\omega_{n_i}) - (\omega/\omega_{n_i})^2]$$

where "arg" stands for the phase or argument. We see that if every term in the right-hand sides of (10-15) and (10-16) is plotted, by adding them up we obtain the total gain and phase. There are basically four different types of factors in (10-15) and (10-16). They are: (1) a constant gain k, (2) poles or zeros at the origin, (3) poles or zeros on the real axis, and (4) complex conjugate poles or zeros. The Bode plots of these factors will be discussed in the following.

A constant gain, k. A constant gain, k, has logarithmic gain $20 \log |k|$ dB and phase $0°$ or $180°$ depending on whether k is positive or negative. The gain and phase are independent of ω, and hence they are horizontal lines on the Bode plot.

Poles or zeros at the origin. The repeated pole $1/s^l$ has a logarithmic magnitude $-20 \log |j\omega|^l = -20l \log \omega$ dB and phase $-l \times 90°$. Since the phase is independent of ω, it is a horizontal line in the Bode plot, as shown in Figure 10-6. The gain plot is a straight line in the Bode plot, because the coordinate of frequency is in the units of $\log \omega$. For a simple pole the slope is -20dB/decade, as shown in Figure 10-6. For a repeated pole with multiplicity l, the slope is $-20l$ dB/decade. The repeated zero s^l has a logarithmic gain $20 \log |j\omega|^l = 20l \log \omega$ dB and phase $l \times 90$. Its Bode plot is similar to that of the repeated pole except that the slope of the gain plot and the phase are now positive rather than negative.

Poles or zeros on the real axis. The real pole $1/(1 + \tau_p s)$ has a logarithmic magnitude $-20 \log |1 + j\tau_p\omega|$ dB. The magnitude can be very well approx-

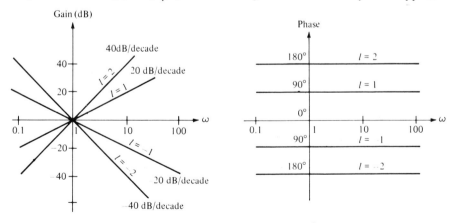

Figure 10-6 The Bode plots of $(j\omega)^2$.

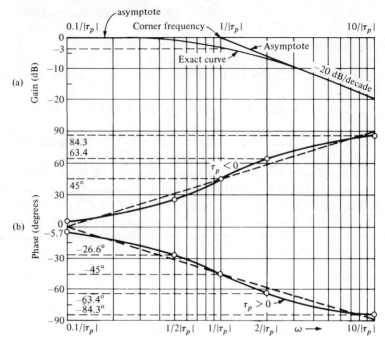

Figure 10-7 The Bode plot of $(1 + j\tau_p\omega)^{-1}$.

imated by two straight lines called *asymptotes*, as shown in Figure 10-7(a). For very small ω, the gain $-20 \log |1 + j\tau_p\omega|$ is approximately equal to $-20 \log 1 = 0$ dB; for very large ω, the gain $-20 \log |1 + j\tau_p\omega|$ is approximately equal to $-20 \log |\tau_p\omega|$ dB. The function $-20 \log |\tau_p\omega|$ is a straight line on the Bode plot, with slope -20 dB/decade and intersecting the ω-axis at $1/|\tau_p|$. It is easy to show that, for $\omega \leq |0.1/\tau_p|$ and $\omega \geq |10/\tau_p|$, the error between the asymptotes and the actual gain plot is less than 0.04 dB. The maximum error occurs at $\omega = 1/\tau_p$ and is equal to 3 dB. The frequency $1/|\tau_p|$ is often called the *corner frequency*. Hence the gain plot of a real pole can be obtained easily from the asymptotes. The gain plot of a real zero is similar to that of a real pole, except that the slope of the asymptote is $+20$ dB/decade.

The phase plot of the real pole, $-\arg(1 + j\tau_p\omega)$, is shown in Figure 10-7(b) for both positive and negative τ_p. A straight line that approximates the phase plot is shown. The straight line is connected from $0.1/|\tau_p|$ at $0°$ to $10/|\tau_p|$ at $-90°$ if $\tau_p > 0$ or $+90°$ if $\tau_p < 0$. The maximum error occurs at $\omega = 0.1/|\tau_p|$ and at $\omega = 10/|\tau_p|$ and is equal to $5.7°$. Hence from the straight line a fairly accurate phase plot can be obtained. The phase plot of a real zero is again similar to that of a real pole.

Complex conjugate poles or zeros. Consider the complex conjugate poles $[1 + (2\delta/\omega_n)s + (s/\omega_n)^2]^{-1}$. Its logarithmic magnitude is

$$-20 \log |1 + j(2\delta/\omega_n)\omega - (\omega/\omega_n)^2| \text{ dB}$$

It is approximately equal to, for $\omega \ll 1$, $-20 \log 1 = 0$ dB and equal to, for very large ω, $-20 \log |-(\omega/\omega_n)^2| = -40 \log(\omega/\omega_n)$ dB. The function

$$-40 \log(\omega/\omega_n)$$

is a straight line on the Bode plot, with slope -40 dB/decade and intersecting the ω-axis at $\omega = \omega_n$, as shown in Figure 10-8. The exact gain plot in the neighborhood of ω_n depends highly on the damping ratio δ and is plotted in Figure 10-8 for several values of δ.

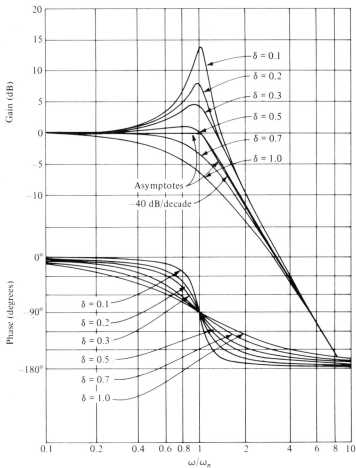

Figure 10-8 Bode plots of $[1 + j(2\delta/\omega_n)\omega - (\omega/\omega_n)^2]^{-1}$.

The phase of $[1 + j(2\delta/\omega_n)\omega - (\omega/\omega_n)^2]^{-1}$ is equal to $-90°$ at $\omega = \omega_n$, approaches $0°$ as $\omega \to 0$, and approaches $-180°$ as $\omega \to \infty$. The exact phase plots for several values of δ are shown in Figure 10-8.

In plotting the Bode plot of a pair of complex conjugate poles, we first compute the corner frequency ω_n and the damping ratio δ. From the corner frequency, two asymptotes can be drawn. Then, by using the family of curves in Figure 10-8, the Bode plot can be obtained. The Bode plot of a pair of complex conjugate zeros can be obtained similarly.

We give two examples to illustrate the plotting of Bode plots.

Example 2

Consider the transfer function

$$\hat{g}_c(s) = \frac{4}{s(s+8)} = \frac{0.5}{s(1+0.125s)}$$

The gain of 0.5 expressed in decibels is equal to $20 \log (0.5) = -6$ dB. It is a horizontal line in the Bode plot. The gain plot of s^{-1} is a straight line with slope -20 dB/decade and passing through 0 dB at $\omega = 1$. The asymptotes of $(1 + 0.125s)^{-1}$ intersect at $\omega = 8$. The sum of these straight lines is shown in Figure 10-9(a) by a heavy dashed line. The Bode plot is then drawn from these straight lines as shown by the solid line.

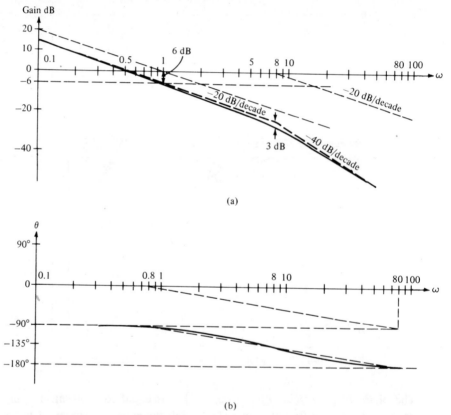

Figure 10-9 The Bode plot of $4/s(s+8)$.

The phase due to gain 0.5 is zero. The phase due to the pole at the origin is $-90°$. The phase plot of $(1 + 0.125s)^{-1}$ is obtained in Figure 10-9(b) from its linear approximation. By shifting this phase plot down 90°, we obtained the total phase plot of the transfer function.

Example 3

Consider the transfer function

$$\hat{g}_c(s) = \frac{50(s+2)}{s(s^2+4s+100)} = \frac{(1+0.5s)}{s\left[1 + 2\left(\dfrac{0.2}{10}\right)s + \left(\dfrac{s}{10}\right)^2\right]}$$

The asymptotes of the zero and the complex conjugate poles are shown in Figure 10-10. By summing these asymptotes with the plot of s^{-1}, we obtain the heavy dashed lines. Since the damping ratio is 0.2, by using Figure 10-8 the final gain plot is obtained as shown by the solid line in Figure 10-10. Though the gain plot is obtained by rough sketch, its degree of accuracy is acceptable in practice.

The phase plots of the factors of $\hat{g}_c(s)$ are shown by light dashed lines in Figure 10-10. The phase plot of $(1 + 0.5s)$ is obtained from the linear approximation; the phase plot of the complex conjugate poles is obtained from Figure 10-8. By adding the three plots algebraically point by point, we obtained the total phase plot of $\hat{g}_c(s)$ as shown by the solid line in Figure 10-10. ∎

We see from these two examples that Bode plots can be easily plotted. A remark is in order concerning the gain and phase plots of a particular class of transfer functions. A transfer function is said to be a *minimum-phase* transfer function if its poles and zeros all lie in the open left-half s-plane. For a minimum-phase transfer function, there is a unique relationship between the gain plot and the phase plot. Hence for this class of transfer functions it is sufficient to plot only gain plots. This fact is not used in this text and will not be discussed further. The interested reader is referred to References [1] and [5].

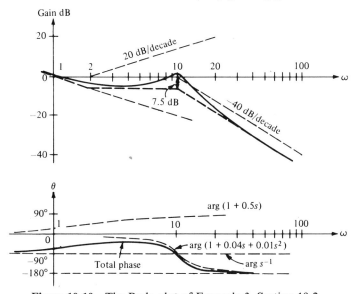

Figure 10-10 The Bode plot of Example 3, Section 10-2.

10-3 Stability Test in the Frequency Domain

Consider the unit feedback system shown in Figure 10-4. The design problem is to find a compensator $C(s)$ so that the overall system is totally stable and satisfies a certain set of specifications. Clearly if the overall transfer function

$$\hat{g}_f(s) = \frac{C(s)\hat{g}(s)}{1 + C(s)\hat{g}(s)} \tag{10-17}$$

is computed, its stability can then be determined by using the Routh-Hurwitz criterion. However in the frequency-domain design, we do not compute $\hat{g}_f(s)$. We plot only the frequency plots of $C(s)$ and $\hat{g}(s)$; hence it is necessary to find the stability of $\hat{g}_f(s)$ from this information. In this section such a stability test will be introduced. Before proceeding, we need to examine the concept of the Nyquist plot.

Definition 10-1

The mapping of the contour C_1 in Figure 10-11(a), or the contour C_1 in Figure 10-11(b) if $\hat{g}_c(s) = C(s)\hat{g}(s)$ has poles on the $j\omega$-axis, by the rational function $\hat{g}_c(s)$ is called the *Nyquist plot* of $\hat{g}_c(s)$. ∎

Note that the radius R in Figure 10-11 should approach infinity. The radius r should approach zero so that the semicircle does not contain any open right-half plane poles of $\hat{g}_c(s)$. Since the contour C_1 is a closed curve, the mapping of C_1 by $\hat{g}_c(s)$ is again a closed curve. The direction of C_1 in Figure 10-11 is chosen arbitrarily as clockwise.

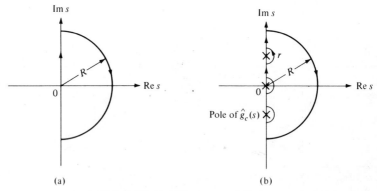

Figure 10-11 The contour C_1.

Every rational function in this text is assumed to have real coefficients. This assumption implies that the real part of $\hat{g}_c(j\omega)$ is an even function of ω and the imaginary part of $\hat{g}_c(j\omega)$ is an odd function of ω; hence we have

$$\hat{g}_c(j\omega) = \overline{\hat{g}}_c(-j\omega)$$

where the overbar denotes the complex conjugate. Hence the plot of $\hat{g}_c(j\omega)$ for $0 > \omega > -\infty$ is just the mirror image, with respect to the real axis, of the plot of $\hat{g}_c(j\omega)$ for $\infty > \omega > 0$. If $\hat{g}_c(s)$ is strictly proper, as is often the case in practice, then the plot of $\hat{g}_c(s)$ along the infinite semicircle is zero. Hence once the polar plot of $\hat{g}_c(s)$ is plotted, the entire Nyquist plot can be obtained immediately.

Theorem 10-1 (Nyquist Stability Criterion)

The unit feedback system shown in Figure 10-4 is BIBO stable if and only if the Nyquist plot of $\hat{g}_c(s)$ does not go through the point $(-1, 0)$ and does encircle it P times in the counterclockwise direction,[2] where P is the number of the open right-half plane poles of $\hat{g}_c(s)$. ∎

We first give an example to illustrate the application of the Nyquist criterion.

Example

Consider the system shown in Figure 10-4. Let

$$\hat{g}_c(s) = \frac{2s + 1}{s(s - 1)}$$

Determine the stability of the overall system by using the Nyquist criterion.

The Nyquist plot of $\hat{g}_c(s)$ is shown in Figure 10-12(a). It can be obtained by computing $\hat{g}_c(j\omega)$, or from the Bode plot, or from measurements as shown in Figure 10-12(b). As $\omega \to \infty$, $\hat{g}_c(j\omega)$ is approximately equal to $2/j\omega$; hence the plot approaches zero with phase $-90°$. For s in the neighborhood of the origin, $\hat{g}_c(s)$ is approximately equal to $1/(-s)$. Hence as $\omega \to 0$, the plot approaches infinity, with phase $90°$. The mapping of the small arc AB in Figure 10-12(b) by $\hat{g}_c(s) \doteq 1/(-s)$ is shown in Figure 10-12(a) by $A'B'$. The complete Nyquist plot is obtained by taking the mirror image, with respect to the real axis, of $\hat{g}_c(j\omega)$ for $\omega > 0$.

The transfer function $\hat{g}_c(s)$ has one open right-half plane pole. Hence the overall system is BIBO stable if the Nyquist plot encircles the point $(-1, 0)$ once in the counterclockwise direction. Since this is the case, the overall system with the transfer function $\hat{g}_f(s) = \hat{g}_c(s)/[1 + \hat{g}_c(s)]$ is BIBO stable. ∎

We now offer a few comments concerning the Nyquist criterion. First, a proof of the Nyquist criterion can be found in any of the references listed at the end of this chapter. Second, the Nyquist plot is essentially a polar plot, and since Bode, polar, and log magnitude-phase plots are all equivalent, the stability of $\hat{g}_f(s)$ can also be determined from the Bode plot and log magnitude-phase

---

[2] If C_1 in Figure 10-11 is chosen as counterclockwise, then the Nyquist plot should encircle the point in the clockwise direction.

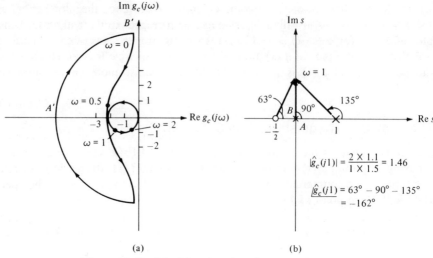

Figure 10-12 The Nyquist plot of $(2s + 1)/s(s - 1)$.

plot. However they are not as convenient as the polar plot. Third, if the plant and the compensator are completely characterized by $\hat{g}(s)$ and $\hat{C}(s)$, respectively, and if there is no pole-zero cancellation between $\hat{g}(s)$ and $\hat{C}(s)$, or the canceled poles are stable poles, then the Nyquist criterion implies total stability of the overall system. Fourth, in counting the number of encirclements in Theorem 10-1, we count the net encirclements. For example, if the Nyquist plot encircles the point $(-1, 0)$ twice in the counterclockwise direction and once in the clockwise direction, then the net encirclement is one in the counterclockwise direction. Finally, the Nyquist criterion is very important in the study of nonlinear systems by using the so-called describing-function method. See Reference [1].

We shall now present a special case of Theorem 10-1 to conclude this section.

Corollary 10-1

Consider the unit feedback system shown in Figure 10-4. If the transfer function $\hat{g}_c(s) = \hat{C}(s)\hat{g}(s)$ has no open right-half s-plane poles, then the overall system is BIBO stable if and only if the Nyquist plot of $\hat{g}_c(s)$ does not go through the point $(-1, 0)$ and does not encircle it. ∎

10-4 Specifications in Terms of Open-Loop Transfer Functions

Consider the unit feedback system shown in Figure 10-4. In this section the frequency-domain specifications of the overall system will be translated into the Bode plot of $\hat{g}_c(s)$. The reason for choosing the Bode plot rather than the polar or the log magnitude-phase plot is twofold. First, the Bode plot is the easiest to obtain. Second, and more important, in the process of searching a compensator, the Bode plot of the compensator can be just added to that of the plant. Hence the Bode plot is most convenient in the design. First, we discuss the steady-state performance and then, the transient performance.

Steady-state performance. Consider the transfer function

$$\hat{g}_c(s) = \frac{k \prod_q (1 + \tau_q s) \prod_j [1 + 2(\delta_j/\omega_{n_j})s + s^2/\omega_{n_j}^2]}{s^l \prod_p (1 + \tau_p s) \prod_i [1 + 2(\delta_i/\omega_{n_i})s + s^2/\omega_{n_i}^2]} \quad (10\text{-}18)$$

We shall call $\hat{g}_c(s)$ a type l transfer function, according to the number of poles at the origin. Hence a type 0 transfer function has no pole at the origin; a type 1 transfer function has one pole at the origin; and so forth. We define

$$K_p = \lim_{s \to 0} \hat{g}_c(s) = \hat{g}_c(0) \quad (10\text{-}19)$$

$$K_v = \lim_{s \to 0} s\hat{g}_c(s) \quad (10\text{-}20)$$

The constant K_p is called the *position-error constant*; and K_v, the *velocity-error constant*. Clearly for a type 0 transfer function, we have

$$K_p = \lim_{s \to 0} \hat{g}_c(s) = k$$

and $K_v = 0$. For a type 1 transfer function we have $K_p = \infty$ and

$$K_v = \lim_{s \to 0} s\hat{g}_c(s) = k$$

We see that the constant k in Equation (10-18) can be K_p or K_v, depending on l. We establish now the relationship between the steady-state error and K_p, K_v. As in (10-2), the steady-state error due to a step function is equal to

$$e_1 = |1 - \hat{g}_f(0)| \times 100 \text{ percent} \quad (10\text{-}21)$$

The substitution of $\hat{g}_f(s) = \hat{g}_c(s)/[1 + \hat{g}_c(s)]$ into Equation (10-21) yields

$$e_1 = \left|1 - \frac{\hat{g}_c(0)}{1 + \hat{g}_c(0)}\right| \times 100 \text{ percent} = \left|\frac{1}{1 + K_p}\right| \times 100 \text{ percent} \quad (10\text{-}22)$$

Hence if the steady-state error due to a step function is specified, the required range of K_p can be obtained immediately from Equation (10-22). As in (10-3), the steady-state error due to a ramp function is equal to

$$e_2 = |[1 - \hat{g}_f(0)]t - \hat{g}_f'(0)| \times 100 \text{ percent} \quad (10\text{-}23)$$

In order to have a finite e_2, it is required that $[1 - \hat{g}_f(0)] = 0$ or, equivalently, $1/(1 + K_p) = 0$. This implies that in order to have a finite e_2, it is required that $K_p = \infty$ or, correspondingly, the transfer function $\hat{g}_c(s)$ has at least one pole at the origin. Under this assumption, $\hat{g}_c(s)$ can be written as $\hat{g}_c(s) = N_c(s)/s\bar{D}_c(s)$. From this, it is straightforward to verify that

$$\hat{g}_f'(0) = \frac{d}{ds}\hat{g}_f(s)\bigg|_{s=0} = \frac{d}{ds}\left[\frac{\hat{g}_c(s)}{1 + \hat{g}_c(s)}\right]\bigg|_{s=0} = \frac{\bar{D}_c(0)}{N_c(0)} = \frac{1}{K_v} \quad (10\text{-}24)$$

Hence if $K_v \neq 0$ [this implies that $g_c(s)$ has at least one pole at the origin], then the steady-state error is given by[3]

$$e_2 = \left|\frac{1}{K_v}\right| \times 100 \text{ percent} \qquad (10\text{-}25)$$

Thus if the steady-state error due to a ramp function is specified, then the required range of K_v can be obtained immediately from (10-25).

We show in the following that the constants K_p and K_v are directly obtainable from the Bode plot, as shown in Figure 10-13. For a type 0 transfer function, we have, at very low frequencies,

$$20 \log |\hat{g}_c(j\omega)| \doteq 20 \log |k| = 20 \log |K_p| \qquad (10\text{-}26)$$

For a type 1 transfer function, we have, at very low frequencies,

$$20 \log |\hat{g}_c(j\omega)| \doteq 20 \log \left|\frac{k}{j\omega}\right| = 20 \log \left|\frac{k}{\omega}\right| = 20 \log \left|\frac{K_v}{\omega}\right| \qquad (10\text{-}27)$$

This is a straight line with slope -20 dB/decade. The gain of this asymptote at $\omega = 1$ is equal to $20 \log K_v$. The intersection of this asymptote with the ω-axis also gives $\omega = K_v$. Hence from the leftmost asymptote of a Bode plot, the constants K_p and K_v can be obtained immediately. For type 2 or higher transfer functions, K_p and K_v are both equal to infinity.

We summarize what we have discussed up to this point. Consider the unit feedback system shown in Figure 10-4. If the steady-state error due to a step

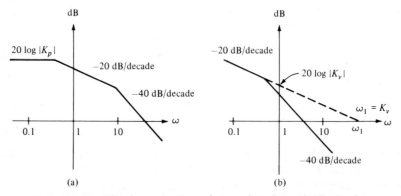

Figure 10-13 The determination of K_p and K_v from the Bode plot.

---

[3] If it is known beforehand that e_2 is bounded and has a limit, then this formula can also be obtained by applying the final-value theorem as follows:

$$\lim_{t \to 0} e_2(t) = \lim_{s \to 0} s e_2(s) = \lim_{s \to 0} s[\hat{y}_d(s) - \hat{y}(s)] = \lim_{s \to 0} s\left[1 - \frac{g_c(s)}{1 + g_c(s)}\right]\frac{1}{s^2}$$

$$= \lim_{s \to 0} \frac{1}{s + sg_c(s)} = \frac{1}{K_v}$$

function is required to be less than $\gamma \times 100$ percent, then Equation (10-22) implies that the position-error constant of the open-loop transfer function $\hat{g}_c(s)$ should be

$$\left|\frac{1}{1+K_p}\right| < \gamma$$

or

$$K_p > \frac{1-\gamma}{\gamma}$$

by considering only positive K_p. If $\hat{g}_c(s)$ is a type 0 transfer function, then the gain at low frequency is required to be larger than

$$20 \log\left(\frac{1-\gamma}{\gamma}\right) \text{ dB}$$

in order to satisfy the steady-state error specification. If $\hat{g}_c(s)$ is a type 1 or higher transfer function, then $K_p = \infty$, and the specification on the steady-state error due to a step function is always met.

Now if the steady-state error due to a ramp function of the overall system is required to be less than $\gamma \times 100$ percent, then (10-25) implies that

$$\left|\frac{1}{K_v}\right| < \gamma$$

or

$$K_v > \frac{1}{\gamma}$$

by considering only positive K_v. It means that if $\hat{g}_c(s)$ is a type 0 transfer function, then this condition can never be met. If $\hat{g}_c(s)$ is a type 1 transfer function, then the gain of the leftmost asymptote should be at least $20 \log (1/\gamma)$ dB at $\omega = 1$ in order to satisfy the steady-state performance specification. If $\hat{g}_c(s)$ is a type 2 or higher transfer function, then this condition is always satisfied.

Transient performance. The frequency-domain specifications on the transient performance of an overall system are stated in terms of the peak resonance M_p, bandwidth, and the high-frequency gain of $\hat{g}_f(s)$. In this subsection the constraints imposed by these specifications on the open-loop transfer function $\hat{g}_c(s)$ will be studied.

Consider the unit feedback system shown in Figure 10-4. The overall transfer function is

$$\hat{g}_f(s) = \frac{\hat{g}_c(s)}{1 + \hat{g}_c(s)} \qquad (10\text{-}28)$$

Consider the polar plot of $\hat{g}_c(s)$ shown in Figure 10-14. Let $x + jy$ be a point in

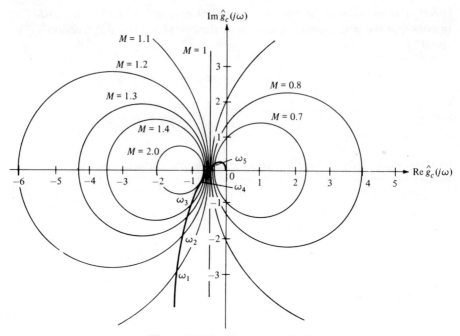

Figure 10-14 Constant-M loci.

the plot. We compute in the following the points that have a constant $|\hat{g}_f(j\omega)|$; that is,

$$M = |g_f(j\omega)| = \left|\frac{x + jy}{1 + x + jy}\right| = \left[\frac{x^2 + y^2}{(1 + x)^2 + y^2}\right]^{1/2}$$

or

$$(1 + x)^2 M^2 + y^2 M^2 = x^2 + y^2 \qquad (10\text{-}29)$$

By algebraic manipulation, Equation (10-29) can be written as

$$\left(x - \frac{M^2}{1 - M^2}\right)^2 + y^2 = \left(\frac{M}{1 - M^2}\right)^2 \qquad (10\text{-}30)$$

This is the equation of a circle with its center at $x = M^2/(1 - M^2)$ and $y = 0$ and of radius $M/(1 - M^2)$. A family of constant-M circles is shown in Figure 10-14. For $M = 1$, the circle reduces to a vertical straight line passing through the point $(-0.5, 0)$. Similar to the constant-gain plots of $\hat{g}_f(j\omega)$, it is possible to plot the points that have the same phase of $\hat{g}_f(j\omega)$. This information is not used in our design, and hence its discussion is omitted.

With the constant-M loci, the gain plot of $\hat{g}_f(j\omega)$ can be easily obtained from the polar plot of $\hat{g}_c(j\omega)$ as shown in Figure 10-15. The peak resonance is given by the M value of the circle tangent to the polar plot of $\hat{g}_c(j\omega)$. Since the bandwidth is defined as the frequency at which the gain of $\hat{g}_f(j\omega)$ is equal to

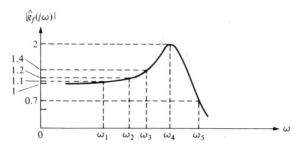

Figure 10-15 The gain plot of $\hat{g}_f(j\omega)$ obtained from the polar plot of $\hat{g}_c(j\omega)$ in Figure 10-14.

$0.707\hat{g}_f(0)$, the bandwidth is given by the frequency at which the polar plot of $\hat{g}_c(j\omega)$ intersects the circle with $M = 0.707|\hat{g}_c(0)/[1 + \hat{g}_c(0)]|$, or $M = 0.707$ if $\hat{g}_c(s)$ is a type 1 or higher-order transfer function.

With the constant-M loci, a part of the design problem can now be stated as follows: Given a plant with transfer function $\hat{g}(s)$, and given the allowable maximum peak resonance M_p, and the desired bandwidth ω_b, find $C(s)$ so that the polar plot of $\hat{g}_c(s) = C(s)\hat{g}(s)$ will intersect the circle with $M = 0.707 \times |\hat{g}_c(0)/[1 + \hat{g}_c(0)]|$ at $\omega = \omega_b$ and will not cut into the circle with $M = M_p$. Furthermore if $C(s)\hat{g}(s)$ has P number of open right-half s-plane poles, the Nyquist plot of $C(s)\hat{g}(s)$ is required to encircle the point $(-1, 0)$ and, consequently, the M_p circle, P times in the counterclockwise direction in order to insure BIBO stability of the overall system. Unfortunately the design is very difficult to carry out by using the polar plot, because of the lack of correlation between the polar plot of $\hat{g}(s)$ and that of $C(s)\hat{g}(s)$. Hence it is desirable to translate the specifications on the polar plot into the Bode plot. Although it is possible to translate the constant-M_p loci into the Bode plot, the loci are rather complicated. Therefore it is desirable to introduce some other specifications to replace the M loci.

Consider the polar plot of $\hat{g}_c(s)$ shown in Figure 10-16. The frequency, denoted as ω_p, at which the phase of $\hat{g}_c(j\omega)$ is equal to $-180°$ is called a *phase crossover frequency*. The frequency, denoted as ω_c, at which the gain of $\hat{g}_c(j\omega)$ is equal to 1 [that is, $|\hat{g}_c(j\omega_c)| = 1$] is called a *gain crossover frequency*. We define

$$\text{gain margin} \triangleq 20 \log |-1| - 20 \log |\hat{g}_c(j\omega_p)| = -20 \log |\hat{g}_c(j\omega_p)|$$

(10-31)

and

$$\text{phase margin} \triangleq |\arg(-1)| - |\arg[g_c(j\omega_c)]| = 180° - |\arg[g_c(j\omega_c)]|$$

(10-32)[4]

---

[4] By convention, an angle is negative if measured in the clockwise direction and positive if measured in the counterclockwise direction.

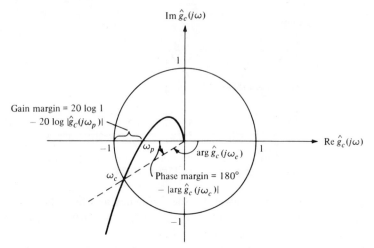

Figure 10-16 Phase margin and gain margin.

In this definition, $\arg\left[g_c(j\omega_c)\right]$ must be measured in the clockwise direction. Note that a gain margin and a phase margin can be negative or positive. If a gain margin or a phase margin is equal to zero, then the polar plot and consequently the Nyquist plot of $\hat{g}_c(s)$ will pass through the point $(-1, 0)$, and the overall system shown in Figure 10-4 is unstable. From Figure 10-17 we can see easily that the peak resonance imposes some constraints on the phase margin and the gain margin. For example, if the peak resonance is required to be less than 1.2, then the gain margin should be at least 5.3 dB, and the phase margin at least 50°; otherwise the peak resonance will be larger than 1.2.

In order to specify the peak resonance from the phase margin and the gain margin, the polar plot of $\hat{g}_c(j\omega)$ should be of the form shown in Figures 10-17 or 10-18(a). We see from Figure 10-17 that, if the gain margin is larger than 10 dB and the phase margin is larger than 45°, and if the polar plot of $\hat{g}_c(j\omega)$ is of the form shown, then the peak resonance of the overall system $\hat{g}_f(s)$ is very possibly in the range of 1.3. If the gain margin is larger than 12 dB, and the phase margin is larger than 60°, then the peak resonance may be in the range of 1. Therefore for this type of polar plots, the specification on the peak resonance can be reduced to the specifications on the phase margin and the gain margin.

The foregoing discussion however cannot be applied to every polar plot. For example, although the phase and gain margins of the polar plot in Figure 10-18(b) are very large, the peak resonance of the overall system is almost infinity. The polar plots in Figure 10-18(c) and (d) have two or three gain margins; the specifications of peak resonance from these gain margins are rather complicated, and hence the gain and phase margins are not used for this case.

Although the phase and gain margins are useful in specifying the peak resonance only for a special class of polar plots, it turns out that a great number of polar plots of control systems are of the form shown in Figure 10-17; hence they are widely applicable in the design of control systems. The transformation

10-4 SPECIFICATIONS IN TERMS OF OPEN-LOOP TRANSFER FUNCTIONS 277

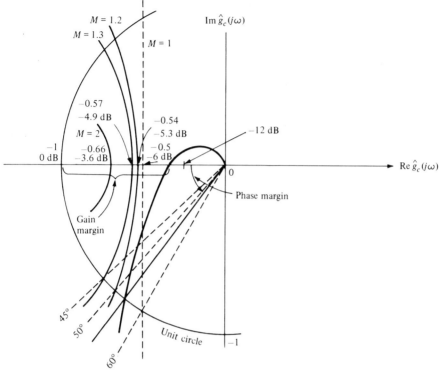

Figure 10-17 Constant-M loci and phase and gain margins.

of these two margins into the Bode plot is also very simple. At the gain crossover frequency ω_c, we have $\hat{g}_c(j\omega_c) = 1$, or 0 dB; hence the frequency ω_c can be easily determined from the gain plot. The phase crossover frequency ω_p is the frequency at which the phase is equal to $-180°$. The gain at that frequency is the gain margin as shown in Figure 10-19.

The specification on the peak resonance of an overall system has been translated into the phase margin and the gain margin of the Bode plot of the open-loop transfer function. Can the bandwidth of an overall system be also translated into the Bode plot of the open-loop system? Unfortunately the answer is negative. Therefore if we use the Bode plot in the design, the bandwidth of the overall system is not directly considered. Instead, the gain crossover frequency ω_c is often specified. Although there is no direct relationship between the gain crossover frequency of an open-loop system and the bandwidth of its overall system, the *bandwidth is* however *generally larger than the gain crossover frequency*. Hence if the bandwidth is required to be, say, at least ω_b, then by assigning the gain crossover frequency at ω_b, the bandwidth of the resulting system will satisfy the bandwidth specification. On the other hand, if the bandwidth is required to be not greater than ω_b because of the presence of high-frequency noise, then the Bode plot alone cannot handle this specification; the constant-M loci on the polar plot must be used.

278 FREQUENCY-DOMAIN TECHNIQUE

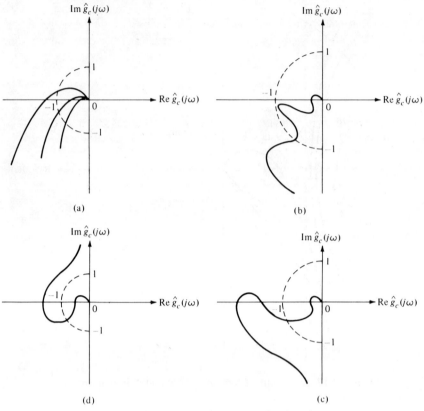

Figure 10-18 Various types of polar plots.

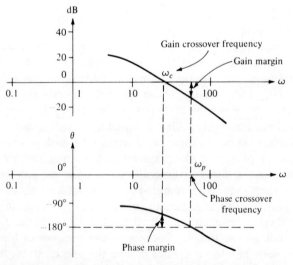

Figure 10-19 Gain and phase margins in the Bode plot.

The high-frequency specification on $\hat{g}_f(s)$ can be easily reduced to the Bode plot of $\hat{g}_c(s)$. If $\hat{g}_c(s)$ is strictly proper, as is almost always the case, then in the high-frequency region we have $|\hat{g}_c(j\omega)| \ll 1$ and, consequently,

$$|\hat{g}_f(j\omega)| = \left|\frac{\hat{g}_c(j\omega)}{1 + \hat{g}_c(j\omega)}\right| \doteq |\hat{g}_c(j\omega)|$$

Hence the translation of $|\hat{g}_f(j\omega)| < \varepsilon$ for large ω to the Bode plot of $\hat{g}_c(s)$ is straightforward.

We summarize what we have discussed in this subsection. The transient performance of an overall system is specified in terms of the peak resonance M_p, bandwidth ω_b, and high-frequency gain. In order to carry out the design on the Bode plot of an open-loop system, the specifications are translated into the gain margin, the phase margin, the gain crossover frequency, and high-frequency gain. Although the phase and gain margins may not necessarily yield the specified peak resonance, and the gain crossover frequency may not yield the required bandwidth, these specifications on the Bode plot can be easily obtained and thus are widely used in practice.

We make the following remark to conclude this section: In the design the Bode plot of an open loop system not only should satisfy the given specifications but also should ensure BIBO stability of the overall system.[5] From the Bode plot, the Nyquist plot can be easily sketched, and the stability of the overall system can then be determined by employing Theorem 10-1. If the open-loop system $\hat{g}_c(s)$ is known to have no open right-half s-plane poles and if $\hat{g}_c(s)$ is either a type 0 or type 1 transfer function with a positive K_p or K_v, then the overall system will be BIBO stable if the Bode plot of $\hat{g}_c(s)$ has only one phase margin and only one gain margin, and these margins are both positive. In this case the Nyquist plot will not circle the point $(-1, 0)$, and the overall system is BIBO stable according to Corollary 10-1. The systems that will be studied in the next section are all of this type; hence by requiring the phase and gain margins to be positive, the stability of overall systems is automatically insured.

10-5 Design on the Bode Plots

The relationships between closed-loop frequency plots and open-loop frequency plots are developed for the unit feedback configuration shown in Figure 10-4; hence the design technique that will be introduced is applicable to this configuration only. We restate the design problem: Given a plant with transfer function $\hat{g}(s)$, find a compensator $C(s)$ so that the Bode plot of $\hat{g}_c(s) = C(s)\hat{g}(s)$ satisfies the specification on the position- or velocity-error constant, the phase margin, the gain margin, the gain crossover frequency, and the high-frequency gain. The search of $C(s)$ is essentially a trial-and-error process; therefore we like

---

[5] Note that BIBO stability implies total stability if missing poles, if there are any, are also stable.

to proceed from a simple compensator to a more complicated one. The compensations used in this approach are almost exclusively of the following four types: (a) gain adjustments (amplification or attenuation), (b) phase-lead compensation, (c) phase-lag compensation, and (d) lag-lead compensation. These compensations will be introduced successively, together with design examples.

Gain adjustments. The compensator $C(s)$ in this compensation is just a gain, k. If k is positive, then the introduction of this compensator will not affect the phase plot of a plant. It will just shift the gain plot of a plant up or down depending upon whether k is greater or smaller than 1.

Example 1

Consider a plant with the transfer function $\hat{g}(s) = 1/s(s + 2)$, as shown in Figure 10-20. Find a compensator $C(s)$ such that the overall system satisfies the following specifications: (a) the steady-state error due to a step function is less than 10 percent, (b) the percentage overshoot is less than 5 percent, and (c) the response is as fast as possible.

We first translate these specifications on the overall system to the open-loop transfer function $C(s)\hat{g}(s)$. The steady-state error requires that the position-error constant of $C(s)\hat{g}(s)$ satisfies

$$\left| \frac{1}{1 + K_p} \right| < 0.1$$

where $K_p = C(0)\hat{g}(0)$. From Figure 10-2 we see that if the peak resonance is equal to 1, then the specification on the overshoot can be met. From Figure 10-17 we see that if the gain margin is larger than 12 dB and the phase margin is larger than 60°, then the peak resonance of the overall system may be equal to or less than 1. Hence the specification on the overshoot imposes the requirement that the gain margin be larger than 12 dB and the phase margin be larger than 60°. The speed of response is inversely proportional to the bandwidth of the overall system. Although there is no simple relation between the bandwidth of the overall system and the Bode plot of $\hat{g}(s)C(s)$, the bandwidth is always larger than the gain crossover frequency. Hence if the gain crossover frequency is required to be as large as possible, the specification on the speed of response can be met.

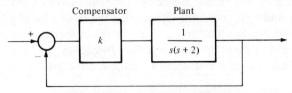

Figure 10-20 A design problem.

The Bode plot of $\hat{g}(s) = 1/s(s + 2)$, or of

$$\hat{g}(s) = \frac{0.5}{s(1 + 0.5s)}$$

is shown in Figure 10-21. Its gain crossover frequency is $\omega_c = 0.5$ rad/s; the phase margin is measured as 76°. The phase crossover frequency is $\omega = \infty$; the gain margin is infinity. Hence even without introducing any compensation, the gain and phase margins are satisfied. The plant $\hat{g}(s)$ is a type 1 system; hence $K_p = \infty$, and the specification on the steady-state error is also met. Hence if no specification is imposed on the speed of response, the compensator $C(s)$ can be chosen as 1, and the design is completed. Now the specification on the speed of response requires that the gain crossover frequency be as large as possible. This can be achieved by simply moving the gain plot upward; hence we choose $C(s) = k$. Now as k increases from 1, the gain crossover frequency increases, and the phase margin decreases, as can be seen from Figure 10-21. Note that the gain margin remains to be infinity as k changes. Since the phase margin is required to be larger than 60°, from the phase plot we see that the permissible largest gain crossover frequency of the compensated plant is $\omega_c' = 1.15$. Thus the required k is, from the gain plot,

$$20 \log k = 8$$

or $k = 2.5$. Hence if we choose $C(s) = 2.5$, the design is completed.

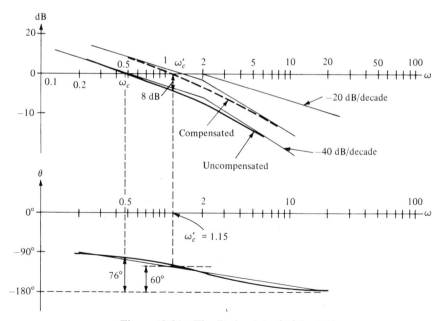

Figure 10-21 The Bode plot of $1/s(s + 2)$.

This design problem was also solved by using the root-locus technique in Section 9-4. The solutions obtained in both methods are fairly close. This might be due to the fact that the overall transfer function of the control problem is a second-order transfer function with a constant numerator, and hence the relations among overshoot, pole locations, and peak resonance are exact. Thus the design results for this problem by using the root-locus method and the Bode plot are very close. This however cannot be expected for more complicated systems.

It is not always possible to design a control system by merely adjusting a gain. This can be seen from the following example.

Example 2

Consider a plant with the transfer function $\hat{g}(s) = 1/s(s + 2)$. Design an overall system so that the steady-state error due to a ramp function is smaller than 10 percent, the gain margin is larger than 12 dB, and the phase margin is larger than 60°.

In order to have the steady-state error due to a ramp function smaller than 10 percent, the velocity error constant of $\hat{g}(s)C(s)$, that is,

$$K_v = \lim_{s \to 0} s\hat{g}(s)C(s)$$

should be larger than 10; see Equation (10-25). If $C(s) = kC_1(s)$ with $C_1(0) = 1$, then

$$K_v = \lim_{s \to 0} sg(s)k = \frac{k}{2} \geq 10$$

or $k \geq 20$. The Bode plot of $20/s(s + 2) = 10/s(1 + 0.5s)$ is shown in Figure 10-22. The phase crossover frequency is infinity, and the gain margin is also infinity. The gain crossover frequency is $\omega_c = 4.2$ rad/s, and the phase margin is 26°, as measured from Figure 10-22. Now if $C(s)$ is just a gain, say k, in order to meet the specification on the steady-state error we need $k \geq 20$. Clearly there is no $k \geq 20$ that will yield a phase margin larger than 60°. Hence it is not possible to design a satisfactory system by choosing $C(s) = k$; a more complicated compensator is required. ∎

Phase-lead compensation. In the foregoing example, if we can find a compensator that renders some positive phase at the gain crossover frequency, then the design may be achieved. In this subsection we introduce such a compensator. Consider the network shown in Figure 10-23. Its transfer function can be found as

$$C_1(s) = \frac{\hat{e}_2(s)}{\hat{e}_1(s)} = \frac{R_1 + R_2}{R_2} \cdot \frac{R_2 + R_1 R_2 Cs}{R_1 + R_2 + R_1 R_2 Cs} = \frac{1 + aT_1 s}{1 + T_1 s} \quad \textbf{(10-33)}$$

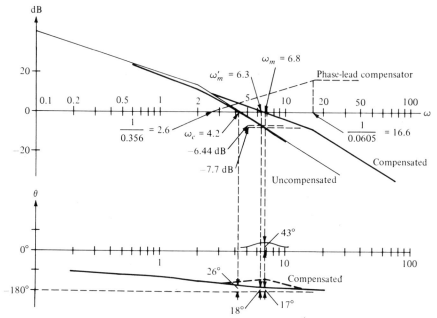

Figure 10-22 The Bode plot of $20/s(s+2)$.

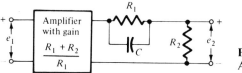

Figure 10-23 A phase-lead network.

with

$$a = \frac{R_1 + R_2}{R_2} > 1 \tag{10-34}$$

$$T_1 = \frac{R_1 R_2}{R_1 + R_2} \cdot C \tag{10-35}$$

The constant a is always larger than 1, and hence the Bode plot of $C_1(s)$ is of the form shown in Figure 10-24. The gain plot is obtained by adding the asymptotes of $(1 + aT_1 s)$ and the asymptotes of $(1 + T_1 s)^{-1}$. Since the phase of this network is positive for all ω, the network is called a *phase-lead network*. The phase of $(1 + aT_1 s)/(1 + T_1 s)$ can be computed as

$$\phi = \tan^{-1} aT_1 \omega - \tan^{-1} T_1 \omega$$

or

$$\tan \phi = \frac{a_1 T_1 \omega - T_1 \omega}{1 + aT_1^2 \omega^2}$$

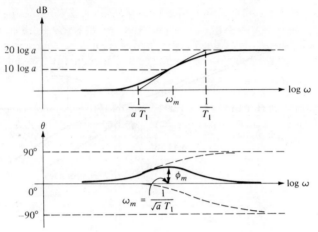

Figure 10-24 The Bode plot of $C_1(s)$.

Since the phase plot is symmetric with respect to the midpoint of $1/aT_1$ and $1/T_1$, as shown in Figure 10-24, the maximum phase occurs at

$$\log \omega_m = \frac{1}{2}\left(\log \frac{1}{aT_1} + \log \frac{1}{T_1}\right) = \log \frac{1}{\sqrt{a}T_1}$$

or

$$\omega_m = \frac{1}{\sqrt{a}T_1} \tag{10-36}$$

and the maximum phase is equal to

$$\tan \phi_m = \frac{(a-1)T_1\omega_m}{1 + aT_1^2\omega_m^2} = \frac{a-1}{2\sqrt{a}}$$

which implies that

$$\sin \phi_m = \frac{a-1}{a+1} \quad \text{or} \quad a = \frac{1 + \sin \phi_m}{1 - \sin \phi_m} \tag{10-37}$$

We see that the larger the constant a is, the larger the phase ϕ_m. However the network in Figure 10-23 also requires a larger amplification. In practice the constant a is seldom chosen greater than 15. Equations (10-36) and (10-37) are two key relations used to choose a phase-lead compensator in the design.

With the introduction of a phase-lead compensator, we should be able to complete the design problem in Example 2. Before proceeding, the procedure of designing a phase-lead compensator is listed in the following:

1. Compute the required position-error or velocity-error constant from the specification on the steady-state error.

2. Plot the Bode plot of the plant with the new position- or velocity-error constant. Determine the phase margin and gain margin from the plot.

3. If it is decided to use a phase-lead compensator, determine the necessary phase lead ψ to be added. The introduction of a phase-lead compensator generally results in a shift of the gain crossover frequency and, consequently, a reduction of the phase margin. To compensate this reduction, we add θ, say 5° to ψ. Compute $\phi_m = \psi + \theta$.

4. Compute the constant a from Equation (10-37).

5. Determine the frequency at which the gain of the uncompensated plant is equal to $-10 \log a$. Call this frequency ω_m, and compute the phase margin of the uncompensated plant at this frequency. If the reduction of the phase margin is larger than θ in step 3, then choose a larger θ and repeat steps 3 through 5 until the reduction of the phase margin is approximately equal to θ.

6. Compute T_1 from Equation (10-36). If the resulting system satisfies all other specifications, then synthesize the network by using Equations (10-33) through (10-35).

Example 2 (continued)

The Bode plot of the plant with the required velocity-error constant is shown in Figure 10-22. The phase margin is 26°; the required phase margin however is 60°. Hence we have $\psi = 34°$. Rather arbitrarily, we choose $\theta = 5°$; hence $\phi_m = 34° + 5° = 39°$. The constant a is computed from Equation (10-37) as $a = (1 + \sin 39°)/(1 - \sin 39°) = 1.63/0.37 = 4.4$. Next compute $-10 \log a = -6.44$ dB. The new gain crossover frequency ω_m' is then found from Figure 10-22 as shown. The corresponding phase margin is measured as 18°. The reduction of the phase margin is $26° - 18° = 8°$, which is greater than $\theta = 5°$; hence the design is unsatisfactory. As a next try, we choose $\theta = 9°$, and hence $\phi_m = 34° + 9° = 43°$. The constant a is then computed as 5.9. Next compute $-10 \log a = -7.7$ dB. The new ω_m is then found from the plot as $\omega_m = 6.8$ rad/s. The reduction in the phase margin is $26° - 17° = 9°$. Hence the choice of $\theta = 9°$ is satisfactory. Next we compute

$$T_1 = \frac{1}{\sqrt{a}\omega_m} = \frac{1}{\sqrt{5.9} \times 6.8} = \frac{1}{16.5} = 0.06$$

The required phase-lead compensator is then

$$C_1(s) = \frac{1 + aT_1 s}{1 + T_1 s} = \frac{1 + 0.356s}{1 + 0.0605s}$$

The completed control system is shown in Figure 10-25. The asymptotes of the Bode plot of $kC_1(s)\hat{g}(s)$ are shown in Figure 10-22. ∎

The introduction of a phase-lead compensator will always increase the gain crossover frequency and, consequently, the bandwidth of the overall system. Hence the introduction of a phase-lead compensator generally increases the speed of response.

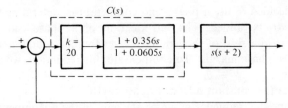

Figure 10-25 A system with phase-lead compensation.

The phase-lead compensation is rather ineffective if the rate of change of the phase in the neighborhood of the gain crossover frequency is very large, because the reduction of the phase margin due to the shift of the crossover frequency will offset the phase introduced by the compensator. In this case a different compensation has to be used.

Phase-lag compensation. Consider the network shown in Figure 10-26. Its transfer function is

$$C_2(s) = \frac{\hat{e}_2(s)}{\hat{e}_1(s)} = \frac{1 + R_2Cs}{1 + (R_1 + R_2)Cs} = \frac{1 + bT_2s}{1 + T_2s} \qquad (10\text{-}38)$$

where

$$b = \frac{R_2}{R_1 + R_2} < 1 \qquad (10\text{-}39)$$

$$T_2 = (R_1 + R_2)C \qquad (10\text{-}40)$$

Since the constant b is smaller than 1, the Bode plot of $C_2(s)$ is of the form shown in Figure 10-27. The phase of the network is negative for all ω, and hence it is called a *phase-lag network*.

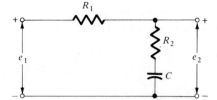

Figure 10-26
A phase-lag network.

In a phase-lead network, we use mainly its phase property in the design. By contrast, we use only the gain property of a phase-lag network in the design. The main idea in using a phase-lag network is to shift the gain crossover frequency to a desired location. The procedure of designing a phase-lag network is listed in the following:

1. Compute the required position- or velocity-error constant from the specification on the steady-state error.

2. Plot the Bode plot of the plant with the new error constant. Determine the phase margin and gain margin from the plot.

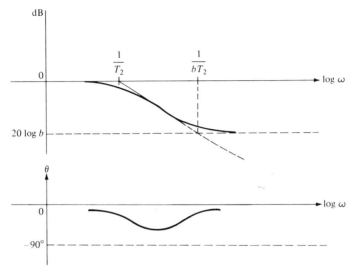

Figure 10-27 The Bode plot of $C_2(s)$.

3. If it is decided to use a phase-lag compensator, determine the frequency at which the uncompensated plant has the required phase margin plus 6°. This frequency is designated as the new gain crossover frequency ω_c.

4. Measure the attentuation necessary so that the gain plot will have the new gain crossover frequency. This attentuation will be provided by a phase-lag network. Let this attenuation be α. Then compute b from $\alpha = -20 \log b$.

5. Compute T_2 from

$$\frac{1}{bT_2} = \frac{\omega_c}{10} \qquad \text{(10-41)}$$

that is, the corner frequency $1/bT_2$ is placed one decade below the new crossover frequency ω_c. The phase lag introduced by this phase-lag network at ω_c is at most 5.7°, as can be seen from Figure 10-7. This is the reason we add 6° to the required phase margin in step 3.

6. If the resulting system satisfies all other specifications, then the design is completed. The compensating network can be built by using Equations (10-38) through (10-40).

Example 3

Consider a plant with the transfer function $\hat{g}(s) = 2/s(s + 1)(s + 5)$. Design an overall system to satisfy the following specifications: (a) the steady-state error due to a ramp function is less than 15 percent, (b) the phase margin is at least 50° and the gain margin is at least 12 dB, and (c) signals (or noise) with frequency higher than 10 rad/s should be reduced at the output to less than 1 percent of their values at the input.

The configuration of compensator is chosen as shown in Figure 10-28.

288 FREQUENCY-DOMAIN TECHNIQUE

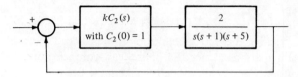

Figure 10-28 A design problem.

The compensator is assumed to have transfer function $kC_2(s)$ with $C_2(0) = 1$. The specification on the steady-state error requires that

$$\frac{1}{0.15} = 6.67 \leq k_v = \lim_{s \to 0} skC_2(s)\hat{g}(s) = \frac{2k}{5}$$

which implies that $k \geq 16.675$. The Bode plot of

$$\frac{16.675 \times 2}{s(s+1)(s+5)} = \frac{6.67}{s(1+s)(1+0.2s)} \tag{10-42}$$

is shown in Figure 10-29. The specification on high-frequency signals implies that $|\hat{g}_f(j\omega)| \leq 0.01$ for $\omega \geq 10$ rad/s. This, in turn, implies that the gain of (10-42) should be smaller than $20 \log 0.01 = -40$ dB for all $\omega \geq 10$ rad/s. That is, the gain plot for $\omega \geq 10$ rad/s should be lower than the point denoted by the asterisk in Figure 10-29. This is not the case. Furthermore the phase margin is $-6°$, and the gain margin is -3 dB. Hence if $C_2(s) = 1$, the system is entirely unsatisfactory. In fact, the system is unstable, and therefore a more complicated compensator is required.

A phase-lead compensator cannot be used in this problem for the following three reasons: First, the rate of change of the phase is large in the neighborhood of the gain crossover frequency; hence a phase-lead compensation is not effective.

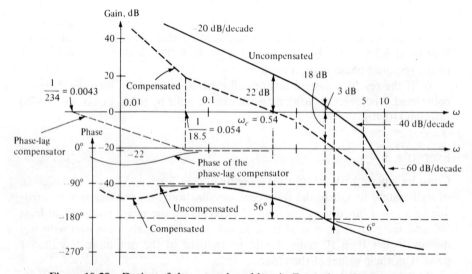

Figure 10-29 Design of the control problem in Example 3, Section 10-5.

Second, the introduction of a phase-lead network will increase the gain at high frequencies. Finally, the introduction of a phase-lead network will further reduce the gain margin. Hence the phase-lead compensation cannot be used. Clearly the adjustment of gain alone is inadequate for this problem, and hence it is decided to use a phase-lag compensator.

From the phase plot in Figure 10-29, we find that the phase margin at ω_c = 0.54 rad/s is equal to $50° + 6° = 56°$. In order to have the new gain crossover frequency, the attentuation needed is measured as 22 dB. Since $22 = -20 \log b$, we compute b as 0.079. The constant T_2 is computed from (10-41) as $T_2 = 234$. Hence the required compensator is

$$kC_2(s) = 16.675 \cdot \frac{1 + 18.5s}{1 + 234s}$$

Note that after the introduction of this compensator, the gain margin becomes 18 dB, and the gain for $\omega \geq 10$ rad/s is less than -40 dB. Hence all the specifications are met, and the design is completed. ∎

Three remarks are in order at this point. First, the introduction of a phase-lag network always reduces the gain crossover frequency; hence the bandwidth of the overall system may be reduced. Consequently the speed of response of the overall system may also be reduced. Second, the phase introduced at ω_c due to a phase-lag network is at most equal to 5.7°, and hence the resulting system usually has a phase margin larger than the specified value. Finally, the problem in Example 3 was also solved by using the root-locus method in Section 9-4. The resulting compensator was a phase-lead network. The reason for this difference is that the specifications in both cases are not exactly the same. If the specification on the steady-state error in Example 3 of this section is relaxed, the required compensator may become a phase-lead network.

Lag-lead compensation. For some control problems it is not possible to use either a phase-lead network or a phase-lag network alone to design a satisfactory system. In this case we may try to use both networks in the design. This is illustrated by an example.

Example 4

Consider the tracking antenna problem studied in Example 2 of Section 8-2. The transfer function of the plant is, as derived in (8-23),

$$\hat{g}(s) = \frac{300}{s(s + 0.225)(s + 3.997)(s + 179.77)} \tag{10-43}$$

Design an overall system to meet the following specifications: (a) The steady-state error due to a step function is zero, (b) the phase margin is 55° and the gain margin is 6 dB, and (c) the gain crossover frequency is not smaller than that of the uncompensated plant.

Since the transfer function of the plant is type 1, the position-error constant is infinity, and the specification on the steady-state error is always satisfied. The Bode plot of Equation (10-43), or of

$$\hat{g}(s) = \frac{1.85}{s(1 + 4.45s)(1 + 0.25s)(1 + 0.00556s)}$$

is shown in Figure 10-30. Although the gain margin (8 dB) is satisfactory, the phase margin (11°) is smaller than the specified value. Because of the requirement on the gain crossover frequency, we rule out the use of a phase-lag network. The adjustment of gain alone cannot yield a satisfactory system, because the increase of gain will reduce the gain margin and the decrease of gain will decrease the gain crossover frequency. We discuss now the use of a phase-lead network. The required phase lead is $55° - 11° = 44°$ plus $\theta = 5°$. Hence we have $\phi_m = 49°$, and $a = (1 + \sin \phi_m)/(1 - \sin \phi_m) = (1 + 0.75)/(1 - 0.75) = 7$. After the introduction of this phase-lead network, the new gain crossover frequency becomes $\omega_m = 1.05$ rad/s, which is determined by the gain $-10 \log a = 8.45$ dB. We see that the reduction of the phase margin, $11° -(-1°) = 12°$, is larger than $\theta = 5°$. Hence the chosen θ is not satisfactory. If we choose $\theta = 12°$, similarly, it can be shown that the reduction of the phase margin is 20°. In other words, a phase-lead network alone cannot provide the necessary phase margin because the reduction of the phase margin due to the shift of the gain crossover frequency nullifies the effect of the network.

From the above discussion we see that if it is possible to hold the gain crossover frequency at $\omega_c = 0.64$ rad/s after the introduction of a phase-lead network, then the design may be accomplished. Recall that the introduction

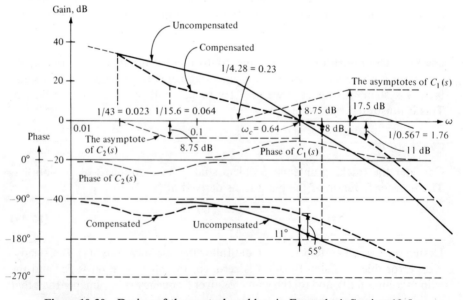

Figure 10-30 Design of the control problem in Example 4, Section 10-5.

10-5 DESIGN ON THE BODE PLOTS

of a phase-lead network will increase the gain crossover frequency, whereas the introduction of a phase-lag network will decrease the gain crossover frequency. Hence by introducing both networks, the gain crossover frequency can be kept unchanged. In order to have a phase margin of 55° at $\omega_c = 0.64$ rad/s, an additional 44° phase lead is required. In order to compensate the phase lag introduced by the phase-lag network, yet to be designed, we need an additional 6°. Hence we shall design a phase-lead network that provides a phase lead of $44° + 6° = 50°$. This can be obtained from a phase-lead network with the constant a given by, from (10-37),

$$a = \frac{1 + \sin \phi_m}{1 - \sin \phi_m} = \frac{1 + \sin 50°}{1 - \sin 50°} = \frac{1 + 0.766}{1 - 0.766} = 7.55$$

Since we like to put the maximum phase at $\omega_c = 0.64$ rad/s, we choose $\omega_m = \omega_c$. From (10-36), we compute

$$T_1 = \frac{1}{\sqrt{a}\omega_m} = \frac{1}{\sqrt{7.55} \times 0.64} = \frac{1}{1.76} = 0.567$$

With a and T_1, the required phase-lead network is

$$C_1(s) = \frac{1 + aT_1 s}{1 + T_1 s} = \frac{1 + 4.28s}{1 + 0.567s}$$

Note that the synthesis of $C_1(s)$ requires an amplification of gain a as shown in Figure 10-23. The asymptotes of $C_1(s)$ are shown in Figure 10-30. After the introduction of this network, the gain at ω_c is no longer 0 dB; it is equal to $20 \log \sqrt{a} = 10 \log a = 8.75$ dB, as shown in Figure 10-30. In order to bring the gain at ω_c back to 0 dB, a phase-lag network is required. The constant b in the phase-lag network can be computed from

$$-20 \log b = 8.75$$

or

$$b = 0.363$$

By placing the corner frequency $1/bT_2$ one decade below ω_c, from (10-41) we obtain

$$T_2 = \frac{10}{b\omega_c} = \frac{10}{0.363 \times 0.64} = \frac{1}{0.0232} = 43$$

Hence the required phase-lag compensator is, from (10-38),

$$C_2(s) = \frac{1 + bT_2 s}{1 + T_2 s} = \frac{1 + 15.6s}{1 + 43s}$$

This network can be synthesized as shown in Figure 10-26. The Bode plot of the compensated plant is shown in Figure 10-30 by using the dashed lines. We see that the phase margin is 55° and the gain margin is 11 dB. Hence the design is completed. The overall compensator is

$$C(s) = C_1(s)C_2(s) = \frac{1 + 4.28s}{1 + 0.567s} \cdot \frac{1 + 15.6s}{1 + 43s} \qquad (10\text{-}44)$$

The unit step response of the resulting system is shown in Figure 12-3. The design is rather satisfactory. ∎

We see that the design on the Bode plot can be very easily carried out by using the asymptotes. Although the asymptotes yield only approximate results, they are often acceptable in practice. After all, linear equations including transfer functions are often obtained by approximations. We also see that the pole $(1 + 0.00556s)$ is automatically neglected in this design. It is important to note that the design is not unique; for example, if $C(s)$ in (4-44) is replaced by

$$C'(s) = 3.775 \frac{1 + 4.28s}{1 + 0.567s} \cdot \frac{1 + 15.6s}{1 + 118s} \qquad (10\text{-}45)$$

the obtained phase margin, gain margin, and gain crossover frequency will remain to be the same. In fact, the asymptotes of $C'(s)\hat{g}(s)$ and the asymptotes of $C(s)\hat{g}(s)$ are identical for $\omega \geq 1/43$. The reader is advised to verify this assertion.

The transfer function in (4-45), excluding the gain 3.775, is of the form

$$C_3(s) = \frac{1 + aT_1s}{1 + T_1s} \cdot \frac{1 + bT_2s}{1 + T_2s} \qquad (10\text{-}46)$$

with $a = 1/b = 7.55$, $T_1 = 0.567$, and $T_2 = 118$. This type of transfer function can be synthesized directly, as shown in Figure 10-31(a), rather than by con-

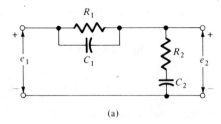

(a)

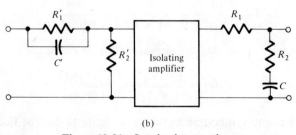

(b)

Figure 10-31 Lag-lead network.

necting a phase-lag network, a phase-lead network, and an isolating amplifier, as shown in Figure 10-31(b). The transfer function of the network in Figure 10-31(a) can be computed as

$$\frac{\hat{e}_2(s)}{\hat{e}_1(s)} = \frac{(1 + R_1C_1s)(1 + R_2C_2s)}{1 + (R_1C_1 + R_1C_2 + R_2C_2)s + R_1R_2C_1C_2s} \quad \text{(10-47)}$$

By identifying $T_1T_2 = R_1R_2C_1C_2$, $T_1 + T_2 = R_1C_1 + R_2C_2 + R_1C_2$, $a = R_1C_1/T_1$, and $b = R_2C_2/T_2 = 1/a$, Equation (10-47) reduces immediately to (10-46).

10-6 Remarks and Review Questions

What is the peak resonance? What is its relationship to the overshoot? Is this relationship established for a general transfer function?

What is the bandwidth? What is the relationship between the bandwidth and the response time?

It is often said that the speed of response of a system is limited by high-frequency noise. Why?

As $\omega \to 0$ and $\omega \to \infty$, the parts of the polar plot (and the Nyquist plot) can be obtained almost by inspection. How?

Is it possible to obtain the polar plot of $\hat{g}(s)C(s)$ from that of $\hat{g}(s)$?

Is it possible to obtain the Bode plot of $\hat{g}(s)C(s)$ from that of $\hat{g}(s)$?

Why, in the Bode plot, must a transfer function be factored as a product of terms $1 + \tau s$ rather than of terms $s + (1/\tau)$, as we did in the plot of root loci?

What is the Nyquist stability criterion? Do the right-half s-plane zeros of a transfer function play any role in the criterion? Is the criterion applicable to any configuration of compensators?

What is the position-error constant? What is the velocity-error constant? How do they relate to the steady-state error? How can they be determined from a Bode plot?

How do you determine phase margin and gain margin from a Nyquist plot?

How do you determine phase margin and gain margin from a Bode plot?

Are phase margin and gain margin always positive?

State the type of transfer functions for which, if phase margin and gain margin are both positive, then a negative unit feedback system consisting of such a transfer function on its forward path is BIBO stable.

What is the gain crossover frequency? In most of control systems why is the bandwidth larger than the gain crossover frequency?

Table 10-1 Specifications of Control Systems

	Time domain (overall systems)	Frequency domain	
		Overall system	Open-loop system (Unit feedback system)
Steady-state performance	Steady-state error due to step input or ramp input	$\hat{g}_f(0)$ (Eq. (10-2)) $\hat{g}_f'(0)$ (Eq. (10-3))	K_p (Eq. (10-22)) K_v (Eq. (10-25))
Transient performance	Overshoot Settling time Rise time	Peak resonance M_p Bandwidth High-frequency gain	Gain margin, phase margin Gain crossover frequency High-frequency gain

The specifications introduced in this chapter are summarized in Table 10-1. Can you establish their relationships?

Is the frequency-domain design carried out on a closed-loop or open-loop system? How about the root-locus method and the optimal design method?

If the polar plot, Bode plot, and log magnitude-phase plot of a transfer function are all equivalent, why is the design most often carried out in the Bode plot?

Which property, gain or phase, of a phase-lead compensator do we use in the compensation?

Which two equations are the key relationships needed in the determination of a phase-lead compensator?

For what type of Bode plot is the phase-lead compensation not effective?

Which property, gain or phase, of a phase-lag compensator do we use in the compensations?

Why will the introduction of a phase-lead compensator generally increase the speed of response?

Why will the introduction of a phase-lag compensator generally decrease the speed of response?

The Bode-plot design, as introduced, is applicable only to the unit feedback configuration. It can be extended to the configuration shown in Figure 8-17. The interested reader is referred to References [1], [2], and [3].

The frequency-domain design can also be carried out on a Nichols chart, which is a log magnitude-phase plot with the constant-M loci superimposed on it. The design on a Nichols chart yields more accurate information on the overall system (because of the M loci); however its plotting is not as convenient as the Bode plot. The interested reader is referred to Reference [2].

Chronologically the frequency-domain method was the first method developed to design control systems. Even with the availability of the optimal design technique and the root-locus method, the frequency-domain method is still widely used in practice. There are several reasons. First, the Bode plot of a system can be obtained by direct measurements. By connecting the plant to a special instrument, such as the Model 911A frequency-response analyzer, manufactured by BAFCO, Inc., the gain and phase can be read out immediately over a wide frequency range. Hence the development of exact mathematical descriptions is not needed in this design. Second, the method is independent of the complexity of the system. No matter what degree a transfer function has, the method is equally applicable. Third, the method is very easy to carry out; it requires only simple calculations. Finally, the method is applicable to systems that consist of time-delay elements. The frequency-domain method however is not without handicaps. It is a trial-and-error technique. It is applicable mainly to the unit feedback configuration. Moreover it does not consider the saturation problem. Finally, the applicability of the open-loop specifications, such as phase margin and gain margin, is questionable for some control systems.

References

[1] Bower, J. L., and P. M. Schultheiss, *Introduction to the Design of Servomechanisms*. New York: Wiley, 1958.
[2] Chestnut, H., and P. W. Mayer, *Servomechanisms and Regulating System Design*, vol. 1. New York: Wiley, 1951.
[3] D'Azzo, J. J., and C. H. Houpis, *Feedback Control System Analysis and Synthesis*, 2d ed. New York: McGraw-Hill, 1965.
[4] Kuo, B. C., *Automatic Control Systems*, 2d ed. Englewood Cliffs, N. J.: Prentice-Hall, 1967.
[5] Melsa, J. L., and D. G. Schultz, *Linear Control Systems*. New York: McGraw-Hill, 1969.

Problems

10-1 Given the Bode plots shown in Figure P10-1, what are their transfer functions?

10-2 Plot the polar plot, log magnitude-phase plot, and Bode plot of $20/s^2(s + 1)$.

10-3 Plot the Bode plots of the following transfer functions. What are their gain margins, phase margins, K_p, and K_v?

a. $\dfrac{s + 5}{s(s + 1)(s + 10)}$

b. $\dfrac{s - 5}{s(s + 1)(s + 10)}$

c. $\dfrac{10}{(s + 2)(s^2 + 8s + 25)}$

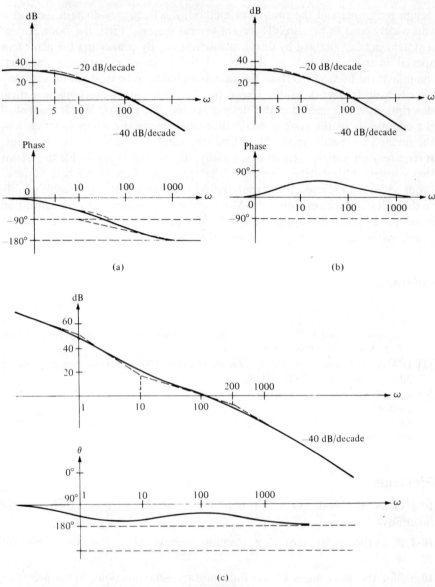

Figure P10-1

10-4 Determine, by using the Nyquist criterion, the BIBO stability of the system shown in Figure P10-2.

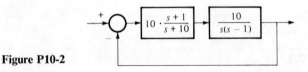

Figure P10-2

10-5 Establish the following assertion from Theorem 10-1: The unit feedback system shown in Figure P10-3 is BIBO stable if and only if the Nyquist plot of $\hat{g}_c(s)$ does not go through the point $(-1/k, 0)$ and does encircle it P times in the counterclockwise direction, where P is the number of the open right-half plane poles of $\hat{g}_c(s)$.

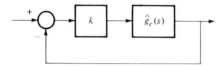

Figure P10-3

10-6 Consider the system shown in Figure P10-3. The polar plot of $\hat{g}_c(s)$ is assumed to be of the form shown in Figure P10-4. Determine the stability range of k for the following cases:

a. $\hat{g}_c(s)$ has no open right-half plane (RHP) pole and zero.
b. $\hat{g}_c(s)$ has one open RHP zero, no open RHP pole.
c. $\hat{g}_c(s)$ has one open RHP pole.
d. $\hat{g}_c(s)$ has two open RHP poles, one open RHP zero.
e. $\hat{g}_c(s)$ has three open RHP poles.

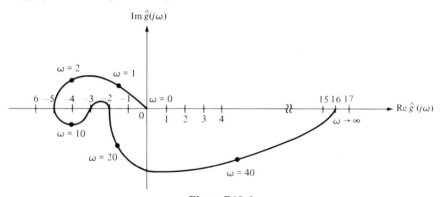

Figure P10-4

10-7 Find the stability range of the system shown in Figure P10-3 with

$$\hat{g}_c(s) = \frac{s + 2}{(s^2 + 3s + 6.25)(s - 1)}$$

by using
a. The Routh-Hurwitz criterion.
b. The root-locus method.
c. The Nyquist stability criterion.

10-8 Consider the tracking antenna problem studied in Problem 8-3. Design an overall system to satisfy the following specifications:
a. The steady-state error due to a step function is zero.
b. The phase margin is at least 50°.
c. The gain margin is at least 6 dB.

10-9 Consider the problem of controlling the yaw of the airplane discussed in Problem 8-8. Design an overall system, by using the Bode plot, to satisfy the following specifications:
a. The steady-state error due to a ramp function is less than 10 percent.
b. The overshoot is smaller than 20 percent.
c. The response is as fast as possible.

10-10 Consider the ship stabilization problem discussed in Problem 8-6. Design an overall system to satisfy the following specifications:
a. The steady-state error due to a step input is smaller than 15 percent.
b. The phase margin is at least 60°.
c. The margin is at least 10 dB.
d. The bandwidth of the overall system is at least 10 rad/s.

10-11 Consider the control of the depth of a submarine discussed in Problem 9-9. Design a system to satisfy the following specifications:
a. The steady-state error due to a step function is less than 10 percent.
b. The phase margin is at least 60°.
c. The gain margin is at least 10 dB. Compare the result with the one obtained in Problem 9-9.

CHAPTER

11

Parameter Optimization

In this chapter we shall introduce a different design technique called *parameter optimization*. In the design of a control system, the designer may or may not have complete freedom in using compensators. If he has complete freedom in using compensators, he may employ the optimization technique introduced in Chapter 8. If he does not have complete freedom, because of cost, parts availability, or other reasons, he has to start with a fixed configuration and then try to make the best use of the configuration. In this case he may employ the root-locus method, the frequency-domain technique, or the parameter optimization technique. Hence the parameter optimization problem can be stated as follows: Given a plant and a configuration of compensator, with some or all of the parameters of the compensators yet to be determined, find the parameters so that the resulting system is optimal with respect to some performance criterion. The difference between this problem and that of optimal design, discussed in Chapter 8, is that no constraint on compensators is imposed on the latter. In optimal design a designer is permitted to use compensators of any degree and any configuration, whereas in parameter optimization the configuration and the type of compensators are predetermined. Since there is no constraint imposed on compensators, optimal design generally results in a better system. However if there is a sufficient number of parameters to be adjusted, the system resulting from parameter optimization will be as good as the one from optimal design.

The performance criterion used in this chapter is assumed, as in Chapter 8, of the form

$$J = \int_0^\infty \{q[y(t) - y_d(t)]^2 + u^2(t)\} \, dt \tag{11-1}$$

where q is a positive number, $y(t)$ is the output, $y_d(t)$ is the desired, or command, signal, and $u(t)$ is the actuating signal. The parameters that yield the smallest J are the optimal parameters. For the criterion chosen in Equation (11-1), it is possible to study the parameter optimization problem analytically and numerically. We study first the analytical method and then the numerical method.

11-1 Analytical Method

In the analytical approach parameter optimization consists of two steps: the computation of the performance criterion J as a function of parameters $p_1, p_2, \ldots, p_m$, and the determination of optimal p_i by solving the equations

$$\frac{\partial J}{\partial p_i} = 0 \quad i = 1, 2, \ldots, m \tag{11-2}$$

Equation (11-2) generally consists of nonlinear algebraic equations and may have many sets of solutions. Among these solutions we have to find the sets of solutions that ensure the stability of the overall system and have the property

$$\frac{\partial^2 J}{\partial p_i^2} > 0 \quad i = 1, 2, \ldots, m$$

If there is only one solution, then it must be the optimal parameters. If there are two or more such sets of solutions, then we have to compute the corresponding J. The set that has the smallest J is the optimal parameters.

In the analytical method the computation of J will be carried out in the frequency domain. By using Parseval's theorem (see Appendix A), Equation (11-1) can be written as

$$J = \frac{1}{2\pi j} \int_{-j\infty}^{+j\infty} \{q[\hat{y}(s) - \hat{y}_d(s)][\hat{y}(-s) - \hat{y}_d(-s)] + \hat{u}(s)\hat{u}(-s)\} \, ds \tag{11-3}$$

The computation of J from Equation (11-3) is not a simple task. Fortunately there is a table available for computing J (see Table 11-1). After J is computed, the optimum parameter can then be obtained by solving Equation (11-2). The procedure is illustrated by examples.

Table 11-1 Table of Integration

$$I_n = \frac{1}{2\pi j}\int_{-j\infty}^{j\infty} \frac{D_n(s)}{A_n(s)A_n(-s)}\,ds$$

where

$$A_n(s) = a_0 s^n + a_1 s^{n-1} + \cdots + a_n$$

$$D_n(s) = d_0 s^{2n-2} + d_1 s^{2n-4} + \cdots + d_{n-1}$$

and $A_n(s)$ is a Hurwitz polynomial.

$$I_1 = \frac{d_0}{2a_0 a_1}$$

$$I_2 = \frac{a_0 d_1 - d_0 a_2}{2a_0 a_1 a_2}$$

$$I_3 = \frac{a_3(a_0 d_1 - d_0 a_2) - a_0 a_1 d_2}{2a_0 a_3(a_0 a_3 - a_1 a_2)}$$

$$I_4 = \frac{a_4[d_0(a_2 a_3 - a_1 a_4) - a_0 a_3 d_1 + a_0 a_1 d_2] + a_0 d_3(a_0 a_3 - a_1 a_2)}{2a_0 a_4(a_0 a_3{}^2 + a_1{}^2 a_4 - a_1 a_2 a_3)}$$

Example 1

Given a plant with the transfer function $20/s(s+2)$. The configuration of compensator is chosen as shown in Figure 11-1. Find the gain k so that the resulting system minimizes the performance criterion

$$J = \int_0^\infty \{q[y(t) - y_d(t)]^2 + u^2(t)\}\,dt$$

where $q = 1$ and y is a unit step function.

In order to apply Table 11-1, the performance criterion is transformed into the frequency domain as

$$J = \frac{1}{2\pi j}\int_{-j\infty}^{j\infty} \{[\hat{y}(s) - \hat{y}_d(s)][\hat{y}(-s) - \hat{y}_d(-s)] + \hat{u}(s)\hat{u}(-s)\}\,ds \quad (11\text{-}4)$$

Hence it is necessary to compute $\hat{y}(s) - \hat{y}_d(s)$ and $\hat{u}(s)$. The overall transfer function of the system in Figure 11-1 is

$$\hat{g}_f(s) = \frac{20k}{s^2 + 2s + 20k} \triangleq \frac{20k}{D_f(s)}$$

Hence we have

$$\hat{y}(s) - \hat{y}_d(s) = \hat{g}_f(s)\hat{y}_d(s) - \hat{y}_d(s) = \left(\frac{20k}{s^2 + 2s + 20k} - 1\right)\frac{1}{s} = \frac{-s-2}{D_f(s)}$$

302 PARAMETER OPTIMIZATION

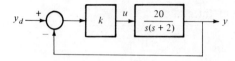

Figure 11-1
A design problem.

and

$$\hat{u}(s) = \frac{\hat{y}(s)}{\hat{g}(s)} = \frac{\hat{g}_f(s)\hat{y}_d(s)}{\hat{g}(s)} = \frac{ks + 2k}{D_f(s)}$$

Simple manipulation yields

$$[\hat{y}(s) - \hat{y}_d(s)][\hat{y}(-s) - \hat{y}_d(-s)] + \hat{u}(s)\hat{u}(-s)$$

$$= \frac{(-s-2)(s-2) + (ks+2k)(-ks+2k)}{D_f(s)D_f(-s)}$$

$$= \frac{-(1+k^2)s^2 + (4k^2+4)}{D_f(s)D_f(-s)} \quad (11\text{-}5)$$

We see that the integrand of Equation (11-4) has the form shown in Table 11-1, and hence the table can be directly applied. Before the application, the polynomial $D_f(s) = s^2 + 2s + 20k$ is required to be a Hurwitz polynomial; this is the case if k is positive. The application of I_2 in Table 11-1 yields

$$J = \frac{(4k^2+4) + (1+k^2) \times 20k}{2 \times 2 \times 20k} = \frac{5k^3 + k^2 + 5k + 1}{20k}$$

To find the optimal k, we take the derivative of J with respect to k; that is,

$$\frac{dJ}{dk} = \frac{10k^3 + k^2 - 1}{20k^2}$$

The solutions of $dJ/dk = 0$ or, equivalently, of $10k^3 + k^2 - 1 = 0$ are -4.875, -0.455, and 0.43. Since the Hurwitz condition requires k to be positive, the gains -4.875 and -0.455 can be discarded. To check that $k = 0.43$ is indeed the optimum gain, we compute

$$\frac{d^2J}{dk^2} = \frac{20k^3 + k^2 + 1}{20k^3}$$

which is positive at $k = 0.43$. Hence we conclude that if k is chosen as 0.43, then the resulting system minimizes J in Equation (11-4).

It is of interest to note that if the performance index J does not contain u^2, then it can be shown that the optimum k is infinity, and the overall transfer function in Figure 11-1 is essentially equal to 1. In this case the plant will always saturate, and the linear model will no longer be valid. Therefore it is important to include the actuating signal u in the performance criterion.

Example 2

Consider a plant with the transfer function $20/s(s + 2)$. The configuration of the compensator is chosen as shown in Figure 11-2. It can be synthesized by the use

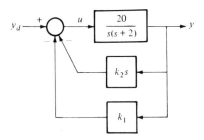

Figure 11-2
A design problem.

of a potentiometer and a tachometer. Find the gains k_1 and k_2 so that the resulting system minimizes

$$J = \int_0^\infty \{[y(t) - y_d(t)]^2 + u^2(t)\} \, dt \tag{11-6}$$

where y_d is a unit step function.

The overall transfer function of the system shown in Figure 11-2 is, by Mason's formula,

$$\hat{g}_f(s) = \frac{20}{s^2 + (2 + 20k_2)s + 20k_1}$$

As in Example 1, we compute

$$\hat{y}(s) - \hat{y}_d(s) = (\hat{g}_f(s) - 1)\hat{y}_d(s) = \frac{-s^2 - (2 + 20k_2)s - 20k_1 + 20}{s^2 + (2 + 20k_2)s + 20k_1} \cdot \frac{1}{s} \tag{11-7}$$

and

$$\hat{u}(s) = \frac{\hat{y}(s)}{\hat{g}(s)} = \frac{s+2}{s^2 + (2 + 20k_2)s + 20k_1} \tag{11-8}$$

We note that the denominator of $\hat{y}(s) - \hat{y}_d(s)$ has a root at $s = 0$; hence it is not a Hurwitz polynomial, and Table 11-1 cannot be employed. However if we choose $k_1 = 1$, then Equation (11-7) reduces to

$$\hat{y}(s) - \hat{y}_d(s) = \frac{-s - (2 + 20k_2)}{s^2 + (2 + 20k_2)s + 20}$$

and the denominator of $\hat{y}(s) - \hat{y}_d(s)$ can be Hurwitz for a certain range of k_2. Hence Table 11-1 can now be used. The necessity of choosing $k_1 = 1$ can also be seen directly from $\hat{g}_f(s)$. If $20k_1 \neq 20$, or $k_1 \neq 1$, then the overall system has a nonzero steady-state error due to a step function; hence the performance index (11-6) will be infinity for any k_2. Therefore it is necessary to choose $k_1 = 1$. With $k_1 = 1$, we have

$$[\hat{y}(s) - \hat{y}_d(s)][\hat{y}(-s) - \hat{y}_d(-s)] + \hat{u}(s)\hat{u}(-s)$$

$$= \frac{-2s^2 + 4 + (2 + 20k_2)^2}{[s^2 + (2 + 20k_2)s + 20][s^2 - (2 + 20k_2)s + 20]}$$

Again from Table 11-1, the performance index J in (11-6) can be computed as

$$J = \frac{4 + (2 + 20k_2)^2 - 20(-2)}{2 \times 20 \times (2 + 20k_2)} = \frac{50k_2^2 + 10k_2 + 6}{10(1 + 10k_2)} \qquad (11\text{-}9)$$

Taking the derivative of J with respect to k_2 yields

$$\frac{dJ}{dk_2} = \frac{10(1 + 10k_2)(100k_2 + 10) - (50k_2^2 + 10k_2 + 6) \times 100}{100(1 + 10k_2)^2}$$

$$= \frac{5(10k_2^2 + 2k_2 - 1)}{(1 + 10k_2)^2}$$

The solutions of $dJ/dk_2 = 0$ or of $10k_2^2 + 2k_2 - 1 = 0$ are $-8.63/20$ and $4.63/20$. If Table 11-1 is to be employed, it is required that the polynomial $s^2 + (2 + 20k_2)s + 20$ be a Hurwitz polynomial. Hence we have, from the Routh-Hurwitz criterion, $2 + 20k_2 > 0$. The solution $k_2 = -8.63/20$ does not satisfy this condition, and hence it cannot be an optimum gain. The solution $k_2 = 4.63/20$ satisfies the condition; furthermore the second derivative of J with respect to k_2

$$\frac{d^2J}{dk_2^2} = \frac{5 \times 22}{(1 + 10k_2)^3}$$

is positive at $k_2 = 4.63/20$. Hence $k_2 = 4.63/20$ does yield a minimum J in Equation (11-9). Thus we conclude that if $k_1 = 1$ and $k_2 = 4.63/20$, then the overall system in Figure 11-2 minimizes the performance criterion in (11-6).

Comparisons of this result with the ones in Example 1 of Section 8-2 and the Example in Section 8-3 are in order. With $k_1 = 1$ and $k_2 = 4.63/20$, the overall transfer function of the system in Figure 11-2 is

$$\hat{g}_f(s) = \frac{20}{s^2 + 6.63s + 20}$$

This is identical to Equation (8-21). Again, the gains $k_1 = 1$ and $k_2 = 4.63/20$ are identical to the feedback gains in Equation (8-40). Hence we conclude that if there is a sufficient number of parameters to be adjusted, the result obtained by the parameter optimization will be the same as the one obtained by the optimal design.

Example 3

Consider the tracking antenna problem studied in Example 2 of Section 8-2. The transfer function of the plant, as derived in (8-23), is given by

$$\hat{g}(s) = \frac{300}{s(s^3 + 184s^2 + 760.5s + 162)} \qquad (11\text{-}10)$$

11-1 ANALYTICAL METHOD

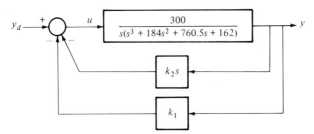

Figure 11-3 Design of a tracking antenna problem.

The configuration of compensators is chosen as shown in Figure 11-3. Find the gains k_1 and k_2 so that the resulting system minimizes the performance index

$$J = \int_0^\infty \{25[y(t) - y_d(t)]^2 + u^2(t)\} \, dt \qquad (11\text{-}11)$$

where y_d is a unit step function.

The configuration of compensators can be implemented by using a potentiometer and a tachometer. It is assumed that the moments of inertia of the transducers are negligible compared with that of the antenna; hence the transfer function $\hat{g}(s)$ in Equation (11-10) need not be modified. The overall transfer function of the system in Figure 11-3 is, by Mason's formula,

$$\hat{g}_f(s) = \frac{300}{s^4 + 184s^3 + 760.5s^2 + (162 + 300k_2)s + 300k_1} \qquad (11\text{-}12)$$

If $k_1 \neq 1$, then the steady-state error due to a step function is different from zero, and the performance index in Equation (11-11) will become infinity. Hence in order to have a finite performance index, it is required that $k_1 = 1$. With $k_1 = 1$, Equation (11-12) becomes

$$\hat{g}_f(s) = \frac{300}{s^4 + 184s^3 + 760.5s^2 + (162 + 300k_2)s + 300} \qquad (11\text{-}13)$$

In order to use Table 11-1, we compute

$$\hat{y}(s) - \hat{y}_d(s) = [\hat{g}_f(s) - 1]\hat{y}_d(s)$$

$$= \frac{-(s^3 + 184s^2 + 760.5s + 162 + 300k_2)}{s^4 + 184s^3 + 760.5s^2 + (162 + 300k_2)s + 300}$$

and

$$\hat{u}(s) = \frac{\hat{y}(s)}{\hat{g}(s)} = \frac{s^3 + 184s^2 + 760.5s + 162}{s^4 + 184s^3 + 760.5s^2 + (162 + 300k_2)s + 300}$$

Next we compute

$$25[\hat{y}(s) - \hat{y}_d(s)][\hat{y}(-s) - \hat{y}_d(-s)] + \hat{u}(s)\hat{u}(-s)$$

$$= \frac{-26s^6 + 8.4 \times 10^5 s^4 + (2.76 \times 10^6 k_2 - 1.35 \times 10^7)s^2 + 6.82 \times 10^5 + 2.43 \times 10^6 k_2 + 2.25 \times 10^6 k_2^2}{D_f(s)D_f(-s)}$$

where

$$D_f(s) \triangleq s^4 + 184s^3 + 760.5s^2 + (162 + 300k_2)s + 300 \qquad (11\text{-}14)$$

Now by using Table 11-1, the performance criterion J in (11-11) can be computed as

$$J = 12.5 \times \frac{k_2^3 - 464k_2^2 - 392k_2 - 1305}{k_2^2 - 466k_2 - 139} \qquad (11\text{-}15)$$

In order to find the k_2 that yields the minimum J, we take the derivative of J with respect to k_2:

$$\frac{dJ}{dk_2} = \frac{12.5(k_2^4 - 932k_2^3 + 21.7 \times 10^4 k_2^2 - 13.2 \times 10^4 k_2 - 55.35 \times 10^4)}{(k_2^2 - 466k_2 - 139)^2}$$

There are four solutions of k_2 in $dJ/dk_2 = 0$. A positive real solution is found, by trial and error, as

$$k_2 = 1.93$$

It is easy to show that $D_f(s)$ in Equation (11-14) with $k_2 = 1.93$ is a Hurwitz polynomial (see Problem 4-11b); hence Table 11-1 is indeed applicable. It can also be shown that $d^2 J/dk_2^2$ is positive at $k_2 = 1.93$; hence this k_2 yields a minimum J. It can be shown that no other positive real k_2 yields a minimum J, and thus we conclude that if $k_1 = 1$ and $k_2 = 1.93$, then the resulting system in Figure 11-3 minimizes the performance index in Equation (11-11). ∎

The response of the system designed in this example due to a unit step function is shown in Figure 12-4 in Chapter 12. We see that the response is extremely slow. Hence even though the system in Figure 11-3 with $k_1 = 1$ and $k_2 = 1.93$ optimizes the performance index in Equation (11-11), the system is unacceptable. There are two ways to remedy this problem. The first one is to change the performance index—for example, the increase of the weighting factor q in (11-11) can speed up the response. The second one is to choose a different configuration of compensators; this is done in the following example.

Example 4

Consider the design problem discussed in Example 3. The configuration of compensators is now chosen as shown in Figure 11-4. Find k_1 and k_2 so that the resulting system minimizes the performance index

$$J = \int_0^\infty \{q[y(t) - y_d(t)]^2 + u^2(t)\}\, dt$$

$$= \frac{1}{2\pi j} \int_{-j\infty}^{j\infty} \{q[\hat{y}(s) - \hat{y}_d(s)][\hat{y}(-s) - \hat{y}_d(-s)] + \hat{u}(s)\hat{u}(-s)\}\, ds$$

where $q = 25$ and y_d is a unit step function.

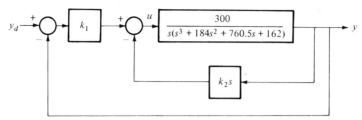

Figure 11-4 A configuration of compensators.

The overall transfer function of the system in Figure 11-4 can be easily computed as

$$\hat{g}_f(s) = \frac{300k_1}{s^4 + 184s^3 + 760.5s^2 + (162 + 300k_2)s + 300k_1} \triangleq \frac{N_f(s)}{D_f(s)}$$

We then compute

$$\hat{y}(s) - \hat{y}_d(s) = [\hat{g}_f(s) - 1]\frac{1}{s} = \frac{s^3 + 184s^2 + 760.5s + (162 + 300k_2)}{D_f(s)}$$

and

$$\hat{u}(s) = \frac{\hat{y}(s)}{\hat{g}(s)} = \frac{\hat{g}_f(s)s^{-1}}{\hat{g}(s)} = \frac{(s^3 + 184s^2 + 760.5s + 162)k_1}{D_f(s)}$$

where $\hat{g}(s)$ is given in Equation (11-10). In order to apply Table 11-1, we compute

$$q[\hat{y}(s) - \hat{y}_d(s)][\hat{y}(-s) - \hat{y}_d(-s)] + \hat{u}(s)\hat{u}(-s) = \frac{d_0 s^6 + d_1 s^4 + d_2 s^2 + d_3}{D_f(s)D_f(-s)}$$

where

$$d_0 = k_1^2 - 25$$

$$d_1 = 80.75 \times 10^4 + 3.23 \times 10^4 k_1^2$$

$$d_2 = -1296 \times 10^4 + 275 \times 10^4 k_2 - 51.84 \times 10^4 k_1^2$$

$$d_3 = 65.5 \times 10^4 + 243 \times 10^4 k_2 + 225 \times 10^4 k_2^2 + 2.62 \times 10^4 k_1^2$$

With these the performance index J can be computed as a function of k_1 and k_2. In order to find the optimal k_1 and k_2, we solve the simultaneous equations $\partial J(k_1, k_2)/\partial k_1 = 0$ and $\partial J(k_1, k_2)/\partial k_2 = 0$. This is, as can be seen from Example 3, an extremely complicated computation. Hence, rather than solve it analytically, we use a digital computer to compute $J(k_1, k_2)$ for $k_1 = 1, 2, \ldots, 20$ and $k_2 = 1, 2, \ldots, 20$. The optimal k_1 and k_2 are then found as $k_1 = 4$ and $k_2 = 5$, with $J = 30.53$. Thus the design is completed.

The response of the system in Figure 11-4 with $k_1 = 4$ and $k_2 = 5$ due to a unit step function is shown in Figure 12-4. The response is very satisfactory. ∎

11-2 Numerical Method

Parameter optimization involves two separate problems: the computation of the performance criterion and the search of the optimal parameters. In the analytical approach the performance index must be in a closed form and expressed as a function of the parameters. This can be achieved by using Table 11-1. The table however becomes very unwieldy for n larger than 4. Therefore it is desirable to introduce a different method of computing the performance index. The method to be introduced will compute the performance index iteratively and is most convenient for digital computer programming. The method yields a numerical value however; hence the search of the optimal parameters also must be carried out numerically. Before proceeding we introduce some notations.

Consider the polynomials

$$A_n(s) = a_0(n)s^n + a_1(n)s^{n-1} + \cdots + a_{n-1}(n)s + a_n(n) \quad \text{(11-16)}$$

$$B_n(s) = b_1(n)s^{n-1} + \cdots + b_{n-1}(n)s + b_n(n) \quad \text{(11-17)}$$

Define

$$\tilde{A}_n(s) = a_1(n)s^{n-1} + a_3(n)s^{n-3} + \cdots \quad \text{(11-18)}$$

The polynomial $\tilde{A}_n(s)$ consists of the even part of $A_n(s)$ if n is odd, and the odd part of $A_n(s)$ if n is even. Now we define recursively the following polynomials, for $k = n, n-1, \ldots, 2, 1$,

$$A_{k-1}(s) = A_k(s) - \alpha_k s \tilde{A}_k(s) \quad \text{(11-19)}$$

$$B_{k-1}(s) = B_k(s) - \beta_k s \tilde{A}_k(s) \quad \text{(11-20)}$$

where

$$\alpha_k = \frac{a_0(k)}{a_1(k)} \quad \text{(11-21)}$$

$$\beta_k = \frac{b_1(k)}{a_1(k)} \quad \text{(11-22)}$$

Note that the polynomials $A_{k-1}(s)$ and $B_{k-1}(s)$ have degree $k-1$ and $k-2$, respectively. The computation of $A_k(s)$ can be carried out in a table form as follows:

$A_n(s)$	$a_0(n)$	$a_1(n)$	$a_2(n)$	$a_3(n)$	$a_4(n)$ ...
$\tilde{A}_n(s)$	$a_1(n)$	0	$a_3(n)$	0	$a_5(n)$ ...
$A_{n-1}(s)$	$a_0(n-1)$	$a_1(n-1)$	$a_2(n-1)$	$a_3(n-1)$	$a_4(n-1)$ ...
$\tilde{A}_{n-1}(s)$	$a_1(n-1)$	0	$a_3(n-1)$	0	$a_5(n-1)$ ...
$A_{n-2}(s)$	$a_0(n-2)$	$a_1(n-2)$	$a_2(n-2)$	$a_3(n-2)$ ...	
$\tilde{A}_{n-2}(s)$	$a_1(n-2)$	0	$a_3(n-1)$	0 ...	

⋮

where, for $k = n, n-1, \ldots,$

$$a_0(k-1) = \frac{a_1(k)a_1(k) - a_0(k) \cdot 0}{a_1(k)} = a_1(k)$$

$$a_1(k-1) = \frac{a_1(k)a_2(k) - a_0(k)a_3(k)}{a_1(k)}$$

$$a_2(k-1) = \frac{a_1(k)a_3(k) - a_0(k) \cdot 0}{a_1(k)} = a_3(k)$$

$$\vdots$$

We see that the computations of the coefficients of $A_{k-1}(s)$ are identical to the formation of the Routh-Hurwitz table in Chapter 4. Similarly the polynomial $B_k(s)$ can be computed as follows:

$$\begin{cases} B_n(s) & b_1(n) & b_2(n) & b_3(n) & b_4(n) & \cdots \\ \tilde{A}_n(s) & a_1(n) & 0 & a_3(n) & 0 & \cdots \end{cases}$$

$$\begin{cases} B_{n-1}(s) & b_1(n-1) & b_2(n-1) & b_3(n-1) & b_4(n-1) & \cdots \\ \tilde{A}_{n-1}(s) & a_1(n-1) & 0 & a_3(n-1) & 0 & \cdots \end{cases}$$

$$\begin{cases} B_{n-2}(s) & b_1(n-2) & b_2(n-2) & b_3(n-2) & \cdots \\ \tilde{A}_{n-2}(s) & a_1(n-2) & 0 & a_3(n-2) & \cdots \end{cases}$$

$$\vdots$$

where, for $k = n, n-1, \ldots,$

$$b_1(k-1) = \frac{a_1(k)b_2(k) - b_1(k) \cdot 0}{a_1(k)} = b_2(k)$$

$$b_2(k-1) = \frac{a_1(k)b_3(k) - b_1(k)a_3(k)}{a_1(k)}$$

$$b_3(k-1) = \frac{a_1(k)b_4(k) - b_1(k) \cdot 0}{a_1(k)} = b_4(k)$$

$$\vdots$$

After computing $A_k(s)$ and $B_k(s)$, the constants α_k and β_k can then be easily obtained. With α_k and β_k, we introduce the following theorem.

Theorem 11-1

Consider the integral

$$I_n = \frac{1}{2\pi j} \int_{-j\infty}^{j\infty} \frac{B_n(s)B_n(-s)}{A_n(s)A_n(-s)} ds$$

If $A_n(s)$ is a Hurwitz polynomial or, equivalently, if all the roots of $A_n(s)$ have negative real parts, then I_n is given by

$$I_n = \sum_{k=1}^{n} \frac{\beta_k^2}{2\alpha_k} = \sum_{k=1}^{n} \frac{b_1^2(k)}{2a_0(k)a_1(k)}$$

where α_k, β_k, $a_0(k)$, $a_1(k)$, and $b_1(k)$ are defined as in Equations (11-16) through (11-22); or

$$I_k = I_{k-1} + \frac{\beta_k^2}{2\alpha_k} \quad k = 1, 2, \ldots, n$$

with $I_0 = 0$. ∎

The proof of this theorem can be found in Reference [2]. We give a simple example to illustrate the application of this theorem.

Example

Compute the integral

$$I_2 = \frac{1}{2\pi j} \int_{-j\infty}^{j\infty} \frac{(4s + 5)(-4s + 5)}{(s^3 + 2s^2 + 3s + 2)(-s^3 + 2s^2 - 3s + 2)} \, ds$$

It is easy to check that the polynomial $(s^3 + 2s^2 + 3s + 2)$ is Hurwitz. We form

$$\begin{cases} A_3(s) \\ \tilde{A}_3(s) \end{cases} \quad \begin{matrix} 1 & 2 & 3 & 2 \\ 2 & 0 & 2 & 0 \end{matrix}$$

$$\begin{cases} A_2(s) \\ \tilde{A}_2(s) \end{cases} \quad \begin{matrix} 2 & 2 & 2 \\ 2 & 0 & \end{matrix}$$

$$\begin{cases} A_1(s) \\ \tilde{A}_1(s) \end{cases} \quad \begin{matrix} 2 & 2 \\ 2 & 0 \end{matrix}$$

and

$$\begin{cases} B_3(s) \\ \tilde{A}_3(s) \end{cases} \quad \begin{matrix} 0 & 4 & 5 \\ 2 & 0 & 2 \end{matrix}$$

$$\begin{cases} B_2(s) \\ \tilde{A}_2(s) \end{cases} \quad \begin{matrix} 4 & 5 \\ 2 & 0 \end{matrix}$$

$$\begin{cases} B_1(s) \\ \tilde{A}_1(s) \end{cases} \quad \begin{matrix} 5 & 0 \\ 2 & 0 \end{matrix}$$

Hence we have $\alpha_3 = 1/2$, $\alpha_2 = 2/2 = 1$, $\alpha_1 = 1$, $\beta_3 = 0$, $\beta_2 = 4/2 = 2$, and $\beta_1 = 5/2 = 2.5$. Thus the integral I_2 is

$$I_2 = \frac{\beta_1^2}{2\alpha_1} + \frac{\beta_2^2}{2\alpha_2} + \frac{\beta_3^2}{2\alpha_3} = \frac{(2.5)^2}{2} + \frac{4}{2} + \frac{0}{1} = 5.125$$

This result can be checked to be the same as the one obtained by using Table 11-1. ∎

With the introduced algorithm, the performance criterion of the type shown in (11-1) or (11-3) for a given set of parameters can be easily obtained in a digital computer. We discuss now the search of the parameters that minimize the performance criterion. Clearly in the search of the optimal parameters, we search only the ranges of the parameters in which the system is totally stable. For convenience of discussion, we study first the one-parameter case.

The performance criterion, $J(k)$, as a function of parameter k is assumed to be of the form shown in Figure 11-5. To start a search, we pick an arbitrary k_0 and compute $J(k_0)$. Next we compute $J(k_1) = J(k_0 + h)$, where h is a positive constant. If $J(k_0 + h)$ is larger than $J(k_0)$, then we proceed in the opposite direction. We continue the process with the same step size h in the direction of decreasing J until three points are found for which the middle point has the least-performance index, as shown in Figure 11-5(a). Let the three points be k_2, k_3, and k_4. Now we know that the minimum point must occur somewhere between k_2 and k_4; that is, the optimal parameter k^* is bounded by

$$k_3 - h \leq k^* \leq k_3 + h$$

Next we compute the performance indices at the center points, k_5 and k_6, of (k_2, k_3) and (k_3, k_4), respectively. Now search from k_3, k_5, and k_6 the point with the least performance index. Let k_6 be the point as shown in Figure 11-5(b). Then the optimal parameter k^* must lie between k_3 and k_4 or, equivalently, must be bounded by

$$k_6 - \frac{h}{2} \leq k^* \leq k_6 + \frac{h}{2}$$

Repeating the process once again, we obtain

$$k_6 - \frac{h}{4} \leq k^* \leq k_6 + \frac{h}{4}$$

Proceeding successively, after n steps we obtain

$$k_j - \frac{h}{2^n} \leq k^* \leq k_j + \frac{h}{2^n}$$

Hence a fairly accurate k^* can be obtained after a number of steps. Generally the searching process is automatically stopped if

$$|J(k_j) - J(k_{j-1})| \leq \varepsilon$$

where ε is a preassigned number.

We now discuss the two-parameter case. If there are two parameters, k_1 and k_2, we may first pick an arbitrary k_1 and then carry out the searching for the optimal k_2, as shown in Figure 11-6. After obtaining the optimal k_2^* for the chosen k_1, we keep k_2^* constant and then carry out the searching for k_1. Proceeding successively, we can obtain the optimal parameters. The process can be extended to cases involving three or more parameters.

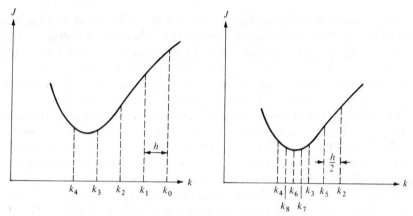

Figure 11-5 The search for the optimal parameter.

The algorithm introduced in this section can be easily programmed in a digital computer. The problem in Example 4 of Section 11-1 was run on an IBM 360/67 computer by using this algorithm, and the optimal parameters were found to be $k_1 = 3.68$ and $k_2 = 4.15$, with $J = 30.51$. The execution time used for this problem is 0.08 minute.

The discussion of searching in this section is not complete. In addition to the search method discussed, there are many others. The interested reader is referred to References [1], [4], [5], and [6]. For existing subroutines the reader is referred to the IBM System/360 Scientific Subroutine Package.

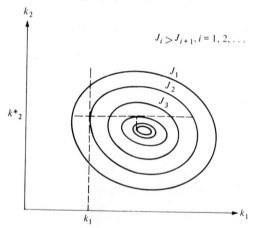

Figure 11-6 The search for the optimal parameters.

11-3 Concluding Remarks

In the parameter optimization with the performance index (11-1), although the obtained system is optimal for the chosen configuration, the system may still be completely unsatisfactory according to the specification on rise time, overshoot, and settling time. However the larger the number of parameters to be optimized, the better chance we have to obtain a satisfactory system.

The weighting factor q in (11-1) is chosen, as in Chapter 8, by trial and error. If the response of the system is too slow, we may increase q; if it has too much overshoot, then we may decrease q. If it is impossible to obtain a satisfactory system by changing q alone, then we have to choose a different compensator configuration.

The search procedure introduced in Section 11-2 may encounter difficulty in multiparameter cases. For example, the search along one parameter may decrease J, while at the same time the search is progressing in a direction where the contribution due to another parameter causes J to increase. In this case the search will break down. See References [1], [4], [5], and [6].

The computation required in the parameter optimization is comparable with the one in the optimal design. If the parameter optimization and the optimal design use the same performance criterion, the system obtained by the former can never be better than the one obtained by the latter method. However the compensator used in the former is generally simpler than the one used in the latter.

Comparison of the parameter optimization with the root-locus method and the Bode-plot method are in order. In the root-locus method, from the trajectory of the root loci, it is possible to obtain a rough estimate of the time response. A rough estimate of the time response can also be obtained from the phase margin, gain margin, and gain crossover frequency of the Bode plot. However in the parameter optimization, it seems impossible to predict what kind of time response will be obtained in the resulting optimal system. The time response is obtainable by simulation only after the design is completed. The parameter optimization, on the other hand, is more systematic than the root-locus and the frequency-domain methods.

References

[1] Aoki, M., *Introduction to Optimization Techniques*. New York: Macmillan, 1971.
[2] Astrom, K. J., *Introduction to Stochastic Control Theory*. New York: Academic Press, 1970.
[3] Chang, S. S. L., *Synthesis of Optimum Control Systems*. New York: McGraw-Hill, 1961.
[4] Hamming, R. W., *Introduction to Applied Numerical Analysis*. New York: McGraw-Hill, 1971.
[5] Himmelblau, D. M., *Applied Nonlinear Programming*. New York: McGraw-Hill, 1972.
[6] Kowalik, J., and M. R. Osborne, *Methods for Unconstrained Optimization Problems*. New York: Elsevier, 1968.
[7] Newton, G. C., L. A. Gould, and J. F. Kaiser, *Analytical Design of Linear Feedback Controls*. New York: Wiley, 1957.

Problems

11-1 Consider the problem of controlling the yaw of an airplane discussed in Problem 8-8. The configuration of compensator is chosen as shown in Figure P11-1, where $\hat{g}(s) = -2/s^2$. Find, by using the analytical method, the gain k that minimizes the performance index

$$J = \int_0^\infty [10\theta^2(t) + \phi^2(t)]\, dt$$

Find also the response of the resulting system due to a unit step input. What are the percentage overshoot, settling time, and rise time? How do you compare this result with the one obtained in Problem 9-4?

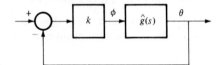

Figure P11-1

11-2 Consider the problem studied in Problem 9-3. Find, by using the analytical method, the gains k_1 and k_2 so that the resulting system minimizes

$$J = \int_0^\infty \{[\theta(t) - 1]^2 + \phi^2(t)\}\, dt$$

11-3 Consider the tracking antenna problem studied in Problem 8-3. The configuration of compensators of the form shown in Figure 11-4 is chosen. Find the gains k_1 and k_2 so that the resulting system minimizes

$$J = \int_0^\infty \{25[y(t) - 1]^2 + u^2(t)\}\, dt$$

Compare this result with the one obtained in Example 4 of Section 11-1.

11-4 Study Problems 11-1 and 11-2 by using the numerical method.

11-5 Consider the control of the depth of a submarine studied in Problem 9-9. Find the k to minimize the performance criterion

$$J = \int_0^\infty \{10[y(t) - 1]^2 + u^2(t)\}\, dt$$

Compare the result with those obtained in Problems 9-9 and 10-11.

11-6 Consider the control system studied in Problem 9-10. Find the k_1 and k_2 to minimize the performance criterion

$$J = \int_0^\infty \{100[y(t) - 1]^2 + u^2(t)\}\, dt$$

Do you think you may obtain a better system if no constraint is imposed on the compensator configuration?

CHAPTER 12
Epilogue

12-1 Comparisons of Various Design Results of the Tracking Antenna Problem

We have introduced in this text four different methods for designing linear control systems. The methods are the optimal design with respect to a quadratic criterion, the root-locus method, the Bode-plot method, and parameter optimization. In optimal design, it is possible to obtain many different optimal systems by employing output feedback, state feedback, or partial state feedback. In the following the design results of the tracking antenna problem in the text will be summarized and compared. Before proceeding, we remark that all the responses are obtained by employing IBM System/360 CSMP on a digital computer.

Consider the tracking antenna problem studied in Example 2 of Section 8-2. The transfer function of the plant is

$$\hat{g}(s) = \frac{300}{s(s^3 + 184s^2 + 760.5s + 162)} \quad (12\text{-}1)$$

This problem is solved twice by the optimal design technique, once by using state feedback in Example 3 of Section 8-7 and once by using partial state feedback in the Example in Section 8-8. When we use state feedback, because of the insertion of resistors into the armature and field circuits, the transfer function of the plant is modified to

$$\hat{g}(s) = \frac{300}{s(s^3 + 194.2s^2 + 838.9s + 171)} \qquad (12\text{-}2)$$

The performance index in both cases is chosen as

$$J = \int_0^\infty \{q[y(t) - y_d(t)]^2 + u^2(t)\}\, dt \qquad (12\text{-}3)$$

with $q = 25$ and y_d a unit step function. The unit step responses of the resulting systems are shown in Figure 12-1. We see that, although the coefficients of the transfer functions in (12-1) and (12-2) are quite different, the results are extremely close. In Figure 12-2 we also show the unit step responses of the optimal systems for $q = 9$ and $q = 100$.

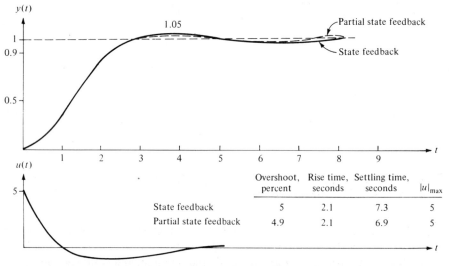

Figure 12-1 Optimal design with $q = 25$.

The tracking antenna problem is designed by using the Bode plot in Example 4 of Section 10-5. The resulting compensator is a lag-lead network, given in Equation (10-44). The unit step response of the resulting system is shown in Figure 12-3. The overshoot is 13.8 percent, the rise time is 2.6 seconds, and the settling time is 18.9 seconds. The steady-state error due to a step function is zero.

In Figure 12-3 we also show the response of the system with the compensator $C(s)$ replaced by the compensator $C'(s)$ in Equation (10-45). Although the gain margin, phase margin, and gain crossover frequency of the Bode plot with compensator $C(s)$ and those of the Bode plot with $C'(s)$ are identical, the step responses of the two systems are quite different. This is conceivable because the specifications on phase margin, gain margin, and gain crossover frequency are derived by the use of an argument that is not very vigorous.

12-1 COMPARISONS OF THE TRACKING ANTENNA PROBLEM 317

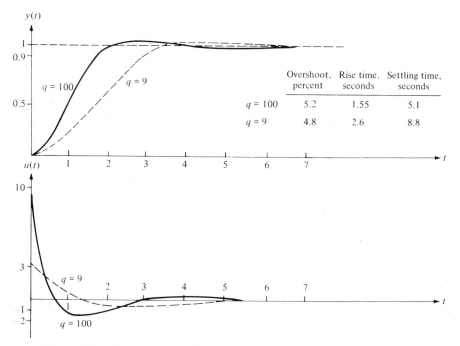

Figure 12-2 Step responses of the optimal systems with $q = 9$ and $q = 100$.

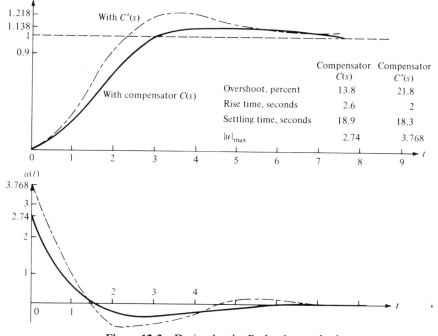

Figure 12-3 Design by the Bode-plot method.

We compare now the results obtained by the optimal design and the Bode-plot method. We compare the response denoted by the solid line in Figure 12-2 and the one denoted by the dashed line in Figure 12-3. The responses have the same rise time. The one obtained by the optimal design has a smaller overshoot and a smaller settling time; the one obtained by the Bode-plot method however has a smaller $|u(t)|_{max}$. Hence the system obtained by the Bode-plot method is less susceptible to saturation; furthermore its compensator is of a degree one less than that used in the optimal design.

The procedure in the optimal design is systematic. If the response at the output is found to be too slow, or the magnitude of the actuating signal is too large, the design can be improved by changing the weighting factor q. This however is not the case in the Bode-plot method. If the design is not satisfactory, it seems that the Bode-plot method does not provide any direction for improvement. It does however have the distinctive feature that the design can be carried out from the measured data without first obtaining the exact mathematical description.

The tracking antenna problem is designed in Examples 3 and 4 of Section 11-1 by the use of parameter optimization. Although the system in Figure 11-3 with $k_1 = 1$ and $k_2 = 1.93$ is optimal with respect to the performance index in Equation (12-3), the response, as shown in Figure 12-4, is totally unacceptable. The optimal system in Figure 11-4 with $k_1 = 4$ and $k_2 = 5$ however has an excellent step response, as shown in Figure 12-4. The rise time is 2.3 seconds; the settling time is 5.7 seconds. There is no overshoot in the step response. We see that the response compared very well with the one obtained by the optimal design shown in Figure 12-1.

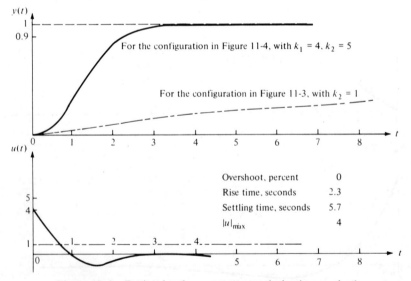

Figure 12-4 Design by the parameter optimization method.

The same problem is solved in Example 3 of Section 9-4 by using the root-locus method. The unit step response of the resulting system is shown in Figure 12-5. It is again very satisfactory and compares very well with the ones obtained by the optimal design and parameter optimization techniques. The method is however a trial-and-error process.

We see in this Epilogue that, if properly employed, the optimal design technique, the Bode-plot method, the root-locus method, and parameter optimization all yield good systems. Hence they all can be used in the design of linear time-invariant control systems.

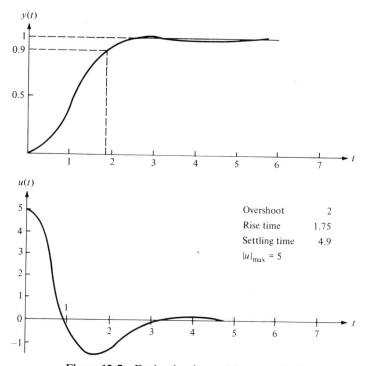

Figure 12-5 Design by the root-locus method.

12-2 Further Topics

The control problems studied in this text are modeled exclusively as linear time-invariant single-variable systems. Although we have given a fairly complete treatment of this class of systems, the study of control theory is far from complete. We introduce briefly in the following some topics that are not discussed in the text.

Multivariable systems. A system is said to be a *multivariable system* if it has two or more inputs and/or two or more outputs. A large number of physical systems are of this type. For example, the economic system discussed in Section 1-1 has government spending, interest rate, and monetary policy as controlling factors (inputs) to control the rate of unemployment, average hourly wage, gross national product, and consumer price index (outputs). Hence a properly chosen model of this economic system should have three inputs and four outputs. The power plant shown in Figure 1-1 is another example of a multivariable system. Although it can be modeled as a single-variable system by neglecting steam pressure and temperature, and by considering the valve position as input and the generated voltage as output, a more realistic model should have valve position, control rod position, and the rate of water flow to the boiler as inputs and the generated voltage, steam temperature, and pressure as outputs. Traffic control of a downtown area is again a multivariable system. The inputs are the traffic lights at various intersections; the outputs are the traffic flows of the major streets.

As discussed in Section 2-6, if a multivariable system is lumped (that is, has a finite number of state variables), linear, and time-invariant, then it can be described by the input-output description

$$\hat{\mathbf{y}}(s) = \hat{\mathbf{G}}(s)\hat{\mathbf{u}}(s) \qquad (12\text{-}4)$$

and the state-variable description

$$\dot{\mathbf{x}} = \mathbf{A}\mathbf{x} + \mathbf{B}\mathbf{u} \qquad (12\text{-}5a)$$

$$\mathbf{y} = \mathbf{C}\mathbf{x} + \mathbf{D}\mathbf{u} \qquad (12\text{-}5b)$$

The analysis techniques used in the single-variable case are directly applicable to these multivariable equations. For example, the response of Equation (12-4) can be obtained by considering each output at a time; the solution of (12-5) is readily obtainable after computing $e^{\mathbf{A}t}$. The system described by (12-4) is BIBO stable if all poles of each element of $\hat{\mathbf{G}}(s)$ have negative real parts; the zero-input response of (12-5) is asymptotically stable if all eigenvalues of $\mathbf{A}$ have negative real parts.

The design of multivariable systems is much more complicated than the single-variable cases. We discuss first the designs by using transfer-function matrices. There have been many attempts to extend the root-locus method and Bode-plot design to the multivariable cases. Since these two design techniques are essentially methods of adjusting parameters by trial and error, and since multivariable systems generally have large numbers of parameters, the root-locus and Bode-plot methods are rather ineffective in the design of multivariable systems. The design of optimal multivariable systems with respect to the quadratic performance index can also be carried out by using transfer-function matrices. The computations required in the single-input, multiple-output or the multiple-input, single-output cases are comparable with the single-variable cases. The computations required in the multiple-input, multiple-output cases are very involved however. Hence the design of optimal multivariable systems by the use of transfer-function matrices has not received much attention.

The introduction of the state-variable approach in control theory was often deemed as a breakthrough in the design of multivariable systems. Indeed the equations and computations required in the design of optimal single-variable systems and those in the design of optimal multivariable systems are almost identical. Hence most of the texts on optimal design use the state-variable approach almost exclusively. The interested reader is referred to Reference [1].

Linear discrete-time systems. The inputs, outputs, and state variables of the systems studied in the text and the above subsection are defined for all time in the interval $[0, \infty)$; that is, the signals are available continuously for all t in $[0, \infty)$. This class of systems is called *continuous-time systems*. In this subsection we introduce a different class of systems called *discrete-time systems*. A discrete-time system is one in which one or more signals (inputs, outputs, or state variables) of the system are defined only at discrete instants of time. This kind of signal can be represented by a sequence of numbers. A digital computer reads in and prints out sequences of numbers; hence it is a discrete-time system. A radar sends out pulses at fixed intervals of time; hence it is another example of a discrete-time system. A continuous-time physical system can be modeled as a discrete-time system if its response is of interest or measurable only at certain instants of time. For example, an economic system is generally modeled as a discrete-time system because the rate of unemployment and consumer price index are censored monthly and the gross national product is compiled quarterly.

In this subsection the discrete instants of time at which the inputs, outputs, and states of discrete-time systems appear will be assumed to be equally spaced and will be denoted by the set of positive integers $n = 0, 1, 2, \ldots$. Thus the input, output, and state are sequences of numbers and will be denoted by $\{u(n)\}$, $\{y(n)\}$, and $\{x(n)\}$, respectively. If the functions $u(t)$, $y(t)$, and $x(t)$ are replaced by $u(n)$, $y(n)$, and $x(n)$, then all the concepts introduced in Section 2-2 and 2-3 are directly applicable to discrete-time systems. If a discrete-time system is single-variable, linear, time-invariant, and relaxed at $n = 0$, then the input and output sequences can be related by

$$y(n) = \sum_{m=0}^{n} g(n - m)u(m) \qquad n = 0, 1, 2, \ldots \qquad (12\text{-}6)$$

where $g(n)$ is called the *weighting sequence* and is the response of the system due to the input

$$u(0) = 1$$
$$u(n) = 0 \qquad n = 1, 2, 3, \ldots$$

The weighting sequence has the property $g(n) = 0$ for negative integer n, for it is the response due to an input applied at $n = 0$. The derivation of Equation (12-6) is quite similar to that of (2-13), except that it is simpler. We see that the integration in the continuous-time case is replaced by the summation in the discrete-time case.

In the continuous-time case, an integral equation as shown in (2-13) can be transformed into an algebraic equation by the application of the Laplace transform. We have a similar situation here; the transformation used is called the z-transform.

Definition 12-1

The z-transform of a sequence $\{u(n)\}$, for $n = 0, 1, 2, \ldots\}$ is defined as

$$\hat{u}(z) = \sum_{n=0}^{\infty} u(n) z^{-n}$$

where z is a complex variable. ∎

Example 1

If $u(n) = 1$, for $n = 0, 1, 2, \ldots$, then

$$\hat{u}(z) = \sum_{n=0}^{\infty} z^{-n} = \frac{1}{1 - z^{-1}} = \frac{z}{z - 1}$$

Example 2

If $u(n) = e^{-2n}$, for $n = 0, 1, 2, \ldots$, then

$$\hat{u}(z) = \sum_{n=0}^{\infty} e^{-2n} = \frac{1}{1 - e^{-2} z^{-1}} = \frac{z}{z - e^{-2}}$$ ∎

Given a z-transform $\hat{u}(z)$, the computation of $\{u(n)\}$ from $\hat{u}(z)$ is called the inverse z-transformation. There are many ways to carry out the inverse z-transformation. The simplest is just to expand $\hat{u}(z)$ into a descending power series of z^{-1}. Then the coefficients of the series immediately give the sequence. For example, if

$$\hat{u}(z) = \frac{3z - 1}{z^2 + 2z + 1} = 3z^{-1} - 5z^{-2} + 7z^{-3} - 9z^{-4} + 11z^{-5} + \cdots$$

then its inverse z-transform is $\{0, 3, -5, 7, -9, 11, \ldots\}$.

The application of the z-transform to (12-6) yields[1]

$$\hat{y}(z) = \hat{g}(z) \hat{u}(z) \tag{12-7}$$

where $\hat{y}(z)$, $\hat{g}(z)$, and $\hat{u}(z)$ are, respectively, the z-transforms of $\{y(n)\}$, $\{g(n)\}$, and $\{u(n)\}$. The function $\hat{g}(z)$ is called the *pulse transfer function*, or *sampled transfer function*. It is, by definition, the z-transform of the weighting sequence. We see that the z-transform converts a summation into an algebraic equation.

---

[1] See References [2] and [6].

This is similar to the continuous case in which the Laplace transform converts an integral equation into an algebraic equation. We give an example to show the application of the z-transform.

Example 3

Given a discrete-time system with the sampled transfer function

$$\hat{g}(z) = \frac{z-1}{(z^2 + 2z + 1)}$$

What is the output due to the application of $u(n) = 1$, for $n = 0, 1, 2, \ldots$?

The z-transform of $\{u(n)\}$ is, as computed in Example 1, $z/(z-1)$. Hence we have

$$\hat{y}(z) = \hat{g}(z)\hat{u}(z) = \frac{z-1}{z^2 + 2z + 1} \cdot \frac{z}{z-1} = \frac{z}{z^2 + 2z + 1}$$

$$= z^{-1} - 2z^{-2} + 3z^{-3} - 4z^{-4} + \cdots$$

Hence the output sequence is $\{0, 1, -2, 3, -4, \ldots\}$. This output can also be obtained by first computing the weighting sequence and then applying Equation (12-6). Try it!

We introduce now the discrete-time dynamical equation. A linear time-invariant discrete-time dynamical equation is defined as

$$\mathbf{x}(n+1) = \mathbf{A}\mathbf{x}(n) + \mathbf{b}u(n) \tag{12-8a}$$

$$y(n) = \mathbf{c}\mathbf{x}(n) + du(n) \tag{12-8b}$$

where $\mathbf{x}(n)$, $u(n)$, and $y(n)$ are the state, input, and output at time instant n. Note that it is a set of first-order *difference* equations, rather than *differential* equations as in the continuous-time case. Given an initial state $\mathbf{x}(0)$, and an input sequence $\{u(n)$, for $n = 0, 1, 2, \ldots\}$, the solution of (12-8) can be easily obtained by first computing $\mathbf{x}(1)$, then $\mathbf{x}(2)$, and so forth.

The application of the z-transform to (12-8) yields[2]

$$\hat{\mathbf{x}}(z) = (z\mathbf{I} - \mathbf{A})^{-1} z\mathbf{x}(0) + (z\mathbf{I} - \mathbf{A})^{-1} \mathbf{b}\hat{u}(z) \tag{12-9a}$$

$$\hat{y}(z) = \mathbf{c}(z\mathbf{I} - \mathbf{A})^{-1} z\mathbf{x}(0) + [\mathbf{c}(z\mathbf{I} - \mathbf{A})^{-1}\mathbf{b} + d]\hat{u}(z) \tag{12-9b}$$

These equations are algebraic and are similar to (2-27) and (2-28).

The relationship between the input-output description (12-7) and the state-variable description (12-9) can be obtained, by setting $\mathbf{x}(0) = \mathbf{0}$, as

$$\hat{g}(z) = \mathbf{c}(z\mathbf{I} - \mathbf{A})^{-1}\mathbf{b} + d \tag{12-10}$$

This is identical to Equation (2-29) except that the Laplace-transform variable s is replaced by the z-transform variable z.

---

[2] See References [2] and [6].

From the foregoing discussion we see that there is a close analogy between the mathematical descriptions of continuous-time systems and those of discrete-time systems. Therefore it is conceivable that some of the concepts and results in the continuous-time case can be applied to the discrete-time case. Indeed the concepts of controllability and observability and their conditions introduced in Section 2-7 are directly applicable to discrete-time equation (12-8). The realization problem and procedure discussed in Sections 5-3 and 5-4 are also directly applicable to the discrete-time case. Though the concepts of stability introduced in Section 4-4 are still valid, the conditions have to be modified. Similar remarks apply to most of the design techniques.

Discrete-time systems have become increasingly important because of the wide availability of digital computers and microprocessors. Thus their study is essential in the design of control systems. With the background introduced in this text, the study of discrete-time systems should be comparatively easy.

Time-varying systems. Roughly speaking, a system is called a *time-varying* system if its characteristics change with time (see Section 2-3). If a time-varying system is linear, then it can be described by the input-output description

$$y(t) = \int_{t_0}^{t} g(t, \tau)u(\tau) \, d\tau \qquad (12\text{-}11)$$

where $g(t, \tau)$ is the impulse response and is the response due to the application of a δ-function at time τ, or the state-variable description

$$\dot{\mathbf{x}} = \mathbf{A}(t)\mathbf{x} + \mathbf{b}(t)u \qquad (12\text{-}12a)$$

$$y = \mathbf{c}(t)\mathbf{x} + d(t)u \qquad (12\text{-}12b)$$

These equations are quite similar to those of (2-23) and (2-24), except that the matrices $\mathbf{A}$, $\mathbf{b}$, $\mathbf{c}$, and d are now functions of time. Solving these equations by analytical methods is however very complicated; it is best to use numerical methods.

In linear time-invariant systems, Laplace and z-transforms are two important tools in the analysis and design. By applying these transformations, linear time-invariant equations can be transformed into algebraic equations. Although the transformations can still be applied in the linear time-varying cases, the resulting equations are very complicated. Hence the transformation techniques are generally *not* used in the time-varying case.

Depending on the type of problem, either the input-output equation (12-11) or the state-variable equation (12-12) can be used. In the optimal design of time-varying control systems, the state-variable equation (12-12) is used almost exclusively. In the study of the stability problem the two equations appear to be used equally.

Nonlinear systems. Although a large number of physical systems can be modeled as linear systems by linearization and by restricting the operational range to a certain region, this is not always possible for every physical system.

For example, the control of a home heating system is accomplished by automatically opening or closing a relay. This kind of system cannot be modeled adequately by a linear system. Hence it is sometimes necessary to study nonlinear systems.

A system is called a *nonlinear system* if its responses do not satisfy the superposition property (see Section 2-3). The nonlinear systems that have been studied by control engineers are mostly of, or can be reduced to, the form shown in Figure 12-6. The block denoted by N is a nonlinear element such as saturation, backlash, relay, or hysteresis. The block denoted by L is a linear time-invariant system. Depending on the analysis technique used, the block L is described by either a transfer function or a state-variable equation. The methods used in the analysis of nonlinear systems are the graphical method, the describing function method, the Liapunov method, and the Popov theorem. The interested reader is referred to Reference [5].

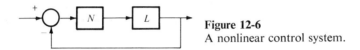

Figure 12-6
A nonlinear control system.

Stochastic systems. A signal is called a *deterministic signal* if it can be defined precisely without any uncertainty or chance. In contrast, a stochastic signal is one that cannot be defined precisely; it is random in nature and can, at best, be described statistically or with certain probability. Noise is one example. A system having only deterministic signals is called a *deterministic system*; a system having one or more stochastic signals is called a *stochastic system*. As we mentioned in the text, noise is always present in physical devices. If the effect of noise can be neglected, then these physical systems can be modeled as deterministic systems. If the noise cannot be neglected, such as in precision navigation systems, then stochastic control theory must be employed. A study of stochastic system requires the knowledge of probability theory and stochastic processes. The interested reader is referred to Reference [7].

Adaptive and learning systems. If the characteristics of a plant do not change with time over the time interval of interest, then the plant can be modeled as a time-invariant system. If the characteristics of a plant change with time, and if the changes are predictable such as the change of the mass of a rocket during blast-off, then the plant must be modeled as a time-varying system. If the characteristics of a plant change with time, and if the changes are not predictable, then the control system having this kind of plant must be modeled as a learning system. A learning system performs two functions: identification and compensation. In the identification part a device, possibly a computer, is used to learn or identify the characteristics of the plant. Then a compensation, based on the updated characteristics of the plant, is generated to control the plant. This is the most sophisticated control system. The interested reader is referred to References [8] and [9].

References

[1] Athans, M., and P. L. Falb, *Optimal Control*. New York: McGraw-Hill, 1966.
[2] Cadzow, J. A., and H. R. Martens, *Discrete-Time and Computer Control Systems*. Englewood Cliffs, N.J.: Prentice-Hall, 1970.
[3] Chen, C. T., "Synthesis of multivariable feedback systems," *Preprints 1968 JACC*, pp. 224–228. (See, especially, the references listed in this paper.)
[4] Chen, C. T., "Design of compensators for a class of multivariable linear optimum regulators," *2d IFAC Symp. on Multivariable Technical Control Systems*, 1971.
[5] Hsu, J. C., and A. U. Meyer, *Modern Control Principles and Applications*. New York: McGraw-Hill, 1968.
[6] Kuo, B. C., *Discrete-Date Control Systems*, Englewood Cliffs, N.J.: Prentice-Hall, 1970.
[7] Meditch, J. S., *Stochastic Optimal Linear Estimation and Control*. New York: McGraw-Hill, 1969.
[8] Mendal, J. M., and K. S. Fu, *Adaptive, Learning, and Pattern Recognition Systems*. New York: Academic Press, 1970.
[9] Tsypkin, Y. Z., *Adaptation and Learning in Automatic Systems*. New York: Academic Press, 1971.

APPENDIX A
The Laplace Transformation

We give a brief introduction of the Laplace transformation in this appendix. The introduction does not intend to be complete; it covers only the material used in the text.

Consider a time function $f(t)$ defined for all $t \geq 0-$. The Laplace transform of $f(t)$, denoted by $\hat{f}(s)$, is defined as

$$\hat{f}(s) \triangleq \mathcal{L}[f(t)] \triangleq \int_{0-}^{\infty} f(t)e^{-st}\, dt \tag{A-1}$$

where s is a complex variable and is often referred to as the *Laplace-transform variable*. The lower limit $0-$ of the integral indicates that the limit approaches zero from a negative value. Hence if $f(t)$ contains a delta function at $t = 0$, the delta function will be included in the integration. If $f(t)$ has no delta function at $t = 0$, then there is no difference between using $0-$ and 0. As a simple example, the Laplace transform of

$$f_1(t) = \begin{cases} e^{-2t} & \text{for } t \geq 0 \\ 0 & \text{for } t < 0 \end{cases} \tag{A-2}$$

is given by

$$\hat{f}_1(s) = \int_0^{\infty} e^{-2t}e^{-st}\, dt = \frac{-1}{s+2} e^{-(2+s)t} \Big|_{t=0}^{\infty} = \frac{1}{s+2}$$

A time function $f(t)$ can be computed from its Laplace transform $\hat{f}(s)$ by using the formula

$$f(t) = \frac{1}{2\pi j} \int_{c-j\infty}^{c+j\infty} \hat{f}(s)e^{st}\, ds \qquad \text{(A-3)}$$

where the integration is carried out in the complex plane along the vertical line passing the real axis at c. The real constant c is required to be larger than the abscissa of absolute convergence of $f(t)$, which is defined as the smallest real number a so that

$$\int_0^\infty |f(t)e^{-at}|\, dt < \infty$$

The computation of $f(t)$ from Equation (A-3) is rather complicated. In fact, the Laplace transform pair, (A-1) and (A-3), are not used in the text. The Laplace transform and its inverse can be easily obtained from a table, such as the one at the end of this appendix.

The Laplace transform has the linearity property:

$$\mathscr{L}[a_1 f_1(t) + a_2 f_2(t)] = a_1 \mathscr{L}[f_1(t)] + a_2 \mathscr{L}[f_2(t)]$$

The Laplace transform of the derivative of a function can be computed as

$$\mathscr{L}\left[\frac{d}{dt}f(t)\right] = \int_{0-}^{\infty}\left[\frac{d}{dt}f(t)\right]e^{-st}\, dt = f(t)e^{-st}\Big|_{t=0-}^{\infty} - \int_{0-}^{\infty} f(t)\left(\frac{d}{dt}e^{-st}\right) dt$$

$$= s\int_{0-}^{\infty} f(t)e^{-st}\, dt - f(0-) = s\hat{f}(s) - f(0-) \qquad \text{(A-4)}$$

where we used $f(t)e^{-st} = 0$ at $t = \infty$. Similarly it can be verified that

$$\mathscr{L}\left[\frac{d^n}{dt^n}f(t)\right] = s^n \hat{f}(s) - s^{n-1}f(0-) - s^{n-2}f'(0-) - \cdots - f^{(n-1)}(0-) \qquad \text{(A-5)}$$

Example 1

Find the Laplace transform of the derivative of the function $f_1(t)$ defined in (A-2).

Since $f_1(0-) = 0$, the Laplace transform of $df_1(t)/dt$ is, from Equation (A-4), $s/(s+2)$. We check this by direct verification. It is clear that

$$\frac{d}{dt}f_1(t) = \delta(t) - 2e^{-2t}$$

where $\delta(t)$ is the delta function (see Section 2-4). Hence we have

$$\mathscr{L}\left[\frac{d}{dt}f_1(t)\right] = \mathscr{L}[\delta(t)] - 2\mathscr{L}[e^{-2t}] = 1 - \frac{2}{s+2} = \frac{s}{s+2}$$

Example 2

If $f_1(t)$ is instead defined as

$$f_1(t) = e^{-2t} \qquad t \geq 0-$$

Then $f_1(0-) = 1$, and

$$\mathscr{L}\left[\frac{d}{dt}f_1(t)\right] = \frac{s}{s+2} - 1 = \frac{-2}{s+2}$$

This can also be directly verified. Since

$$\frac{d}{dt}f_1(t) = -2e^{-2t} \qquad t \geq 0-$$

we have

$$\mathscr{L}\left[\frac{d}{dt}f_1(t)\right] = -2\mathscr{L}[e^{-2t}] = \frac{-2}{s+2} \qquad \blacksquare$$

The Laplace transform of the integral of $f(t)$ is given by

$$\mathscr{L}\left[\int_{0-}^{t} f(t)\,dt\right] = \frac{1}{s}\hat{f}(s) \tag{A-6}$$

This can be verified as follows:

$$\mathscr{L}\left[\int_{0-}^{t} f(t)\,dt\right] = \int_{0-}^{\infty}\left[\int_{0-}^{t} f(t)\,dt\right] e^{-st}\,dt$$

$$= \left[\int_{0-}^{t} f(t)\,dt\right]\left(\frac{-1}{s}e^{-st}\right)\bigg|_{t=0-}^{\infty} - \int_{0-}^{\infty} f(t)\left(\frac{-e^{-st}}{s}\right) dt$$

$$= \frac{1}{s} - \int_{0-}^{\infty} f(t)e^{-st}\,dt = \frac{1}{s}\hat{f}(s)$$

Hence the integration in the time domain is converted into the division by s in the Laplace transform.

The following two theorems are used in the text.

Final-Value Theorem

If

$$\lim_{t \to \infty} f(t)$$

exists, [or equivalently, if $s\hat{f}(s)$ has no pole on the closed right-half s-plane], then

$$\lim_{t \to \infty} f(t) = \lim_{s \to 0} s\hat{f}(s) \tag{A-7}$$

By the existence of

$$\lim_{t \to \infty} f(t)$$

we mean that $f(t)$ approaches a finite constant as $t \to \infty$. If $f(t)$ becomes infinite or remains oscillatory as $t \to \infty$, then the final value theorem cannot be used. For example, the Laplace transform of $f(t) = e^t$ is $\hat{f}(s) = 1/(s-1)$. Although

$$\lim_{s \to 0} s\hat{f}(s) = 0$$

the function $f(t)$ approaches infinity as $t \to \infty$.

The final-value theorem can be proved as follows. Equation (A-4) implies

$$\lim_{s \to 0} (s\hat{f}(s) - f(0-)) = \lim_{s \to 0} \int_{0-}^{\infty} \left[\frac{d}{dt} f(t)\right] e^{-st} dt$$

$$= \int_{0-}^{\infty} \left[\frac{d}{dt} f(t)\right] dt = f(t) \Big|_{t=0-}^{\infty} = f(\infty) - f(0-)$$

This reduces to (A-7) after canceling $f(0-)$ out from both sides.

Parseval's Theorem

If

$$\int_0^{\infty} [f(t)]^2 \, dt$$

is finite, then

$$\int_0^{\infty} [f(t)]^2 \, dt = \frac{1}{2\pi j} \int_{-j\infty}^{j\infty} \hat{f}(s)\hat{f}(-s) \, ds \quad \text{(A-8)}$$

The assumption that

$$\int_0^{\infty} [f(t)]^2 \, dt$$

is finite implies that the c in Equation (A-3) can be chosen as 0. Using (A-3) with $c = 0$, the left-hand side of (A-8) becomes

$$\int_0^{\infty} f(t)f(t) \, dt = \frac{1}{2\pi j} \int_0^{\infty} \left[\int_{-j\infty}^{j\infty} \hat{f}(s) e^{st} \, ds\right] f(t) \, dt$$

which, if the order of integration is changed, becomes

$$\int_0^{\infty} [f(t)]^2 \, dt = \frac{1}{2\pi j} \int_{-j\infty}^{j\infty} \hat{f}(s) \left[\int_0^{\infty} e^{st} f(t) \, dt\right] ds$$

$$= \frac{1}{2\pi j} \int_{-j\infty}^{j\infty} \hat{f}(s)\hat{f}(-s) \, ds$$

This proves Parseval's theorem.

We give a short table of Laplace transforms to conclude this appendix.

Laplace Transforms

$f(t)$ $t \geq 0$	$f(s)$
$\delta(t)$	1
1 (unit step function)	$\dfrac{1}{s}$
t^n (n = positive integer)	$\dfrac{n!}{s^{n+1}}$
e^{-at} (a = real or complex)	$\dfrac{1}{s+a}$
$t^n e^{-at}$	$\dfrac{n!}{(s+a)^{n+1}}$
$\sin \omega t$	$\dfrac{\omega}{s^2 + \omega^2}$
$\cos \omega t$	$\dfrac{s}{s^2 + \omega^2}$
$e^{-at} \sin \omega t$	$\dfrac{\omega}{(s+a)^2 + \omega^2}$
$e^{-at} \cos \omega t$	$\dfrac{s+a}{(s+a)^2 + \omega^2}$
$tf(t)$	$\dfrac{d}{ds} f(s)$

Reference

[1] Cooper, G. R., and C. D. McGillem, *Methods of Signal and System Analysis*, New York: Holt, Rinehart and Winston, 1967.

APPENDIX
B
Matrix Theory

B-1 Matrices

A matrix is a rectangular array of elements such as

$$\mathbf{A} = \begin{bmatrix} a_{11} & a_{12} & \cdots & a_{1m} \\ a_{21} & a_{22} & \cdots & a_{2m} \\ \vdots & & & \\ a_{n1} & a_{n2} & \cdots & a_{nm} \end{bmatrix} = [a_{ij}]$$

The elements a_{ij} can be real numbers, complex numbers, or rational functions. The matrix has n rows and m columns and is called an $n \times m$ matrix. The matrix is called a *square matrix* of order n if $m = n$; a *column vector* if $m = 1$; a *row vector* if $n = 1$. A square matrix is called a *diagonal matrix* if $a_{ij} = 0$ for all $i \neq j$. A diagonal matrix is called a *unit matrix* if $a_{ii} = 1$ for all i. A unit matrix of order n is denoted by $\mathbf{I}_n$.

Two $n \times m$ matrices are said to be equal if and only if all the corresponding elements are the same. The addition and multiplication of matrices are defined as follows:

$$\underset{n \times m}{\mathbf{A}} + \underset{n \times m}{\mathbf{B}} = \underset{n \times m}{[a_{ij} + b_{ij}]}$$

$$\underset{1 \times 1}{c} \underset{n \times m}{\mathbf{A}} = \underset{n \times m}{[c a_{ij}]}$$

$$\mathop{\mathbf{A}}_{n \times m} \mathop{\mathbf{B}}_{m \times p} = \left[\sum_{k=1}^{m} a_{ik} b_{kj} \right]_{n \times p}$$

Matrix multiplication has the following properties:

$$\mathbf{A} = \mathbf{AI}_m = \mathbf{I}_n \mathbf{A}$$

$$(\mathbf{AB})\mathbf{C} = \mathbf{A}(\mathbf{BC})$$

$$(\mathbf{A} + \mathbf{B})\mathbf{C} = \mathbf{AC} + \mathbf{BC}$$

$$\mathbf{A}(\mathbf{B} + \mathbf{C}) = \mathbf{AB} + \mathbf{AC}$$

Generally the equality $\mathbf{AB} = \mathbf{BA}$ does not hold.

The determinant of a square matrix of order n is defined as

$$\det \mathbf{A} = \sum (-1)^j a_{1j_1} a_{2j_2} \cdots a_{nj_n}$$

where $j_1, j_2, \ldots, j_n$ are all possible orderings of the second subscripts $1, 2, \ldots, n$, and the integer j is the number of interchanges of two digits required to bring the ordering $j_1, j_2, \ldots, j_n$ into the natural ordering $1, 2, \ldots, n$. For example, we have

$$\det \begin{bmatrix} a_{11} & a_{12} \\ a_{21} & a_{22} \end{bmatrix} = (-1)^0 a_{11} a_{22} + (-1)^1 a_{12} a_{21} = a_{11} a_{22} - a_{12} a_{21}$$

$$\det \begin{bmatrix} a_{11} & a_{12} & a_{13} \\ a_{21} & a_{22} & a_{23} \\ a_{31} & a_{32} & a_{33} \end{bmatrix} = a_{11} a_{22} a_{33} + a_{12} a_{23} a_{31} + a_{13} a_{21} a_{32} - a_{13} a_{22} a_{31} - a_{12} a_{21} a_{33} - a_{11} a_{23} a_{32}$$

For a square matrix with a *nonzero* determinant, we may define the inverse. The inverse of $\mathbf{A}$, denoted by $\mathbf{A}^{-1}$, has the property $\mathbf{A}^{-1}\mathbf{A} = \mathbf{AA}^{-1} = \mathbf{I}$. It can be computed by using the formula

$$\mathbf{A}^{-1} = \frac{1}{\det \mathbf{A}} [c_{ij}]$$

where

$$c_{ij} = (-1)^{i+j} \cdot \text{(determinant of the array formed by deleting the } j\text{th row and } i\text{th column from } \mathbf{A}\text{)}$$

B-2 The Rank of a Matrix

Consider an $n \times m$ matrix $\mathbf{A}$. The following operations on the matrix $\mathbf{A}$ are called the elementary transformations:

1. The interchange of two rows or columns.
2. The multiplication of a row or column by a *nonzero* constant.
3. The addition of one row or column to another.

Every matrix **A** can be transformed by a sequence of elementary transformations into the form

$$\mathbf{A} \rightarrow \begin{bmatrix} 1 & x & x & \cdots & x & x & x & \cdots & x \\ 0 & 1 & x & \cdots & x & x & x & \cdots & x \\ 0 & 0 & 1 & \cdots & x & x & x & \cdots & x \\ \vdots & \vdots & \vdots & & \vdots & \vdots & \vdots & & \vdots \\ 0 & 0 & 0 & \cdots & 1 & x & x & \cdots & x \\ 0 & 0 & 0 & \cdots & 0 & 0 & 0 & \cdots & 0 \\ \vdots & \vdots & \vdots & & \vdots & \vdots & \vdots & & \vdots \\ 0 & 0 & 0 & \cdots & 0 & 0 & 0 & \cdots & 0 \end{bmatrix}$$

or

$$\begin{bmatrix} 1 & x & x & \cdots & x & x & x & \cdots & x \\ 0 & 1 & x & \cdots & x & x & x & \cdots & x \\ 0 & 0 & 1 & \cdots & x & x & x & \cdots & x \\ \vdots & \vdots & \vdots & & \vdots & \vdots & \vdots & & \vdots \\ 0 & 0 & 0 & \cdots & 1 & x & x & \cdots & x \\ 0 & 0 & 0 & \cdots & 0 & 1 & x & \cdots & x \end{bmatrix}$$

where "x" denotes a possible nonzero element. Then the *rank* of **A**, denoted by $\rho\mathbf{A}$, is defined as the number of 1's on the diagonal, as shown.

Example

Consider the 4 × 4 matrix

$$\begin{bmatrix} 2 & 1 & 4 & 1 \\ 4 & -1 & 2 & 2 \\ 1 & 2 & 5 & -1 \\ 0 & 1 & 2 & 1 \end{bmatrix} \begin{array}{l} \times (-2) \\ \times (-0.5) \end{array}$$

$$\rightarrow \begin{bmatrix} 2 & 1 & 4 & 1 \\ 0 & -3 & -6 & 0 \\ 0 & 1.5 & 3 & -1.5 \\ 0 & 1 & 2 & 1 \end{bmatrix} \begin{array}{l} \times (+0.5) \\ \times (\frac{1}{3}) \end{array}$$

$$\rightarrow \begin{bmatrix} 2 & 1 & 4 & 1 \\ 0 & -3 & -6 & 0 \\ 0 & 0 & 0 & -1.5 \\ 0 & 0 & 0 & 1 \end{bmatrix} \times 1.5 \rightarrow \begin{bmatrix} 2 & 1 & 4 & 1 \\ 0 & -3 & -6 & 0 \\ 0 & 0 & 0 & -1.5 \\ 0 & 0 & 0 & 0 \end{bmatrix}$$

$$\rightarrow \begin{bmatrix} 2 & 1 & 1 & 4 \\ 0 & -3 & 0 & -6 \\ 0 & 0 & -1.5 & 0 \\ 0 & 0 & 0 & 0 \end{bmatrix} \begin{matrix} \times 0.5 \\ \times (-\frac{1}{3}) \\ \times \left(\frac{1}{1.5}\right) \\ \end{matrix} \rightarrow \begin{bmatrix} 1 & 0.5 & 0.5 & 2 \\ 0 & 1 & 0 & 2 \\ 0 & 0 & 1 & 0 \\ 0 & 0 & 0 & 0 \end{bmatrix}$$

Hence the rank of the matrix is 3. ∎

From the definition of rank, it is clear that if $\mathbf{A}$ is an $n \times m$ matrix, then

$$\rho \mathbf{A} \leq \min(n, m)$$

Consider an $n \times m$ matrix $\mathbf{A}$, written as

$$\mathbf{A} = [\mathbf{a}_1 \quad \mathbf{a}_2 \quad \cdots \quad \mathbf{a}_m]$$

where $\mathbf{a}_i$, for $i = 1, 2, \ldots, m$, are $n \times 1$ column vectors. If the rank of $\mathbf{A}$ is k, then there exist k columns in $\mathbf{A}$, say $\mathbf{a}_{j_1}, \mathbf{a}_{j_2}, \ldots, \mathbf{a}_{j_k}$, such that

$$\alpha_1 \mathbf{a}_{j_1} + \alpha_2 \mathbf{a}_{j_2} + \cdots + \alpha_k \mathbf{a}_{j_k} = 0$$

if and only if $\alpha_j = 0$, for $j = 1, 2, \ldots, k$. These k columns are said to be *linearly independent*. Every remaining column of $\mathbf{A}$ can then be written as a linear combination of these k columns as

$$\mathbf{a}_j = \beta_1 \mathbf{a}_{j_1} + \beta_2 \mathbf{a}_{j_2} + \cdots + \beta_k \mathbf{a}_{j_k}$$

and is said to be *linearly dependent* on $\{\mathbf{a}_{j_1}, \mathbf{a}_{j_2}, \ldots, \mathbf{a}_{j_k}\}$. A similar statement is also applicable to the rows of $\mathbf{A}$. For an $n \times m$ matrix of rank k, by deleting the linearly dependent columns and rows, we obtain a $k \times k$ submatrix of $\mathbf{A}$ with rank k.

It can be shown that a square matrix of order k has a rank k if and only if the determinant of the matrix is different from zero. A square matrix of order k is called *nonsingular* if it has a rank k, or equivalently, its determinant is different from zero.

B-3 Linear Algebraic Equations

Consider the set of linear algebraic equations

$$a_{11}x_1 + a_{12}x_2 + \cdots + a_{1m}x_m = y_1$$
$$a_{21}x_1 + a_{22}x_2 + \cdots + a_{2m}x_m = y_2$$
$$\vdots$$
$$a_{n1}x_1 + a_{n2}x_2 + \cdots + a_{nm}x_m = y_n$$

where the a_{ij}'s and y_i's are known, and the x_i's are unknown. This set of equations can be written in matrix form as

$$\mathbf{Ax} = \mathbf{y}$$

where

$$\mathbf{A} = \begin{bmatrix} a_{11} & a_{12} & \cdots & a_{1m} \\ a_{21} & a_{22} & \cdots & a_{2m} \\ \vdots & \vdots & & \vdots \\ a_{n1} & a_{n2} & \cdots & a_{nm} \end{bmatrix} \quad \mathbf{x} = \begin{bmatrix} x_1 \\ x_2 \\ \vdots \\ x_m \end{bmatrix} \quad \mathbf{y} = \begin{bmatrix} y_1 \\ y_2 \\ \vdots \\ y_n \end{bmatrix}$$

It can be shown that, given $\mathbf{A}$ and $\mathbf{y}$, there exists an $\mathbf{x}$ satisfying $\mathbf{Ax} = \mathbf{y}$ if and only if

$$\rho \mathbf{A} = \rho[\mathbf{A} \mid \mathbf{y}]$$

If $n = m$, and if $\rho \mathbf{A} = n$ or, equivalently, $\det \mathbf{A} \neq 0$, then for *any* $\mathbf{y}$ there exists a unique $\mathbf{x}$ satisfying $\mathbf{Ax} = \mathbf{y}$. If the inverse of $\mathbf{A}$ is computed, then the solution $\mathbf{x}$ is given by $\mathbf{A}^{-1}\mathbf{y}$.

The solution of $\mathbf{Ax} = \mathbf{y}$ can be obtained by computating the inverse of $\mathbf{A}$, by using Cramer's rule, or by using the Gaussian elimination method. Among them, the Gaussian elimination method may be the simplest; furthermore the method is still applicable even when the matrix $\mathbf{A}$ is not nonsingular. The method will be illustrated by a simple example.

Example

Find the solutions of

$$x_1 + 2x_2 + 3x_3 = 10 \qquad \text{(B-1)}$$
$$2x_1 + 5x_2 - 2x_3 = 3 \qquad \text{(B-2)}$$
$$x_1 + 3x_2 - x_3 = 0 \qquad \text{(B-3)}$$

The subtraction of twice (B-1) from (B-2), and (B-1) from (B-3), yields

$$x_1 + 2x_2 + 3x_3 = 10 \qquad \text{(B-1')}$$
$$x_2 - 8x_3 = -17 \qquad \text{(B-2')}$$
$$x_2 - 4x_3 = -10 \qquad \text{(B-3')}$$

The subtraction of (B-2') from (B-3') yields

$$x_1 + 2x_2 + 3x_3 = 10 \qquad \text{(B-1'')}$$
$$x_2 - 8x_3 = -17 \qquad \text{(B-2'')}$$
$$4x_3 = 7 \qquad \text{(B-3'')}$$

The solutions x_1, x_2, and x_3 can now be easily obtained by first solving (B-3''), then (B-2''), and finally (B-1'').

References

[1] Hamming, R. W., *Introduction to Applied Numerical Analysis*. New York: McGraw-Hill, 1971.

[2] Pipes, L. A., and S. A. Hovanessian, *Matrix Computer Methods in Engineering*. New York: Wiley, 1969.

Index

AC device, 39
   amplifier, 59
   motor, 43
   potentiometer, 51
   synchros, 56
   tachometer, 53
Accelerating function, 154
Actuating signal, 4
Actuator, 4
Adaptive system, 325
Additivity, 13
Amplidyne, 77
Amplifier, 59
   electronic, 59
   operational, 57
   rotating power, 59
Analog simulation diagram, 125
Angle of arrival, 238
   of departure, 238
Asymptote of root locus, 235
   of Bode plot, 264

Bandwidth, 256
Basic block diagram, 123
Block diagram, 51, 63
   analog simulation, 125
   basic, 123
Bode plot, 262
Breakaway point, 237

Characteristic denominator, 29
Characteristic polynomial, 93
Closed-loop system, 4, 174
Command signal, 4
Comparison sensitivity, 176
Compensation, 160
   feedback, 219
   lag-lead, 289
   network, 60
   phase-lag, 286
   phase-lead, 282
   series, 219
Complete characterization, 27, 29, 100
Composite system, 81, 101
Computer simulation, 122
   analog, 125
   digital, 132
Controllability, 30
   matrix, 30

Convolution integral, 16
Corner frequency, 264

Damping ratio, 227
DC device, 9
    amplifier, 59
    motor, 41
    potentiometer, 51
    tachometer, 53
Degree of rational function, 23
    of vector function, 29
Demodulator, 40
Description of system, 15, 20, 24, 32, 81
    input-output, 15, 24
    state-variable, 20, 24
Desired signal, 4
Digital computer simulation, 132
Disturbance, 174, 179
Dynamical equation, 21, 25, 122, 132
    controllable-form, 139, 144
    Jordan-form, 141, 145
    observable-form, 137, 145

Eigenvalue, 93, 97, 99
Elementary transformation, 333
Error constant, 271
    position, 271
    velocity, 271
Error detector, 56
    operational amplifier, 57
    potentiometer, 56
    synchros, 56

Feedback system, 4, 174
    output, 202
    partial state, 213
    state, 208
Final-value theorem, 329
Forward path, 83
Free speed, 45
Frequency domain, 32, 255, 260, 268
Frequency spectrum, 39
Friction, 14
    Coulomb, 15, 42
    static, 15, 42
    viscous, 15, 42

Gain crossover frequency, 275
Gain margin, 275
Gear train, 47
    backlash of, 48

Homogeneity, 13
Hurwitz polynomial, 104

Impulse function, 16
Impulse response, 16
Induction motor, 43
Instantaneous system, 10
ITAE, 220

Laplace transform, 327
    table of, 331
    variable, 327
Linearity, 12
Loading problem, 52, 63, 67, 163
Log magnitude-phase plot, 262
Loop, 83
Lumped system, 12

Magnitude scaling, 128
Mason's formula, 83, 136
Matrix, 332
    controllability, 30
    nonsingular, 335
    observability, 31
    rank of, 333
Memoryless system, 10
Model, 7, 35
Modulated signal, 39
Modulator, 40
Motor (see Servomotor)
Multivariable system, 9, 319

Network, 60
    ac, 60, 61
    bridge-T, 37
    dc, 60, 61
    passive, 62
    phase-lag, 286
    phase-lead, 283
Noise, 53, 162, 178
Nonlinearity, 12
Numerical control, 77, 252
Nyquist plot, 268
Nyquist stability criterion, 269

Observability, 30
    matrix, 31
Open-loop system, 4, 174
Operational amplifier, 57
Optimal control law, 194
Output equation, 21
Output regulator problem, 198
Overshoot, 158, 227

Parameter optimization, 299
    analytical method, 300
    numerical method, 308

Parseval's theorem, 330
Partial fraction expansion, 85
Peak resonance, 256
Performance criterion, 154
    ITAE, 220
    quadratic, 185, 300
    steady-state, 155, 270
    transient, 157, 273
Phase crossover frequency, 275
Phase margin, 275
Plant, 4, 152
Polar plot, 261
Pole, 85, 172, 226
    closest, 229
    stable, 172
Pole placement by state feedback, 208
Pole-zero cancellation, 102, 245
Pole-zero excess inequality, 169
Position error constant, 271
Potentiometer, 51, 56
    loading of, 52

Ramp function, 111, 154
Rational function, 20
    irreducible, 28
    proper, 20
    strictly proper, 20
Realization, 23, 134
    controllable-form, 139, 144
    irreducible, 134
    Jordan-form, 141, 145
    minimal-dimensional, 134
    observable-form, 137, 145
Regulating system, 121, 197
    voltage, 74
Response time, 258
Rise time, 158, 228
Root locus, 232
    angle of arrival, 238
    angle of departure, 238
    asymptote of, 235
    breakaway point, 237
    properties of, 234
Routh-Hurwitz criterion, 104

Saturation, 160
Seismometer, 112
Sensitivity, 174, 177
    comparison, 176
Servomotor, 40
    ac, 43
    armature-controlled dc, 41
    gain constant of, 43, 45
    hydraulic, 46

    time constant of, 43, 45
Settling time, 158, 228
Signal-to-noise ratio, 163
Simplification, 112
Simulation (*see* Computer simulation)
Space vehicle, 67
Specification, 154
    frequency-domain, 30, 256, 260, 268
    steady-state, 155, 270
    time-domain, 32, 154
    transient, 157, 273
Spectral factorization, 188
Spirule, 236
s-Plane, 89
    closed right-half, 89
    open right-half, 89
Stability, 95, 100, 102, 109
    asymptotic, 97
    BIBO, 95
    total, 98
Stall torque, 45
State, 9
    equation, 21
    initial, 9
    variable, 10, 136, 200
    vector, 10
Steady-state error, 156
Steady-state response, 109
    of ramp function, 111
    of sinusoidal function, 110
    of step function, 111
Step function, 110, 154
Stochastic system, 325
Superposition property, 13
Synchros, 56
    transformer, 56
    transmitter, 56
Synthesizable transfer function, 173
System, 8
    adaptive, 325
    continuous-time, 5
    control, 1
    deterministic, 325
    discrete-time, 321
    distributed, 12
    initially relaxed, 15
    learning, 325
    linear, 5, 12
    lumped, 12
    memoryless, 10
    multivariable, 10
    nonlinear, 12, 324
    optimal control, 185
    physical, 7

System (*continued*)
   single-variable, 9
   stochastic, 325
   time-invariant, 5, 13
   time-varying, 13, 324

Tachometer, 52
Test function, 154
Time constant, 43, 45, 89
Time domain, 32, 154
Time invariance, 13
Time scaling, 130
Tracking antenna problem, 153, 191, 200, 211, 216, 221, 245, 289, 304, 306, 315
Tracking problem, 197
Transducer, 51
Transfer function, 18
   irreducible, 28
   proper, 20
   pulse, 322
   sampled, 322
   strictly proper, 20
   synthesizable, 173
   type l, 271
Transient performance, 157, 270

Velocity-error constant, 271
Viscous friction, 15

Ward-Leonard system, 66
Weighting factor, 187
Weighting sequence, 321

Zero, 85, 172
Zero-input response, 13, 23, 97
Zero-state response, 13, 23
z-Transform, 322